FOREST RESOURCES AND WOOD-BASED BIOMASS ENERGY AS RURAL DEVELOPMENT ASSETS

FOREST RESOURCES AND WOOD-BASED BIOMASS ENERGY AS RURAL DEVELOPMENT ASSETS

Editors

WILLIAM R. BENTLEY
Senior Program Officer,
Winrock International Institute for Agricultural Development, and
Director, Weyerhaeuser Center for Forest Resource Management
and Policy Studies
Yale University

MARCIA M. GOWEN
Director of Economics and Environmental Planning
Alternative Energy Development, Inc.

Winrock International, USA
and
Oxford & IBH Publishing Co. Pvt. Ltd.
New Delhi Bombay Calcutta

ISBN 81-204-0923-X

Published in India by Mohan Primlani, Oxford & IBH Publishing Co. Pvt. Ltd., 66 Janpath, New Delhi 110 001, for the Winrock-Oxford & IBH Series. Photoready copy prepared by Winrock International at the Weyerhaeuser Center for Forest Resource Management and Policy Studies, Yale University, New Haven, CT 06511. Printed at Baba Barkha Nath Printers, 26/7, Najafgarh Road Industrial Area, New Delhi 110 015.

Foreword

Winrock International Institute for Agricultural Development and the Oxford and IBH Publishing Company initiated a book series on agroforestry research and practice in 1990. This is the seventh volume in this series.

Previous volumes in this series have focused on research methods and management, the biological and social underpinnings of agroforestry, and several specific topics and applications. More volumes of this nature are in process for future publication

With this volume, which is edited by myself and my former Winrock colleague, Dr. Marcia Gowen, we move into the broader purposes for agroforestry in development. This volume originated with the papers and discussion during a February 1992 workshop sponsored by Winrock, Yale University, and the U.S. Agency for International Development on *Forest Resources and Wood-Based Biomass Energy as Rural Development Assets*. We added a few more chapters in light of topics raised during the workshop discussions and a broader understanding on our part of the ideas that should be included in this dialogue.

This volume also moves us to a global perspective. We will continue an Asian focus in our series, but the issues raised by biomass energy and forest resource conservation are not limited to Asia. Indeed they are not limited to the tropics or to developing nations, as many recent policy issues in the United States attest.

Our goals remain the same. We want to promote excellent and useful research, including encouraging scholars from all nations to take agroforestry seriously as an area of applied research. We are especially concerned that agroforestry research—and forestry research generally—help produce net benefits for the rural poor.

We also continue to search for funding to help bring the prices of each volume in the series as low as possible, consistent with good editing, layout, and printing, so that students, teachers, and researchers throughout the world will have access to these books. We express our deepest appreciation to the U.S. Agency for International Development, the School of Forestry and Environmental Studies, Yale University, the Rockefeller Foundation, and Winrock International for their financial support and encouragement to produce this volume.

General Series Editor
March 1994

WILLIAM R. BENTLEY

Preface

The purpose of this volume is to bring together the many strands of a week-long workshop in late February 1992. Our theme was *Forest Resources and Wood-Based Biomass Energy as Rural Development Assets*. It was based on collaborative research conducted by Winrock's Biomass Energy Research and Technology (BEST) Project and the School of Forestry and Environmental Studies at Yale University. We were looking at useful ways of integrating promising technologies for generating electrical and other energy from wood residues and other forest based biomass while promoting both environmental sustainability and distribution of new benefits toward the poor.

The workshop participants included professionals with a wide variety of educational backgrounds and work experiences. Our colleagues included energy engineers, silviculturists and forest managers, policy experts, rural development specialists, ecologists, environmental advocates, and several social scientists of quite dissimilar persuasions. Both Winrock and the Yale forestry faculty take considerable pride in their ability to bring ideas together into integrated solutions, but we were stretching all of us in this exercise.

The group was remarkably productive, although each person had struggles with what the problems were and how they might contribute to the identification of solutions. We were impressed with the participants commitment to learning from each other and working together to integrate their information and knowledge. While we reached few solutions during our week together, we did create an excellent testimonial to what can be accomplished through synthesis when coupled to a solid analytical base. Consequently, we begin our acknowledgment and thank yous with a hardy applause for our colleagues who participated in the workshop and did the hard work that proceeded this volume.

We also want to thank our work mates who helped produce this volume. Roberta Britt edited all the original papers, and the ones we added from authors who were not at the workshop. She did yeoman's duty, pulling sense out of our many desperate ideas, and converting them to understandable English. Sonia Varley did the layout and composition for this our first book produced using camera-ready copy. Her patience with both of us, our many changes, and our many delays, deserves a medal. Don Grebner helped throughout with computer problems, computer graphics, and other new tasks as we learned how to convert our ideas into a final manuscript. Karen Seckler, who has edited four other volumes in this series, helped Sonia and the rest of us learn a few of her many skills. Mohan Primlani and Indu Ramchandani, our friends and colleagues with Oxford and IBH, as always were supportive and helpful while occasionally whispering some much needed advice in our ears.

We have acknowledged our several donors in the Foreword, but again we want to thank each of them. Without funding, this kind of book and much of the other activities that make agroforestry research cannot continue.

Our thanks finally to both Yale and Winrock for providing the right kind of environment to create and write such a book. Dr. John Gordon, former dean of forestry and environmental studies at Yale, was most encouraging throughout our planning process and, through the Weyerhaeuser Center for Forest Resource Management and Policy Studies, provided some of the support for Roberta Britt and Sonia Varley. Dr. Enrique Ospina, Director of the Center for Institutional and Human Resource Development at Winrock's headquarters, was a continuous supporter of the Yale/Winrock collaboration, and he also provided much needed financial support for Dr. Bentley's time and for Sonia Varley.

William R. Bentley
Marcia M. Gowen
New Haven, Connecticut

March 1, 1994

CONTENTS

PART FOUR: GLOBAL AND TROPICAL FOREST ECONOMIES

List of Figures

PART ONE: THE ISSUES AND LINKAGES

PART THREE: MANAGING THE TROPICAL FOREST: CAN IT BE DONE?

PART FOUR: GLOBAL AND TROPICAL FOREST ECONOMIES

List of Tables

List of Boxes

PART THREE: MANAGING THE TROPICAL FOREST: CAN IT BE DONE?

Contributing Authors

Ashton, P.M.S., Assistant Professor, Yale School of Forestry and Environmental Studies, 205 Prospect Street, New Haven, CT 06511.

Bell, R.H., Program Assistant, Winrock International, Rt. 3, Morrilton, AR 72110.

Bender, Ozzie, Vice President, W.L. Bender, Inc., Consulting Foresters, 805 North C Street, Tacoma, WA 98403.

Bentley, William, R., Senior Program Officer, Winrock International, Rt. 3, Morrilton, AR 72110; Senior Research Scientist and Lecturer and Director, Weyerhaeuser Center for Forest Resource Management and Policy Studies, Yale School of Forestry and Environmental Studies, 205 Prospect St., New Haven, CT 06511.

Briscoe, C.B., La Suiza, Turrialba 7151, Costa Rica.

Burnham, Margo, Renewable Energy Analyst, DynCorp•Meridien Corporation, 4300 King Street, Suite 400, Alexandria, VA 22303-1508, (formerly with Winrock International).

Cook, James, The National Audubon Society, Scully Science Center, 550 South Bay Avenue, Islip, NY 11751.

Dalla Rosa, Karl, State of Hawaii, Department of Land and Natural Resources, 1151 Punchbowl Street, Rm. 325, Honolulu, HI 96813.

de Albuquerque dos Santos, Antonio C., PhD candidate, Yale School of Forestry and Environmental Studies, 205 Prospect St., New Haven, CT 06511.

del Castillo, Romulo, University of the Philippines, Los Banos (UPLB), P.O. Box 150, College, Laguna 4031, Philippines.

Dominguez, Jules J., Manager, Utility Engineering, Georgia-Pacific Corporation, P.O. Box 105605, Atlanta, GA 30348-5605.

Ducey, M.J., PhD candidate, Yale School of Forestry and Environmental Studies, 205 Prospect Street, New Haven, CT 06511.

Evenson, Robert E., Professor, Economic Growth Center, Department of Economics, Yale University, 27 Hillhouse Avenue, New Haven, CT 06520.

Gowen, Marcia M., Director of Environmental Planning, Alternative Energy Development, Inc., 8455 Colesville Road, Silver Spring, MD 20910, (formerly with Winrock International).

Grebner, Donald L., Forest Resource Analyst, Winrock International, Rt. 3, Morrilton, AR 72110.

Grubbs, Kelton, President, Thermal System Engineering, P.O. Box 421, Hot Springs, AR 71902.

Gupta, Anil K., Centre for Management in Agriculture, Indian Institute of Management, Ahmedabad 380015, India.

Hammett, A.L., Chief of Party and Senior Forestry Advisor, Institute of Forestry Project, Kathmandu, Nepal.

Hyde, William F., Department Head, Department of Forestry, 307 Cheatham Hall, Virginia Tech, Blacksburg, VA 24061-0324.

Jezek, Peter A., Private Power Adviser, Ministry of Mines and Energy, Jakarta, Indonesia.

Jones, Donald W., Energy Division, Oak Ridge National Laboratory, P.O. Box 2008, Oak Ridge, TN 37831.

Knoettner, Gregory, New England Field Assistant, Appalachian Trail Conference, Lyme, New Hampshire 03768.

Larson, B.C., Lecturer in Forest Management and Director of School Forests, Yale School of Forestry and Environmental Studies, 205 Prospect Street, New Haven, CT 06511.

Montagnini, Florencia, Associate Professor and John M. Musser Director of the Tropical Resources Institute, Yale School of Forestry and Environmental Studies, 205 Prospect Street, New Haven, CT 06511.

Morris, Gregory P., Senior Associate, Future Resources Associates, Inc., 2000 Center Street, Suite 418, Berkeley, CA 94704.

Newman, David H., Associate Professor, Daniel B. Warnell School of Forest Resources, University of Georgia, Athens, GA 30602-2152.

Padoch, Christine, Scientist, The New York Botanical Garden, Bronx, NY 10458.

Ramakrishnan, P.S., Professor of Ecology, Jawaharlal Nehru University, School of Environmental Sciences, New Delhi 110067, India.

Sancho, Freddy, University of Costa Rica, School of Agronomy, San Pedro, Costa Rica.

Sedjo, Roger, Senior Fellow and Director, Forest Economics and Policy Program, Resources for the Future, 1616 P Street, NW, Washington, D.C., 20036.

Stijfhoorn, Eirik, Norwegian Centre for International Agricultural Development, Agricultural University of Norway, P.O. Box 5002, N-1432 AS, Norway.

Sutton, W.R.J., Tasman Forestry Limited, Ngahere House, Private Bag 3031, Rotorua 3200, New Zealand.

Tugwell, Frank, Executive Director, The Heinz Endowments, 625 Liberty Avenue, Pittsburgh, PA 15222, (formerly with Winrock International).

Vernegaard, L.M., Associate Director, Planning and Ecology, Trustees of Reservations, 572 Essex Street, Beverly, MA 01915.

Weingart, Jerome M., Jerome Weingart and Associates, Arlington, VA 22209.

Wimberly, Jim, Program Officer, Winrock International, Rt. 3, Morrilton, AR 72110.

Part One

The Issues and Linkages

INTRODUCTION

Wood-based energy, like other renewable energy sources, holds significant economic and environmental opportunities around the world. Well-designed wood-energy projects are constructed in context of sustainable resource production and proper waste management, fully accounting for any positive or negative environmental leakages. This section explores the broad issues that will affect the utilization of renewables, wood in particular, as modern energy sources in developing countries. The first paper focuses on new electricity and cogeneration configurations in the future that may accelerate renewable energy use by leapfrogging historic developed country patterns. In place of large central grids, new distribution utilities with mini-grids powered from a wide variety of low-cost renewable options may provide rural and urban areas with more reliable and cheaper energy than large facilities which use imported fossil fuels.

The second paper demonstrates how wood-based energy could provide a significant portion of the thermal-electric needs of rural areas in many developing countries. The current wood wastes from Asia, Africa, and Latin America, if converted into energy represent large potentials. At present, much of the wood product industries' residues are wasted and the feedstock is not managed on a sustainable basis.

Asia has the largest opportunity for energy production from its wood wastes. A transition to sustainable management through mixed species, multiple use plantations must occur before we can rely on this source over time. Africa has some large tracts of virgin tropical forests that need to be preserved, but only isolated pockets with plantations can economically support long-term wood energy production. Latin America represents the world's largest untapped forest resource. The opportunities for wood energy production in Latin America, however, are mixed because of competing land uses, poor forest management practices, and government policies that discourage resource stewardship.

CHAPTER 1

RENEWABLE ENERGY FOR DEVELOPMENT: THE CHALLENGE TO UTILITY MANAGEMENT[1]

Franklin Tugwell
Margo Burnham
Robert H. Bell

INTRODUCTION

Scarcities of electrical power will significantly impair the economic growth of developing countries in the next few decades. Although substantial investments were made by governments to increase generation capacity and extend grid service to rural populations, decades of poor policies and failing infrastructures have placed many utilities of developing countries in financial straits. Deteriorating power services plague most of these nations. Concern over increasing environmental degradation now limits the development of more large-scale hydroelectric and fossil-fuel power facilities. The current outlook is grim. Existing power infrastructures are already deteriorated to an alarming degree. Financing for future generation-capacity will be limited. Prospects for reversing this downward trend with a conventional approach in the next two decades are poor.

A new approach to energy development shows considerable promise. In the United States, thousands of widely dispersed renewable-energy systems, operated by independent producers, now contribute a significant proportion of the national energy supply. Such systems may be least-cost options for providing power to areas at the outer reaches of and beyond the electric grid in developing countries. Since they are small and depend on local energy resources, these systems are ideally suited to development by the private sector. By allowing private enterprise a role in the power sector, governments in developing countries could increase their power capacity at less cost to the public treasury, particularly in foreign exchange, than comparable development of conventional power. Existing power infrastructures in developing countries are less developed and, hence, more adaptable. Thus, in contrast to utilities in industrialized countries, those in developing countries may find it easier to

1 Research for this chapter was supported by the U.S. Agency for International Development, Bureau for Research and Development, Office of Energy and Infrastructure.

integrate distributed renewable-energy systems with their existing infrastructure. In the near-term, developing countries could even surpass the achievements of industrialized countries in adopting dispersed generation to substantially increase their national energy supplies.

THE DEEPENING POWER CRISIS

Symptoms and Costs of Deteriorating Power Service

Most analysts and decision-makers have been inattentive to the power crisis in developing countries. Electric utilities are increasingly unable to provide the amount and kind of power that is needed to sustain economic development. In country after country, electric-supply systems are falling into disrepair because of poor management and lack of investment in maintenance. The symptoms of this crisis are commonplace in the developing world—frequent blackouts, brownouts, and poor quality power, when it is available. In the late 1980s, many developing nations experienced shortages exceeding 10% of generation capability. To cite but a few examples, power supply in India fell short of national demand by over 10%; in the Dominican Republic, by over 15%; and in Pakistan, by over 25% (USAID 1988).

Less visible is the high cost, in lost economic productivity and duplication of basic infrastructure, that shortages of electricity generate. Sanghvi (1991) has established that insufficient and unreliable power can exceed $1.00/kWh (kWh = kilowatt-hour) in lost economic wealth in most developing economies. Unfortunately, rural communities and industries with the most pressing need for economic productivity must cope with the least reliable and lowest quality power supply that is typical at the end of the transmission line. Developing nations have tried to reduce their reliance on imported fossil fuels, but unreliable power service forces increasing numbers of utility customers to install their own diesel and gas-fired back-up facilities (USAID 1988). In addition to the tariffs charged by the utility, these self-generators willingly incur costs for reliable power that exceed those of the utility itself. For example, Sanghvi (1991) estimates that costs for self-generation in Pakistan range from $0.14/kWh to $0.76/kWh, while the estimated long-run, marginal cost to the utility is $0.076/kWh.

This duplication of infrastructure is costly to the economies of developing countries. Not only does it result in unnecessary investments in generation capacity, it also leads to a dramatic reduction in overall system efficiency. Since the duplication almost always involves small diesel-generator sets, it encourages increased dependence on fossil-fuel imports with their drain on foreign exchange.

Historical Causes

There are several distinct causes of the power crisis:

- Insufficient investments in the power sector
- Continued subsidies of prices for electricity
- Politically motivated government programs to provide rural electricity
- Operational inefficiencies of electric utilities

First, and most basic, the level of investments in generation capacity has not kept pace with the steady growth in demand for electricity. Despite large loans by international banks, bilateral foreign-assistance programs, and private banks, many governments of developing countries have been unable to meet the demand for electric power. Demand has grown at a rate of nearly 7%/yr since the 1970s. The power sector is typically the most capital-intensive in the developing world, and it requires substantial amounts of hard currency to support the purchase of imported equipment.

The widespread practice of subsidizing the electricity price, or tariff, paid by customers is the second cause of the crisis. Because electric utilities are usually owned by the government in developing countries, they must fulfill political mandates to provide cheap electrical service, regardless of supply costs. This would not be possible if utilities were operating as private, profit-making companies. Eager to gain political favor and fearful of public outrage, most governments did not allow utilities to pass on to consumers their increased costs that were caused by OPEC's escalating oil prices in the 1970s and 1980s. Many governments also failed to pass on the increasing costs of imported generating equipment. In the 1980s, while tariffs for electricity in industrialized countries rose by 1.4%/yr, average tariffs in developing countries actually *decreased* by 3.5%/yr (World Bank 1990a). Figure 1.1 shows that these tariffs now are often less than half of the estimated marginal supply-costs (World Bank 1992).

Third, many governments of developing countries have adopted politically motivated rural-electrification programs that include price subsidies to provide benefits to rural populations. Such programs only exacerbated their financial problems. Most of these governments have traditionally regarded electricity as a basic service that should be provided to all. Governments have not regarded the production of electricity as a business. Rural electrification has been an integral part of development programs that provide social benefits. Governments and international development institutions based the design of these programs on the theory that economic growth would, in effect, follow the electric lines to rural communities. Recent analyses have revealed that the assumptions of economic growth as a consequence of electrification in rural areas are seriously flawed. Rural electrification does not, in and of itself, spur rural development, but rather is one of several important factors that result in acceleration of rural development (Pearce and Webb 1987, Jechoutek 1990, Schramm 1991).

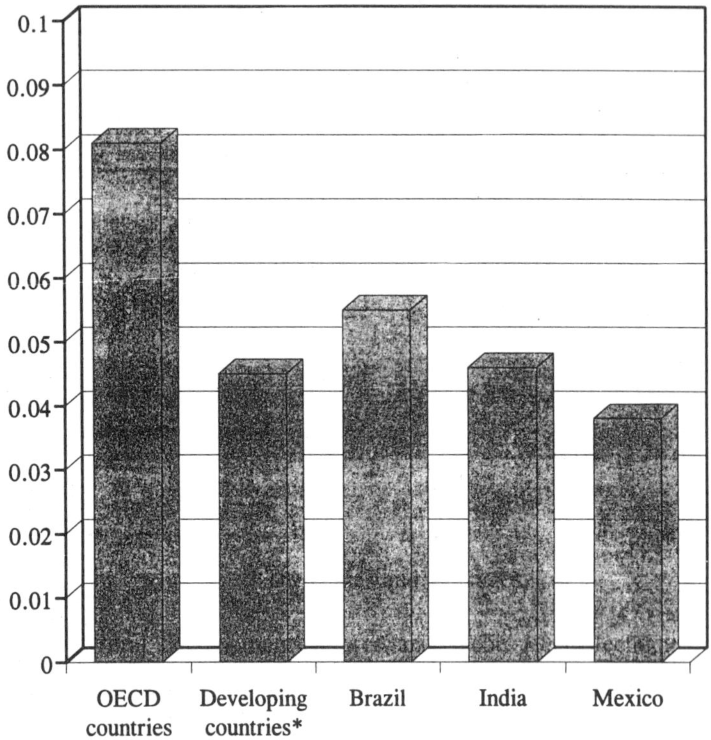

Figure 1.1. Tariffs for Electric Power Compared with Marginal Supply costs ($/kWh, 1988).
* Marginal supply costs are generally $0.10 or more per kWh. Weighted average for 63 countries.
Source: World Bank, World Development Report, 1992.

Thus, many politically motivated programs for rural electrification have actually undermined the financial integrity of power utilities. Indeed, the utilities have been forced to depend increasingly on government subsidies. From the perspective of the public utilities, each new rural household that is added to the grid is, in effect, a hole into which money—via subsidized tariffs—must be poured for many years without a commensurate increase in rural economic growth.

Finally, inefficiency in government-operated utilities is increasing, which compounds all of the problems that we have discussed. Power plants in developing countries typically have capacity factors—measurements indicating the percentage of time that power facilities are available for use—that are well below the standards for modern technologies. Figure 1.2 illustrates the results of a comprehensive survey of utilities in developing countries. The vast majority of these utilities achieved capacity factors that were below 50% (World Bank 1991a). By comparison, utilities in industrialized countries typically achieve capacity factors of 70% to 80%. To make up for this inefficiency, utilities in developing countries must, in effect, overinvest in generation capacity. Making matters worse, transmission and distribution systems in the developing world, as shown in figure 1.3, are in such a poor state that transmission losses commonly exceed 20% (USAID 1988).

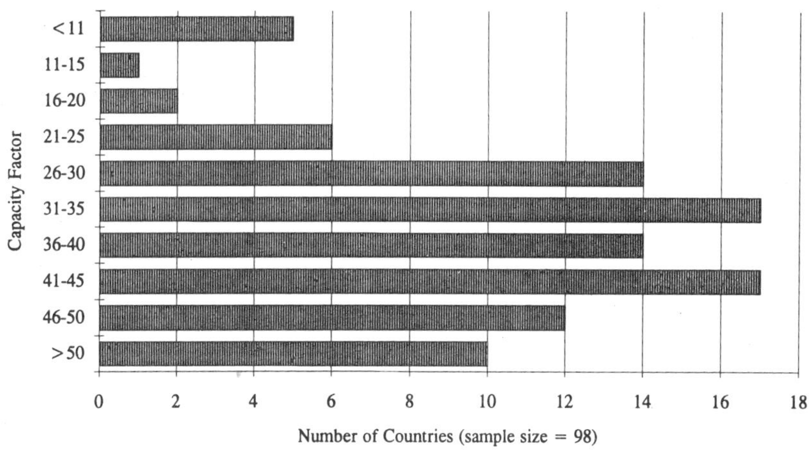

Figure 1.2. Distribution of Public Power Utility Generation Capacity Factor (%). Source: World Bank, Energy Series #40, August 1991.

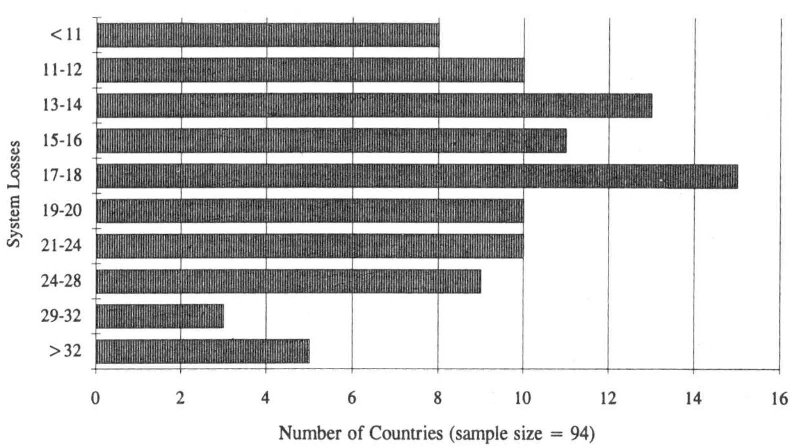

Figure 1.3. Distribution of Public Power Utility Total System Losses (%). Source: World Bank, Energy Series #40, August 1991.

Faced with unreliable electrical power service or the impossibility of affordable access to the electric grid in the near future, suppliers, distributors, and customers have turned to diesel-generator sets for power. Diesel-generators offer the advantage of low up-front capital expenditures, but operational and maintenance costs run quite high, with up to 70%/kWh spent solely for fuel (personal communication with Matthew Mendis, Alternative Energy Development, Inc., January, 1992). Many rural communities, electric cooperatives, and isolated utilities cannot continue to cover such high costs and allow their diesel units to lie idle. A recent appraisal conducted by the World Bank (1991b) of over 60 diesel power plants in developing countries showed characteristically low production levels, low revenues, high costs, and short engine-lifetimes. These results underscore the inadequacies of diesel-powered alternatives to the centralized grid. They leave customers with poor service or none at all.

Poor Prospects for a Conventional Solution

The crisis of inadequate electrical power service in developing countries will only get worse if we continue to use the conventional approach to develop the power sector. Many of the traditionally less-expensive sources of power are becoming scarcer and costlier. Large hydropower facilities provide about one-third of the power supply in developing countries, but new plants are becoming more difficult to site (World Bank 1989). The best and most feasible locations have already been developed. Resistance to new large-scale hydropower projects mounts, and the costs of resettlement and compensation demanded by affected populations become prohibitive. New oil- and coal-fired power plants cannot provide a satisfactory solution for the many countries that have no indigenous fossil-fuel reserves. Heavy reliance on imported fossil fuels exacerbates the problem of insufficient financing to operate existing power infrastructures in most developing countries.

The traditional approach to expand the power supply is the centralized generation of electricity and its transmission by costly grid extensions. Distant and widely dispersed customers cannot pay—and are not required to pay—the high costs of such service. In recent planning, utilities in developing countries could devote 31% of their total capital expenditures to expand the electric system. This represents an enormous drain on their already limited financial resources. The diminishing returns from this approach are clear. For example, utilities in the western region of the United States now pay twice as much for transmission and distribution as they do for generation. In the past, utilities considered these costs insignificant when compared with the cost savings that accrue from the economies of scale of large, central generating plants. In fact, high costs for transmission and distribution are a major obstacle to the large-scale development of localized energy resources, especially hydro, to serve distant customers (World Bank 1989). Until the costs of technology for long-distance transmission are dramatically reduced, the conventional approach to expanding the power supply—the central station and grid extension approach—will prove too costly in many regions.

Lack of financial resources that are needed to meet the growing demand for electricity presents the most serious obstacle to developing countries. Traditional sources of loans to the power sector—international lending institutions and large private banks—will probably not make funds available for public projects available on the concessionary terms that were enjoyed in the past. Lending by private banks has declined dramatically because so many utilities in developing countries have shown a poor financial performance. Funding from traditional sources—governments of developing countries, multilateral development banks, and bilateral assistance agencies—will probably provide only 40% to 60% of the financial needs of developing countries for their power sectors during the remainder of the decade, even if they take aggressive measures to conserve energy (USAID 1988). The potential gap in foreign-exchange financing may be $30 billion/yr.

THE PROMISING PROSPECTS OF RENEWABLE ENERGY

Winrock International (Tugwell 1991) forecasts that generation systems that rely on renewable sources of energy can provide as much as 15% to 20% of the new capacity of the electrical power that is required in developing countries. These facilities would have an equal or lower cost than those that use coal, oil, or nuclear energy. According to the World Bank (1990b), renewable-energy systems could add at least a capacity of 57,000 MW (megawatts), while saving governments over $156 billion, the amount needed to finance an equivalent capacity from conventional-energy systems. Yet another 15% of the costs for additional planned capacity generated from conventional sources of energy can be *avoided entirely* by the introduction of technologies and management techniques that are designed to improve energy efficiency (USAID 1988).

The most promising solutions to the deepening power crisis in developing countries may be found in renewable-energy systems. These systems are cost effective for many applications in rural and remote areas because they rely on local resources. They are well suited to development and financing by the private sector, without recourse to traditional sources of funds, and they are environmentally preferable to conventional sources of energy. Presently, however, these technologies and their applications are poorly understood and cannot be adequately evaluated using the conventional approach to utility planning that is predominant in developing countries.

The Promise of Renewable Energy

Technologies to produce energy from renewable sources already produce power in many locations at costs that are competitive with conventional, centralized thermal technologies. Compared to fossil-fuel generating systems—with their high costs for fuel, including transportation costs—renewable-energy systems are increasingly viable, particularly in rural and remote areas, if they use local resources—biomass, water, wind, sunlight—to generate power. These systems capture local energy values which would otherwise be wasted. The following discussion describes the most promising new technologies and applications for the production of electricity from renewable-energy resources—biomass, light, and wind—their comparatively minor environmental impacts, and the opportunities they offer to overcome financial barriers to energy development. Though not included in this chapter, small hydroelectric systems will continue to present an important opportunity in many countries for supplying electricity to communities that are not connected to the grid.

Biomass

Production of electrical power from biomass has grown rapidly in industrialized countries. The availability of inexpensive biomass resources—primarily residues from agriculture and forestry—has enabled many industries to cogenerate power and steam for internal consumption and to produce surplus power for the utility grid. Generation of energy from biomass in the United States has increased from nearly zero to over 9,000 MW in the past 20 years (U.S. Department of Energy 1990). The wood products industry alone contributes 77% of the total power production from biomass in the United States (Klass 1990) and provides 28% of baseload power in Maine. Hawaii's sugar industry provides 11% of the baseload power in that state. Residues from processing sugarcane—bagasse—provide 40% of the baseload power for the island of Hawaii. This high level of energy production from biomass was achieved within only a few years after oil prices rose in the 1970s (Tugwell 1991).

Figure 1.4 shows that sugarcane residues can play a potentially large role in supplying biomass energy for the developing world (Tugwell 1991). In Mauritius, a single sugar estate exported 16% of the national electricity supply in 1989 and presently provides approximately 10% of power generated for the electric grid (Hall 1991). Current production of bagasse from 80 sugar-producing countries of the developing world could produce 2,800 TWh (terawatt-hours) per year of electricity (1 terawatt = 1,000 gigawatts, and 1 gigawatt = 1,000 megawatts), an amount equivalent to 40% of the current combined demand for electricity in these countries (Hall 1991 and personal communication with Robert H. Williams, February, 1992, Center for Energy and Environmental Studies, Princeton University). To put this estimate in perspective, Indonesia generated a total of 43 TWh of electricity from all energy resources in 1988; Mexico, 110 TWh; and Brazil, 213 TWh (World Bank 1991a). Sugar factories in developing countries have the potential to produce even greater amounts of power for the electric grid by using sugarcane field wastes to supplement bagasse as boiler fuel (USAID 1991).

Adaptation of gasification and gas-turbine technology to the production of electricity from biomass promises to yield greater efficiencies in energy conversion. This improvement in efficiency would make production of electric power from biomass commercially attractive in more locations. A Brazilian utility estimates that the costs of producing power from bagasse could decrease from \$60/MWh, with current combustion technology, to \$31/MWh, with emerging technologies for gasification and gas turbines. In comparison, large hydroelectric plants operated by the utility have generation costs of \$27/MWh; plants powered by natural gas, \$50/MWh to \$60/MWh; and a plant burning imported coal, \$36/MWh to \$43/MWh (presentation by Companhia Hidro Elétrica do Sao Francisco, April, 1992). Other types of biomass residues—such as rice hulls, nut shells, and cotton-gin trash—should become more attractive as energy feedstocks when biomass gasification/gas turbine technology becomes available. Chapters by Bender, Dominguez, Wimberly et al., and Weingart and Jezek in this volume illustrate applications of biomass energy in the forest industries of the United States and developing nations.

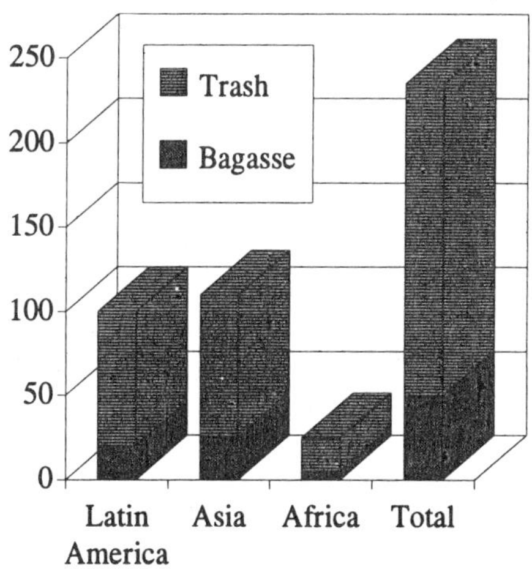

Figure 1.4. Potential for Electricity Production from Sugarcane in 1989 (TWh). Source: Winrock International, 1991.

Solar Photovoltaics

Solar photovoltaic (PV) cells are made from materials such as silicon that produce an electric current in the presence of light. PV installations now power telecommunications systems, lighting, gate controls, and electrical-switch controls throughout the world. The use of PV systems can often provide electricity at lower costs than any conventional option at remote sites. Its use is becoming economical for many applications in remote areas of developing countries.

Solar-powered water pumping, refrigeration, and lighting in individual residences and community centers are the most common applications of PV in developing countries. McNelis (1988) estimates that 2,000 solar water pumps and 800 to 1000 solar refrigerator systems had been installed in developing countries by 1988. Residential PV systems for domestic uses are becoming more popular for lighting and for operating radios, televisions, fans, and cassette players. Researchers are experimenting with larger-scale applications of PV, such as centralized battery recharging, collective water pumping, and remote, hybrid power stations. As of 1989, PV systems worldwide generated an estimated 43 MW of electricity (Foley 1990).

Utilities and governments of developing countries are beginning to evaluate the potential of PV to provide power to isolated communities. In Brazil, the German-funded Eldorado Project will provide 500 PV systems to government-owned electric utility companies in remote communities as an alternative to extending the rural grid. The U.S. Department of Energy has recently announced a cost-shared program in cooperation with government-owned electric companies to install 2,000 PV systems in the Brazilian states of Ceará and Pernambuco. As of 1991, Mexico's PRONASOL program to develop rural infrastructure had funded nearly 9,000 household PV installations (personal communication with Chris Rovero, Meridian Corporation, January, 1992). Zimbabwe, Bolivia, and India are exploring PV applications with funding assistance from international agencies.

The average price for PV panels is $5.00/Wp (Wp = peak watt, the peak or optimum output of a solar panel—at 12 o'clock, noon). Developers of solar energy concur that when the price falls to $1.00/Wp, potentially within a decade, PV technology will become economical for most applications that are powered by electricity.

Wind

Since the 1970s, some 16,000 wind turbines have been installed in the United States. Collective capacity for power from wind turbines was 1,500 MW as of 1990 (Meridian Corporation 1990). Although wind-energy projects were developed with government subsidies under laws that were favorable to the production of independent power, the technology now provides electricity at costs that are competitive with the grid. Utilities have successfully added wind power to their systems by purchasing it from independent producers. For example, in 1991 the Southern California Edison Company purchased about 5% of its peak supply from private producers who were operating wind turbines (Southern California Edison Corporation 1991).

With recent technological advances in wind turbines, several developing countries have become interested in wind-energy systems that are connected to the grid. In India, the federal government has initiated a national program for development of wind energy that includes resource assessments, research and development, and demonstrations. Several Indian states also initiated their own aggressive programs. As in the United States, India has succeeded in stimulating the development of private power by making private wind farms financially attractive. India's total installed capacity in wind energy was 37 MW in 1991, with the average facility producing 200 kW to 250 kW. Two wind farms generate 5 MW and 10 MW. Since 1986, wind turbines have delivered over 57,000 MWh of electricity to the grid. The Tata Energy Research Institute of Tamil Nadu estimates that up to 5,000 MW of wind energy could be generated in the coastal state of Gujarat and that 20,000 MW could be generated throughout India (Alternative Energy Development, Inc. 1991).

A Brazilian utility, now seeking assistance to develop two 1 MW-to-5 MW wind farms, estimates that the costs of wind generation range from $39/MWh to $84/MWh (presentation by Companhia Hidro Elétrica do Sao Francisco, April, 1992). In some cases, wind turbines can produce electricity at costs that are comparable to coal-fired power stations (Asian Development Bank 1991).

Water-pumping systems that are powered by wind have enormous potential for the developing world. Many installed systems are now operating at costs that are lower than those of diesel-powered units (Bergey 1992). By taking advantage of water storage, wind-powered systems are able to minimize the limitations of power production from intermittent energy sources. They are more cost effective (Asian Development Bank 1991).

Offsetting High Costs with Distributed Renewables

Distributed generation with renewable-energy systems has already begun to challenge the conventional approach of centralized power generation and grid extension. The economic competitiveness of the conventional approach depends on the proximity of the customer to the existing grid, the load factor, the number of connections, and the terrain through which transmission lines must pass. Analysts are beginning to take a closer look at options for electrification of rural villages, and some are now evaluating the option of incorporating distributed generation within the grid. New studies reveal that when the full costs of providing power service are included, it is often cheaper to install renewable-energy systems closer to the customer than it is to extend a transmission and distribution line from a large, centralized plant.

Cheatham (1990) conducted an economic analysis of options for electrification in North Tarawa, Kiribati. He found that in a village of 200 or less grid-connected households, PV is the least-expensive option for power supply, followed by stand-alone diesel units, and finally extension of the grid. These results reflect the region's small load and the high fixed costs of extending the grid over long distances (see figure 1.5). Extending transmission lines on the island of Kiribati can cost as little as $7,000/km, but through the highland terrain of North Tarawa, the costs average $12,000/km. According to the criteria of the local commission on electricity, the high incidence of line losses and the low demand for electricity make the extension of transmission lines that are over 5 km to 10 km financially unjustifiable.

Barnes (1988) compares the costs of extending the grid for several villages in India with those for installing distributed biogas, wind, solar thermal, and PV systems, as well as for isolated diesel-generator sets. The costs of extending the grid to supply electricity from a coal-fired thermal plant varied with village characteristics such as population density, distance from substations, and consumer needs. Because consumers were widely dispersed, the average cost of distribution from the substations to villages was three times the cost of both the generation and the transmission of electricity to the substations.

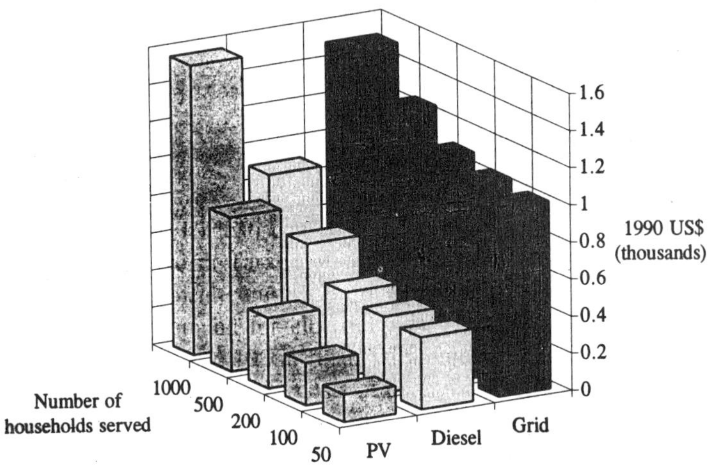

Figure 1.5. Comparison of the Present Value Costs of 3 Electrification Alternatives, North Tarawa, Kiribati.
Source: Cheatham, 1990.

This study acknowledged that the grid would provide comprehensive electrical power services, while alternative sources would provide power for only specific uses that are appropriate to a particular technology of power generation. For instance, diesel engines would provide power specifically for electric pumps, and biogas would provide power for household cooking or lighting. Bearing in mind this variation in quality of service, the study showed that electrification by extending the grid would be preferable where customers require 24-hour service, a variety of end-uses, a large number of connections per mile of electric lines, and a well-managed system load. Decentralized options appeared more attractive in cases requiring only part-time service and single-end uses, where there is a low demand for energy.

A more recent analysis (Alternative Energy Development, Inc. 1991) of Indian villages of varying sizes calculates distances from the grid at which the costs of PV installations become equal to or less than the cost of extending the grid. This study concurs with that made by Cheatham in Kiribati. Electrification with PV systems can be economical even for villages that are located close to the grid, if the villages are small. PV electrification would be the less-expensive option for a village of only 100 people as near as 3 km to 4 km to the grid. A village of 700 inhabitants must be located a minimum of 13 km to 20 km from the grid for PV to be competitive with grid extension.

Two utilities in northeastern Brazil are engaged in an on-going study in cooperation with the Pacific Gas and Electric Company (PG&E) of California and Winrock International to quantify the benefits of installing modular renewable-energy facilities in their distribution systems. Modeled on an analysis by PG&E, the Brazilian study is comparing benefits to the electric system from potential cogeneration from waste biomass at a grid-connected pulp and paper mill with those from proposed, small hydroelectric facilities that are dispersed throughout a region but currently not served by the grid. Earlier PG&E analyses of a proposed 500-kW PV installation in California identified such potential benefits as deferred investments in transmission and distribution, decreased system losses, improved reliability of service, avoided costs for energy and generation capacity, and fewer adverse environmental impacts (Shuger 1990). The Brazilians expect to find similar opportunities to minimize supply costs.

Comparative Environmental Benefits

All energy sources and technologies of power production have some adverse impacts on the environment, but research shows that the known cumulative effects from large-scale, nonrenewable-energy systems are much greater in most cases than those from small-scale, renewable-energy systems on a per-kWh basis (U.S. Department of Energy 1990). Table 1.1 presents estimates of the environmental costs of producing electricity with various renewable-and nonrenewable-energy technologies.

Power produced from biomass, like that from fossil fuels, requires direct combustion, but it results in fewer and less significant environmental impacts. Combustion of most types of biomass feedstocks emits little or no sulfur and less nitrogen oxide than burning fossil fuels (Pace University 1990). If it replaces energy from fossil fuel, biomass-energy can reduce emissions of carbon dioxide because biomass combustion results in no *net* emissions of carbon dioxide. The amount of carbon emitted during combustion is no greater than the amount of carbon sequestered from the atmosphere during the growth of the biomass feedstock. In addition, combustion of biomass does not emit methane, a far more potent greenhouse gas, which could otherwise be produced by the natural decay of biomass in landfills (Klass 1991).

Proposed plantations to produce biomass feedstocks for energy have raised concerns about potential pollution from petroleum-derived fertilizers and herbicides and pesticides, and about the intensity of water use (Pace University 1990). Large-scale biomass plantations may also decrease local biodiversity. Some analysts have suggested, however, that biomass energy plantations could rehabilitate degraded lands or serve as managed buffer zones around protected lands (Cook et al. 1990, Gowen et al. 1993). In addition, use of biomass for energy can solve waste management problems in situations where disposal of processing wastes entails either high economic or environmental costs.

Table 1.1: Estimates of Environmental Costs of Electricity Production[2]

RESOURCE	MIN	MAX
Solar PV	0	0.4
Wind	0	0.1
Biomass	0	0.7
Coal	2.8	6.8
Oil	3.0	7.9
Natural gas	0.8	1.2
Nuclear	2.91	?

Source: Pace University Center for Environmental Legal Studies, *Environmental Costs of Electricity*, 1990.

Photovoltaics are essentially environmentally benign when in operation. Adverse environmental effects of PV may result from the use of hazardous chemicals in the manufacture of system components. Although the actual risks are difficult to quantify, the manufacture of silicon cells involves considerably fewer hazardous materials than the manufacture of cells that use arsenic and cadmium compounds (Pace University 1990).

When in operation, decentralized PV systems present very low risks to users—primarily the possibilities of injury from maintenance on roofs and electrocution from improper handling. Lead-acid batteries, commonly used in conjunction with household PV systems, can pose health hazards if not handled and recycled properly.

Centralized PV installations may require 7 A to 14 A of land per MW. Stations with tracking systems would require considerably greater acreage (Pace University 1990). Planners often consider the land requirements for centralized PV installations prohibitive, but figure 1.6 illustrates that such

2 The values in Table 1 are not definitive values nor do they represent a consensus on environmental costs. These figures were selected as the most representative values from quantitative studies done over several years by numerous researchers. The methodologies and assumptions underlying these figures may not necessarily be camparable. Environmental costs for PV systems consist of only front-end costs because costs for operation and decommissioning are negligible. The costs for producing electricity from other resources include only costs for operation and decommissioning. Front-end costs for some of them are likely to be quite significant.

installations would require less acreage than coal-fired generation if acreage devoted to mining is included (U.S. Department of Energy 1989). In addition, the modularity of PV systems makes the restoration of land a relatively simple process when the installations are disassembled.

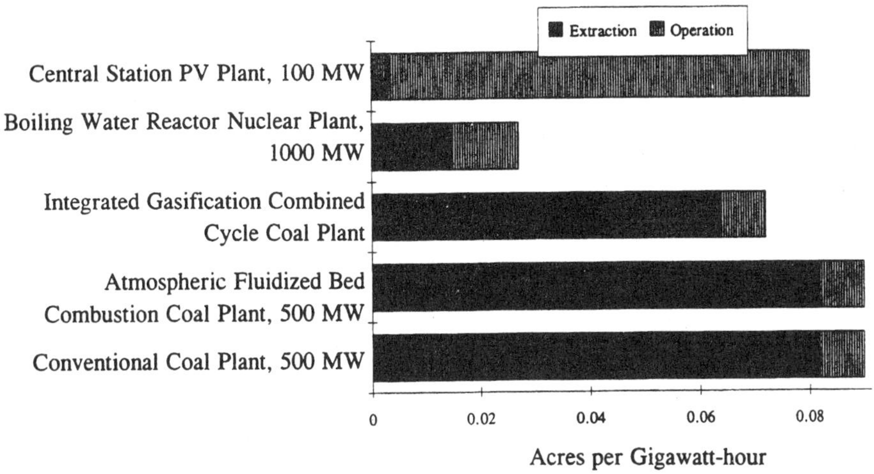

Figure 1.6. PV Land Utilization Compared with Non-renewable Generation Technologies.
Source: U.S. DOE, 1989.

Adverse environmental effects from wind-based energy systems are few and mostly avoidable. Careful site selection can mitigate the greatest potential impacts of power generation with wind turbines—conflicts of land use, noise, visual impacts, and effects on birds. Wind farms can be sited successfully on land used for grazing or crop production, though the best sites are in areas with little vegetation. Excessive noise from wind turbines may be alleviated by making technical adjustments to the equipment or by selecting more remote sites for new projects. Planners should avoid sites in migration paths or habitats of birds (Pace University 1990).

Diversifying the current choices of energy technologies with renewable-energy systems can help minimize the local environmental impacts of energy production and reduce greenhouse gas emissions. However, it is imperative that the development of renewable energy proceed with assertive measures to ensure efficiency and other sustainable development practices. Otherwise, increases in demand for electricity, fueled by population growth, will subsume the overall positive environmental impact of renewable energy, and pressure to exploit local resources will increase its negative environmental impacts. If developed in a sound and sustainable manner, systems based on renewable energy can help developing countries meet their present and future energy needs while improving natural resource management and protection of the environment.

A Solution to Financing Limitations

How can developing countries meet the rising demand for electricity when current projections indicate a financing gap from traditional lending sources of over $30 billion/yr for the remainder of the decade? Part of the answer is to turn to smaller-scale renewable energy technologies and to tap the financial resources and creativity of the private sector. Incorporating dispersed renewable-energy systems throughout the grid, particularly near the end of the transmission lines, would minimize transmission costs for utilities and help reduce financing needs for power generation and delivery expansion in the future.

In many industrialized countries, agricultural and wood processing industries have installed cogeneration plants to produce power from processing wastes for both internal consumption and the electric grid. These companies either have sufficient capital to finance such projects themselves or are sufficiently creditworthy to secure financing from local banks. Developers of independent power projects have found that wind and small-scale hydro technologies are profitable enough to attract investment. They have succeeded in building projects when they are able to obtain reasonable terms from utilities for the purchase of generated power.

Solar-energy technology can also attract private financing. Local entrepreneurs in developing countries have initiated several household PV programs in which rural users pay for installed PV systems at commercial prices. The average wholesale price for PV is around $5.00/Wp. Complete PV systems, with generation capacity ranging from 25 Wp to 48 Wp, cost as little as $400 to $650. Developers of solar energy concur that when the price falls to $1.00/Wp, potentially within a decade, PV technology will become economical for most applications that are powered by electricity. Local groups have developed financing mechanisms to make these systems affordable for many rural families. In one innovative scheme, a local group acquires seed money to establish a revolving fund that makes credit available to community members for the purchase of PV systems, and local entrepreneurs sell and service the systems. These programs have been initially successful on a small scale in the Dominican Republic, Sri Lanka, and Kenya.

Although the quality of PV service is different from grid service, PV and other stand-alone renewable-energy sources may be optimal for serving remote areas. The cost to the consumer will probably be greater than the national electricity tariffs, but the true cost will frequently be lower than the cost of extending the grid. PV users can tailor systems to meet their own requirements and pay the true cost of the power. Such investments by users may be less risky than investments by utilities in grid extension, which require subsidies to cover their cost.

In summary, if developing countries turn to technologies of smaller-scale generation from renewable resources, they will need less funding from traditional financing institutions. These technologies will attract private investment and will ultimately reduce the cost of the public power supply. For renewable-energy technologies to reach their full potential, however, utilities and private companies that seek to develop such energy projects should have access to funds on the same terms as those granted to governments by international lending institutions for larger, conventional energy projects. Conventional-energy projects are often financed on concessionary terms, but without a level playing field for financing projects, the smaller, more environmentally desirable renewable-energy systems will not reach their full economic potential.

THE CHALLENGE TO UTILITIES AND GOVERNMENTS

A New Vision of the Future for Power Utilities

A new pattern of utility growth—*the developmental utility*—will emerge as the optimal choice to expand the power supply in developing countries if utility planners begin to evaluate the benefits of small-scale, distributed renewable-energy installations throughout their national energy systems. Until now, developing countries have reproduced the utility model of industrialized countries—*the central-station utility model*—to strengthen their power sector, but this model relies heavily on presumed economies of scale that are offered by large-scale generation facilities. However, the benefits from the central station decrease as utilities must deliver power to more widely dispersed and distant customers and as the costs of transmission and distribution escalate. This is especially true for rural electrification schemes in developing countries, where poor, rural customers need—and can afford to pay for—only small amounts of power.

The alternative—*the developmental utility model*—is similar to the growth of the power system in the United States. The system was decentralized and expanded with the installation of remote and widely dispersed facilities that used a variety of fuels and technologies. These facilities were eventually linked into a central grid system. Remote and stand-alone systems provide power to areas that cannot be served by the grid in the foreseeable future. Within the grid, utilities use centralized generation for baseload power and rely on distributed generation to meet peak demand. Ultimately, utilities manage electrical power generated by central or distributed sources—both independent and utility-owned—and they control the timing of transitions from remote systems to grid-connected service.

With the *developmental utility model*, utilities in developing countries can create a diversified electrical-power system. The system begins with dispersed power generation that is necessary in areas with large rural populations. Then facilities are built closer together to form the central grid system. This model enables developing countries to provide reliable local service while building a resilient and reliable grid system. The decentralized approach allows developing countries to reap the full benefits of technologies for renewable energy which were not available to the United States during its early stages of utility development.

Integration with the Grid

Utilities in developing countries, which traditionally build large, centralized facilities, may resist exploration of new options for distributed power, even when research reveals that these options are the most cost-effective. Utility planners are experts in achieving efficiency at power plants, but they are often unfamiliar with techniques for maximizing efficiency throughout the system by integrating dispersed generation. Utility planners will need to develop and effectively employ new planning techniques that involve complex analyses of localized demand, resources, and systems characteristics. Governments and utilities in developing countries will need to develop legal and regulatory frameworks that enable utilities to purchase power from independent producers.

Current planning of utilities does not include an evaluation of the costs and benefits of distributed generation. For example, it is standard practice in assessing the costs of service to use aggregated, rather than localized, calculations of the cost of transmission and distribution. As a result, the true costs of transporting power to a specific point are hidden. This method supports the decision to charge flat rates rather than variable rates, which fluctuate with the true costs at the point of service.

Calculating the benefits of distributed generation in a highly interlinked, or webbed, grid system can be complex. However, planning for distributed generation would be more straightforward in developing countries because most of their grid systems are less densely webbed and more radial that those of industrialized countries. Benefits to transmission and distribution systems could be quantified easily for any given transmission line by measuring changes in peak load, voltage, line losses, and operation of electrical-system devices.

In the densely developed grid systems of northern California, micromanagement of individual, distributed power plants to ensure reliability has become unnecessary. There are so many independent producers of electrical power on line in the region, each contributing significant efficiency benefits to the entire system, that the impact on the reliability of the service for the whole system is negligible when any one producer experiences downtime. The systems are so resilient, because of diversification and numerous interlinkages, that only outages at large, central stations pose any real threat to maintaining the high reliability of service.

Combining Private Sector and Utility Initiatives

As utilities begin to modify their planning methods, independent power producers will play a crucial role in introducing technologies for renewable energy in developing countries. Opening the power sector to private enterprise will foster new ideas and resources for developing power projects. For instance, sugar and wood products industries could propose to sell surplus power to the grid when they are adding or upgrading cogeneration plants. A fair price for purchased power will give these and other industries the incentive to produce a new product—electricity—from a variety of energy resources that would otherwise not be available to utilities.

Independent production of power will require changes in laws governing its production in many developing countries. Self-generation is still prohibited in many countries, and power generation for sale to the public is restricted to government-owned institutions and electric companies. In such countries, utilities must lead the effort to change the existing legal and regulatory framework that governs the production of power and to establish a fair pricing-structure for the purchase of power from independent producers.

Finally, as discussed earlier, if private developers are to build new power plants and add capacity to the electric grid, they must gain access to financing at reasonable rates. Renewable-energy projects in developing countries are driven by technology. That is, companies that have the technology form partnerships with governments to finance pilot projects that are designed both to test the applications of the technology and to support rural development. Local developers with small-scale ventures often do not have the resources to find interested investors or to secure low-interest loans from private banks. Governments, international financing agencies, and private banks need to assist these small entrepreneurs by creating financing schemes that are affordable, appropriate, and easily accessible to small-scale ventures.

CONCLUSION

Utilities in developing countries are at a crossroads, where development of conventional electrical power and new alternatives based on renewable energy meet. Many developing countries still have simple grid systems, so they have greater flexibility to expand their power service by incorporating distributed generation. Governments of developing countries, international donors, and utilities must now become leaders in diversifying the power sector by drawing upon the resources of independent developers and alternative energy systems. The more farsighted utility planners have already begun to assemble the tools that are necessary to design and manage such diverse energy systems.

Literature Cited

Alternative Energy Development, Inc. 1991. (Vols. 1 & 2). *India nonconventional energy projects for global environment facility funding*. Silver Spring: Alternative Energy Development, Inc.

Asian Development Bank. 1991. (Vols. 1 & 2). *Environmental considerations in energy development*.

Barnes, Douglas. 1988. *Electric power for rural growth: how electricity affects rural life in developing countries*. Boulder: Westview Press, Inc.

Bergey, Michael L.S. 1992. Sustainable community water supply: a case study from Morocco. In: *Fueling development: energy technologies for developing countries*, OTA-E-516. (U.S. Congress, Office of Technology Assessment). Presented at the American Wind Energy Association National Conference, September, 1990. Washington, D.C.: Government Printing Office.

Cheatham, Chris. 1990. A brief comparison of grid extension, stand-alone diesel generators, and photovoltaic for rural electrification. In: *Proceedings from the Pacific Household and Rural Energy Seminar*, UNDP/World Bank, November 5-9, 1990, Port Vilu, Vanuatu. 2:389–401.

Cook, J.H., J. Beyea, and K.H. Keeler. 1990. Potential impacts of biomass production in the United States on biological diversity. *Annual review of energy and the environment* 16:401–431.

Foley, Gerald. (ed.). 1990. *Electricity for rural people*. London: Panos Publications Ltd.

Gowen, M.M., W.R. Bentley, and E. Stijfhoorn. 1994. Tropical forest management and wood-based biomass energy as development assets. In: *Forest resources and wood-based biomass energy as rural development assets*. (W.R. Bentley and M.M. Gowen, eds.). New Delhi: Oxford & IBH Publishing Company Pvt. Ltd.

Hall, D.O. 1991. Biomass energy. *Energy policy* 19(8):711–37.

Jechoutek, Karl. 1990. The economics of rural electrification. Paper presented at the Rural Electrification Planning Workshop, Manila. February, 1990.

Klass, D.L. 1990. The U.S. biofuels industry. Proceedings of the International Renewable Energy Conference, Honolulu. In: *The potential of renewable energy: an interdisciplinary White Paper*, SERI/TP-260-3674. (U.S. Department of Energy). Golden: Solar Energy Research Institute.

McNelis, B. 1988. Solar electricity for development: an overview. Proceedings of Conference on Solar Energy for Development, UK-ISES, King's College, London. In: *Electricity for rural people*. (G. Foley, ed.). London: Panos Publications Ltd.

Meridian Corporation. 1990. Characterization of U.S. energy resources and reserves. In *Environment costs of electricity*. (Center for Environmental Legal Studies, Pace University). New York: Oceana Publications, Inc. 33:436.

Pace University, Center for Environmental Legal Studies. (ed.). 1990. *Environmental costs of electricity*. New York: Oceana Publications, Inc.

Pearce, D., and M. Webb. 1987. Rural electrification in developing countries—a reappraisal. *Energy policy*.

Sanghvi, Arun J. 1991. Power shortages in developing countries: impacts and policy implications. *Energy Policy* 19(5):433–434.

Schramm, Gunter. 1991. Rural electrification in LDCs as a tool for economic development: facts and fiction. Paper presented at the 30th Annual Meeting of the Western Regional Science Association, February 24-28, 1991, Monterey, California.

Southern California Edison Corporation. 1991. 1991 Annual Report. Rosemead: Southern California Edison Company.

Shugar, Daniel. 1990. Photovoltaics in the distribution system: the evaluation of system and distributed benefits. Pacific Gas & Electric Company paper presented at the Twenty-first IEEE Photovoltaic Specialists Conference. May 22, 1990, Kissimmee, Florida.

Tugwell, Franklin. 1991. *Prospects and problems of biomass energy development in the third world.* Proceedings from the 1990 Conference on Biomass for Utility Applications at the Electric Power Research Institute. Arlington: Winrock International Institute for Agricultural Development.

USAID. 1988. Power shortages in developing countries: magnitude, impacts, solutions, and the role of the private sector. Washington, D.C.: USAID.

USAID. 1991. Baling sugarcane tops and leaves: the Tai experience. Prepared by Winrock International, Biomass Systems and Technology Project (DHR-5737-A-00-9058-00) for USAID Office of Energy and Infrastructure and USAID/Thailand.

U.S. Department of Energy. 1989. Energy systems emissions and material requirements.

U.S. Department of Energy. 1990. The potential of renewable energy: an interdisciplinary White Paper, SERI/TP-260-3674. Golden: Solar Energy Research Institute.

World Bank. 1989. *The future role of hydropower in developing countries.* Industry and Energy Department Working Paper. Energy Series #15.

World Bank. 1990a. *Review of electricity tariffs in developing countries during the 1980s.* Industry and Energy Working Paper. Energy Series #32.

World Bank. 1990b. *Capital expenditures for electric power in the developing countries in the 1990s.* Industry and Energy Department Working Paper. Energy Series #21.

World Bank. 1991a. *Summary 1988 power data sheets for 100 developing countries.* Industry and Energy Working Paper. Energy Series #40.

World Bank. 1991b. *Core report of the electric power utility efficiency improvement Study.* Industry and Energy Department Working Paper. Energy Series #46.

World Bank. 1992. *World Development Report: 1992.*

CHAPTER 2

TROPICAL FOREST MANAGEMENT AND WOOD-BASED BIOMASS ENERGY AS DEVELOPMENT ASSETS

Marcia M. Gowen
William R. Bentley
Eirik Stijfhoorn

INTRODUCTION

Harvesting tropical forests for timber and secondary products, such as wood-based biomass energy, can be an important element in strategies for rural development. We must employ schemes for the sustainable management of resources, however, or the benefits will be short-lived, and serious ecological and social consequences will result. With some wisdom and care, use of wood residues at processing plants to produce energy may positively benefit rural poor people in tropical nations. Benefits may include increased value-added to the raw materials, a more stable wood-processing industry, and a cheaper source of rural electrification. We may also alleviate some of the environmental problems caused by using fossil fuel at the processing plants, while wood residues are dumped in landfills or burned in the open air.

In this chapter, we will focus on the potential for using the residues from wood processing to produce energy in developing countries. Production of energy from wood by the forest products industry, however, is only one component in a broader scheme of forest management. We need to make choices about the use of the forest—both timber and nontimber uses. We need to ensure the participation of local people and the preservation of a sustainable resource base for future generations. We need to evaluate the environmental impacts of our activities. The actual physical potential to produce wood-based energy depends on the sustainability of the forest products industry, alternative options for the use of the forest and wood residues, and the impact that these two factors have on the rural poor in each country.

We will first discuss sustainable forest management and the role of wood-based energy in rural development. Second, we will review the status of forests in developing countries, paying attention to areas that affect the management of tropical forests. Next, we will analyze the patterns of forest use, the production of timber and wood products, and the potential for wood-based energy in Asia,

Latin America, and Africa. Finally, we will place the importance of this potential for rural areas in the context of the countries themselves, highlighting the opportunities and necessary conditions that would allow energy from wood to be a viable commercial option in developing countries.

SUSTAINABLE DEVELOPMENT AND FOREST MANAGEMENT

Rural Development and Forest Use

Rural development is a complex process, and our understanding of potential roles for forest resources within this process is still evolving (Laarman and Sedjo 1991). The "lessons learned" from small-scale irrigation, social forestry, and other strategies for resource development in rural areas suggest that the rural poor must be integrated into all stages of the development process—the planning, the decision-making, and the implementation—not just as target beneficiaries, but as participants (Levine et al. 1986). Increasing the productivity of rural resources on a sustainable basis requires that immediate and well-perceived benefits must accrue to the rural people who are investing their time and meager capital to improve and protect forests and forest-related resources.

The criteria for effective rural development, we and others argue (e.g. Ascher and Healy 1990, Miller et al. 1993), should be based on three general goals:

- **Increased productivity:** The *productivity* of all its resources—its people, forests, soils, and water determines the current income or wealth of a society. Using wood residues for energy makes forest resources more directly productive since each unit of standing timber or harvested logs brings in more net cash.
- **Improved sustainability:** *Sustainability* is both an ecological and an economic concept. It is the relative productivity of natural resources over a long period of time. With well-defined property rights and incentive structures, such as market-price mechanisms, wood-based biomass energy can shift the use of forest resources toward the future while improving current productivity. However, without such safeguards, energy from wood could increase the rate of deforestation.
- **Distribution of benefits:** Effective *distribution* or *equity* requires a process to establish a sense of fairness. Local participation in the planning and implementation of rural development is both an end in itself and a necessary means to broaden social and economic development by ensuring fairness in distribution of new benefits. Wood-based energy, by substituting for or supplementing fossil-

fuel systems, typically lowers the costs of rural electricity and, consequently, improves the livelihoods of the rural poor.

All three goals of rural development allow us to perceive forests as assets. Ciriacy-Wantrup (1963) defined *conservation* as shifting the use of resources towards the future. Our goals stimulate us to look for links between the welfare of today's rural poor and the options available for future generations.

Sustainable management relies on the traditional goal of forest management —*sustained yield*, a concept of soil and forest management developed in feudal Germany. *Sustained yield* is a cultural imperative ensuring that the land passed from one generation to the next will be of equal or better productivity than when it was inherited or received. In the language of forestry, *sustained yield* means that harvest levels cannot exceed forest growth over successive cutting periods (Greeley 1950). In a very real sense, sustainable management is an intergenerational transfer of wealth, where one generation leaves the state of the forest at least the same, if not improved, for the future. In practice, however, sustainable management of natural tropical forests has been elusive. The few known successes are often difficult to replicate in different agro-ecological zones or sociocultural settings (Poore et al. 1989, Bushbacher 1990, World Bank 1991).

Deforestation in Developing Countries

Our global supply of forest resources is vanishing as the twentieth century comes to a close. Historically, rapid population and economic growth has caused sporadic timber scarcity. New sources of wood for commercial development were available with a modest extension of the transportation network. However, the ecological costs—e.g., nutrient degradation, species loss, aesthetic impacts—of converting timberlands to other uses were not included in accounting procedures for development. After millennia of cutting a seemingly inexhaustible global stock of forest resources, clear signals indicate that supplies are now limited (Poore et al. 1989, Repetto and Gillis 1988, Goodland 1990, Sutton 1993, World Bank 1991, WRI 1992). Conservationists are apprehensive about the current rates of deforestation. They correctly link conservation of forests with global climatic change, biological diversity, and complex issues of resource management (Saterson 1990, World Bank 1991). Regardless of their specific viewpoints, virtually all informed observers agree that the current rates of harvest and land conversion are not sustainable in many countries, even in the near-term.

The consensus is that deforestation is occurring widely, particularly in tropical developing countries. Tropical deforestation is now a global issue. Supplies of forest resources from temperate countries are decreasing as old-growth stands are harvested or transferred to parks and other biological reserves. Coupled with increasing demand, this trend creates pressure to harvest more timber from the tropics, affecting the material well-being of resource-poor families in developing nations. Surprisingly, tropical nations, which supply only 13% of the world's current consumption of industrial roundwood (FAO 1991a), have few opportunities to increase harvests significantly.

Actual rates of degradation and deforestation are specific to a country or forest ecotype. Estimating rates of deforestation has been fraught with controversy and difficulty. The FAO (1988) estimated forestlands of developing countries at the end of 1980 using national data. To summarize by region:

	Natural Closed Forest	Plantations	Open Broad-leaved Forest	Shrubs and Fallow Forest	TOTAL
			000 ha/yr		
Asia	317,717	19,690	51,275	166,878	555,560
Latin America	687,509	6,177	207,242	346,555	1,247,483
Africa	217,007	2,983	414,791	589,481	1,224,262

The estimates by country are in table 2.1 (Asia), table 2.2 (Latin America), and table 2.3 (Africa) and are probably biased low because of a variety of reporting errors. Figure 2.1 shows estimates for annual rates of deforestation in tropical developing countries from 1981 to 1985. These rates vary from 5.7 million ha/yr (hectares per year) in Latin America to 3.7 million ha/yr in Africa, and 2 million ha/yr in Asia.

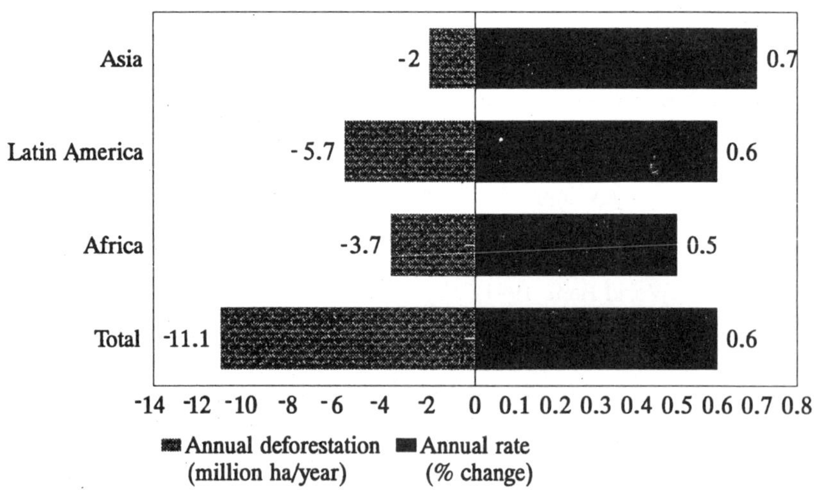

Figure 2.1. Estimated Annual Deforestation Rates of Forest Area in Tropical Developing Countries: 1981-85.
Source: FAO (1988).

Table 2.1. Forest Resource Base in A.I.D.-assisted and non-A.I.D. Countries (000 ha) at end of 1980

ASIA					Natural closed, mature forest (a)			
	Untouched since 1930	Logged since 1930	Managed forests	Parks, protected & unproductive forests	Planta-tions (b)	Open broad-leaved forest (c)	Shrubs and fallow forest (d)	TOTAL
Afghanistan		150	100	485	11	400	690	1836
Bangladesh*	45	10	795	77	128		315	1370
Bahrain								0
Bhutan	1200	625		275	8	40	230	2378
Brunei	270	17		36			237	560
Burma*	14195	5618	3419	8077	15		20700	52024
Cambodia	4614	514		2040	6	5100	625	12899
China				8552	12733	17200	27730	66215
Cyprus								0
Fiji*	176	61		574	40		6	857
French Polynesia							115	115
Hong Kong							13	13
India*	5432	4156	31917	8896	2068	5393	14848	72710
Indonesia*	38915	34780	40	40160	1918	3000	41260	160073
Iran			400	1640	43	1000	14250	17333
Iraq		60			20	1160	300	1540
Jordan					21	50	75	146
Korea, D.P.R.				1800	1628		4200	7628
Korea, Rep.								0
Kuwait								0
Laos	2980			4830	11	5215	5735	18771
Lebanon*					18	20	45	83
Macao								0
Malaysia	7529	5524	2499	5443	26		4825	25846
Mongolia							4335	4335
Nepal*	885	385		670	18	180	340	2478
New Caledonia	288	39		157	7	210	630	1331
New Guinea Papua*	13850	235		20143	22	3945	1530	39725
Oman								0
Pakistan*	425	220	410	1130	160	295	1105	3745
Philippines*	3000	3890		2620	300		3520	13330
Qatar								0
Samoa	99			41	1		31	172
Saudi Arabia				30	1	170	1400	1601
Singapore							3	3
Solomon Isl.	169	85		2169	17	17	40	2497
Sri Lanka*	13	1213		433	112		1068	2839
Syria		26	34		40	90	239	429
Thailand*	4080			4255	114	6440	1300	16189
Tonga							8	8
Turkey								0
Viet Nam	1600	2180		3790	204	1340	11080	20194
Vanuatu	90	3		134				227
Yemen						10	4050	4060
TOTAL	99855	59791	39614	118457	19690	51275	166878	555560

Table 2.2. Forest Resource Base in A.I.D.-assisted and non-A.I.D. Countries (000 ha) at end of 1980

LATIN AMERICA					Natural closed, mature forest (a)			
	Untouched since 1930	Logged since 1930	Managed forests	Parks, protected & unproductive forests	Planta-tions (b)	Open broad-leaved forest (c)	Shrubs and fallow forest (d)	TOTAL
Argentina		1200		11200	600		16500	29500
Bahamas								0
Barbados							5	5
Belize*		891		463	3	92	574	2023
Bolivia*	17760	1290		14160	26	22750	12050	68036
Brazil*	288910	12000		56570	3855	157000	161820	680155
Chile*		6705		845	817		8550	16917
Colombia*	38600	900		6900	95	5300	14400	66195
Costa Rica*	330	770		538	2	160	120	1920
Cuba		795	200	460	157		1005	2617
Dominica	39	137		34			14	224
Dominican Rep.*	242	331		56	6		321	956
Ecuador*	10805	110		3335	43	480	3470	18243
El Salvador*		25		116	1		315	457
French Guiana	6552	132		1148		70	85	7987
Guadeloupe				74	4			78
Guatemala*	1190	1822		1430	15	100	1865	6422
Guyana	12120	1345		5010	1	220	315	19011
Haiti*	9	16		23	1		96	145
Honduras*	815	2165	58	759		200	1900	5897
Jamaica*	41	4		22	12		1	80
Martinique								0
Mexico*	12780	11520		21950	159	2100	85500	134009
Neth. Antilles							7	7
Nicaragua*	3600	226	250	420	1		1580	6077
Panama*	2125	818		1222	3		124	4292
Paraguay*	560	2480		1030	3	15640	12730	32443
Peru*	37500	6000		26180	84	960	8660	79384
Surinam	12075	420		2333	8	170	295	15301
Trin. & Tobago		134	14	60	16		63	287
Uruguay*		70		420	140		120	750
Venezuela	7600	11610		12660	125	2000	14070	48065
TOTAL	453653	63916	522	169418	6177	207242	346555	1247483

Table 2.3. Forest Resource Base in A.I.D.-assisted and non-A.I.D. Countries (000 ha) at end of 1980

AFRICA					Natural closed, mature forest (a)			
	Untouched since 1930	Logged since 1930	Managed forests	Parks, protected & unproductive forests	Plantations (b)	Open broad-leaved forest (c)	Shrubs and fallow forest (d)	TOTAL
Algeria		772		746	431	249	2168	4366
Angola		2450		450	157		28400	31457
Benin	14	33			19	3820	6832	10718
Botswana						32560	20000	52560
Burkina Faso*		271			12	4464	9360	14107
Burundi*		6		9	19	14	24	72
Cameroon*	7000	9940		980	18	7700	15600	41238
Cape Verde					4		1	5
Cent. Afr. Rep.	3120	350		120		32300	21100	56990
Chad*	500				3	13000	10550	24053
Congo	10330	3360		7650	17		2500	23857
Cote d'Ivoire	200	3094	1	1163	45	5376	15390	25269
Djibouti		1		1		68	44	114
Equ. Guinea	785	230		280			1175	2470
Egypt*					40			40
Ethiopia*	650	300		2600	98	22800	35300	61748
Gabon	10655	9250		595	19	75	1500	22094
Gambia*		5		60	1	150	560	776
Ghana*		154	1167	397	75	6975	9480	18248
Guinea*	1150	300		600	3	8600	9900	20553
Guinea-Bissau*	355	70		235		1445	577	2682
Kenya*	240	180	70	450	181	1255	38105	40481
Lesotho					3		16	19
Liberia*	905	425		670	7	40	5640	7687
Libya				134	143	56	446	779
Madagascar*	1600	5070		1200	266	2900	7500	18536
Malawi*		40		146	80	4085	380	4731
Mali*					5	6750	15100	21855
Mauritania				29		525	3980	4534
Mauritius				3	11		32	46
Morocco*		655	421	457	322	1703	1161	4719
Mozambique*	70	380		485	24	14500	42700	58159
Niger		100			9	2450	7880	10439
Nigeria	380	2590		2980	163	8800	49450	64363
Reunion	6		2	69	8		42	127
Rwanda*		56		45	29	110	155	395
Sao Tome & Principe	15	18		23				56
Senegal*	14			206	13	10825	3115	14173
Sierra Leone		219		521	6	1315	4278	6339
Somalia*		90		1450	11	7510	53050	62111
Sudan*		295	50	305	189	47000	98600	146439
Swaziland*				4	101	70		175
Tanzania*	250	580		610	98	40600	17900	60038
Togo*	47	206		51	11	1380	3720	5415
Tunisia*			163	23	127	111		424
Uganda*	100	125	440	85	45	5250	1700	7745

Zaire*	79740	380		25530	22	71840	29700	207212
Zambia*	345	1970	5	690	38	26500	10800	40348
Zimbabwe*				200	110	19620	3570	23500
TOTAL	118471	43965	2319	52252	2983	414791	589481	1224262

The (*) indicates A.I.D. assisted countries.
(a) Closed forests: broadleaved with > 40% crown coverage and coniferous with > 50% crown coverage.
(b) Industrial and nonindustrial hardwood and softwood species.
(c) Open broadleaved forest: 10-40% crown coverage.
(d) Shrubs: 0.5-7m in height; forest fallows: vegetation deriving from clearing of forest land for shifting cultivation
Source: FAO (1988).

Table 2.4. Preliminary Estimates of Forest Area (000 ha) and Rate of Deforestation for Major Tropical Regions

Subregion	Number of countries studied	Total land area (a)	Forest area 1980 (b)	Forest area 1990 (b)	Area deforested annually 1980-90 (b)	Rate of change 1980-90, percent per annum (c)
West Sahelian Africa	8	528000	41900	38000	390	-0.9
East Sahelian Africa	6	489600	92300	85300	700	-0.8
West Africa	8	203200	55200	43400	1180	-2.1
Central Africa	7	406400	230100	215400	1470	-0.6
Tropical Southern Africa	10	557900	217700	206300	1140	-0.5
Insular Africa	1	58200	13200	11700	150	-1.2
Total Africa	40	2243300	650400	600100	5030	-0.8
South Asia	6	445600	70600	66200	440	-0.6
Continental South East Asia	5	192900	83200	69700	1350	-1.62
Insular South East Asia	4	258100	157000	138900	1810	-1.2
Total Asia	15	896600	310800	274800	3600	-1.2
Central America & Mexico	7	245300	77000	63500	1350	-1.8
Caribbean	18	69500	48800	47100	170	-0.4
Tropical South America	7	1360800	797100	729300	6780	-0.9
Total Latin America	32	1675600	922900	839900	8300	-0.9
TOTAL	87	4815500	1884100	1714800	16930	-0.9

(a) Figures may not tally due to rounding.
(b) Forest are defined as vegetal formations with a minimum of 10% crown cover of trees (minimum height 5 m, closed and open forest) and/or bamboos.
(c) Deforestation refers to the change of land use or depletion of crown cover to less than 10%.
Source: FAO (1991b, 1991c).

The latest provisional FAO (1991b, 1991c) estimates of tropical-forest cover and deforestation are based on satellite imagery. As a consequence, they are more accurate, but the data are broken down by continental regions, not countries (Singh 1991). The data do not cover all tropical areas, however, and cannot distinguish between harvested areas that will support growth of secondary forests and those areas that have been converted permanently to nonforest uses. Table 2.4 shows that from 1980 to 1990, the highest rate of deforestation, 1.2%, is observed in Asia, with about 3.6 million ha removed each year. Africa loses more forest area (5 million ha/yr), but the deforestation rate is lower, at 0.8%. Latin America has the highest loss with 8.3 million ha and a 0.9% loss rate each year. Overall, the FAO estimate is for 16.9 million ha/yr, a 0.9% annual deforestation rate. This rate is up from the 9.2 million ha/yr estimated by FAO/UNEP for the period from 1976 to 1980 period (FAO 1990), but differences in the database and procedures suggest some underestimation in 1980 and some overestimation in 1990. Other recent data suggest that tropical deforestation is almost double the rate estimated in the early 1980s—about 17 million ha/yr to 20 million ha/yr (World Bank 1991).

To summarize the FAO (1988, 1991b, 1991c) estimates:

Estimated Annual Deforestation for Tropical Developing Countries:

	1981-85		1980-90	
	percent	ha/yr (000)	percent	ha/yr (000)
Asia	0.70%	1,988	1.20%	3,600
Latin America	0.60%	5,650	0.90%	8,300
Africa	0.50%	3,714	0.80%	5,000
TOTAL	0.60%	11,352	0.90%	16,900

The increase in regional and global rates reflects both more rapid deforestation and better estimates based on satellite imagery.

When examined on a per-country basis, the average annual rates of deforestation from 1981 to 1985 show dramatic losses of forestland in several key countries that produce timber and wood products, such as Brazil, Colombia, Indonesia, Mexico, Côte d'Ivoire, and the Sudan (figure 2.2). Several countries, most notably Brazil, India, and Indonesia, began intensive reforestation efforts before the end of 1980 (figure 2.3). However, these programs do not appear to be regenerating the areas of harvested forests.

Rapid population growth, the need for agricultural expansion, ill-defined or nonexistent property rights, and distorted economic incentives fuel deforestation in most developing countries. While the primary factors for deforestation are different in various regions, the resulting loss of forests and degradation of the land from inappropriate uses of marginal lands have similar consequences. Secondary destruction from poor logging practices and clearcutting for change in land use, coupled with migration onto the land, leads to soil degradation that is difficult to reverse (Poore et al. 1989). In the early 1980s, it was estimated that poor harvesting techniques destroyed 30% to 50% of the residual stands, while, typically, only 10% to 20% of the mature stems were removed for commercial use (Repetto and Gillis 1988). Often little, if any, commercial-timber value was realized from the conversion process (Bushbacher 1990). Lack of clear title or property rights is a fundamental reason for deforestation and conversion of the land to agriculture. Ill-defined property rights or common-access rights on forestlands provide incentives to harvest now and convert to uses where rights are better defined. Many times, the ability of concessionaires to supersede rights of local use creates rapid deforestation or removal of only the highly valued stems. In many cases, no one—public forestry agencies, industry, but especially the rural poor—has incentives to invest in long-term forest management and protection.

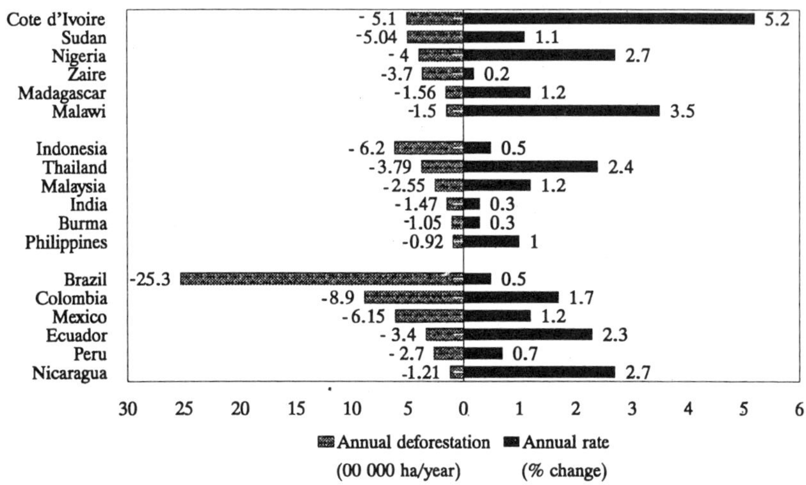

Figure 2.2. Deforestation in Forest Area in Selected Developing Countries: 1981-85.
Source: FAO (1985).

Logging is a major cause of deforestation in Asia, but burning the land to convert it for agricultural use is the primary cause of deforestation in Latin America (Poore et al. 1989). Removal of timber per hectare by loggers in Africa and Latin America is quite small. Leonard (1987) estimated that clearcutting solely for agricultural use in Honduras resulted in an annual $320 million loss of commercial timber. Early in the century, Africa had thriving forest-export industries, especially in Nigeria, but rates of harvest declined after a period of rapid exploitation (Poore et al. 1989). At this point, we need new and innovative forest-policy and management strategies to rehabilitate degraded forests, pastures, and marginal croplands throughout the tropics as well as to protect the remaining forest reserves in Latin America and Asia.

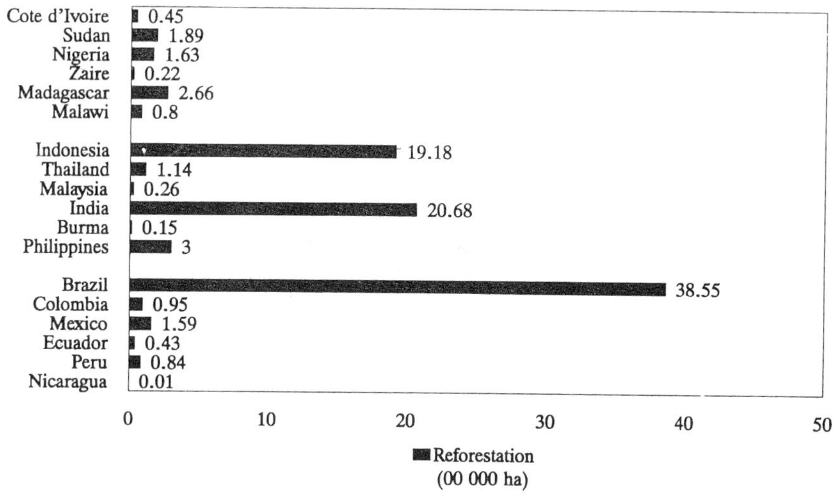

Figure 2.3. Reforestation (plantations) in Selected Developing Countries at the end of 1980.
Source: FAO (1988).

Forest Uses and Sustainable Management

A forest provides a diversity of important monetary and nonmonetary benefits to both a community and a nation (figure 2.4). Wood-based energy from the residues of forest products industries is one option in a variety of possible forest-based income or benefit streams. Use of on-site wood and bark residues as feedstock for energy production can be another step in the complete utilization of harvested timber.

Adding production of biomass energy to the mix of possible uses for timber could encourage deforestation if higher values of timber encourage early harvesting rather than sustainable forest management. Increased value-added to timber leads to higher values of forest resources. If these higher values cannot be protected for a future harvest through assured tenure rights, they simply will encourage yet more rapid exploitation. On the other hand, if values are protected through enforceable property rights and fair market-prices of public timber, higher values will encourage increased investments. Delaying harvest of current timber inventories and adding managerial inputs will increase timber growth and quality. The very act of not harvesting trees today is a decision to invest in the future.

Forest resources and their management, however, are more complex than the simple production of highly valued trees. Food, fodder, medicinals, nuts, and various other extractive products that we often categorize as minor, or nontimber, forest products are often equally, if not more, important to rural communities (Myers 1988). In cases like India, values of nontimber products are considerably greater than values of processed industrial roundwood (Gupta 1983). Peters et al. (1989) and others argue that in some tropical settings, indigenous forest dwellers can earn more by managing natural forests for the values of nontimber products than for the timber harvest itself.

Figure 2.4. Benefits of Forested Ecosystems

Worldwide, forest systems provide such services as aesthetic landscapes, protection of wildlife habitats, biodiversity, protection of the watershed, carbon sequestering, nutrient recycling, and balancing of the global climate. Environmental amenities associated with many types of recreation are probably more valuable than harvested products. The linkages that enable these services to exist, however, have been generally undervalued or simply not recognized by markets and political mechanisms for the allocation of resources. Some argue that loss of carbon storage, increase of CO_2, soil and water runoff, and, especially, loss of genetic materials will result in irreversible negative impacts on the environment (World Bank 1991). Maintenance of synergistic resources or values, such as a basic set of biological, geological, and nutrient cycles, is critical to sustainable forest management (Gowen 1989).

Trends in Forest Management

The short-term picture suggests more exploitation of forests and soil degradation. Timber prices will increase sharply, and so will scarcities of fuelwood. Major political battles over issues concerning biological reserves will occur (Sutton 1993). The consequences for the near future will fall especially hard on the rural poor, who are most dependent on the forest for its value as an open-access area for fuelwood and fodder, and the urban poor, who will pay higher prices for fuelwood and construction materials.

Participation of local people is often advocated as the key to sustainable management (e.g., Levine et al. 1986, Lai and Kahn 1992). Their involvement in developing, implementing, and directly benefiting from plans to manage the local forest will determine success (Wells and Brandon 1992). Without adequate incomes or livelihoods, rural people turn for survival to open-access forestlands, lands that are often fragile and marginal for crops and livestock (Leonard 1987). A new pattern of land use is emerging in many tropical nations. Areas of biological reserves, commercial-timber production, and a broad array of agroforestry practices are located next to each other. Forest preserves must serve the purposes of conservation, but they must also provide livelihoods for the local people through the management of adjacent buffer zones. However, a recent report by Wells and Brandon (1992) suggests that management problems, especially in buffer zones of public forests, are more difficult than widely assumed by advocates of participation.

Possibilities for the long-term include higher production from both natural and plantation forests under sustainable management. These possibilities include more value-added to timber from secondary products—such as energy—and increased market-values, subsidies, or other cash flows from maintaining reserves for biological diversity, recreation, or other services. New opportunities for employment can develop in harvesting, processing, and transportation, especially if rehabilitation of degraded lands contributes to major new sources of timber. Agroforestry and plantation technologies could produce significant new amounts of wood while moving these lands toward sustainable systems. Brazil's experiment with fast-growing species for energy production stands out as an example of intensive management that could reduce pressure on natural forests in the future. If appropriate social and tenurial arrangements can be designed, from individual or village ownership to cooperative agreements with public forestry authorities, the new benefits and incentive structures could favor the poor.

PRODUCTION OF WOOD-BASED ENERGY

Rural Benefits

Use of on-site residues from the forest products industry, "wood wastes," can be an important asset for rural development, provided that the industry practices sustained-yield forestry. Energy from wood provides not only electrical power, but critical income and many environmental benefits. First, the use of wood residues for the production of energy increases the value-added of the primary product to the industry and to the producers of the raw materials. This diversification of the industry into the production of energy results in increased income and jobs. Additionally, since wood generally replaces fossil fuel as a source of energy in rural areas, there would be national and regional benefits, such as savings in foreign exchange and a more reliable, cheaper source of electricity.

The production of energy from wood, if designed well, can benefit both the local and global environment. If forest products industries could use their wood residues for energy, they would alleviate a major waste disposal problem and reduce the associated costs of water, air, and land pollution to the community. Impacts from the use of wood-based energy are more benign to the global climate than those from the use of fossil fuel (Ferrero et al. 1987). However, such environmental benefits can be reduced if a processing plant does not integrate equipment to abate pollution as a standard capital-need in the design and operation of the energy system.

Specific benefits to rural development from the production of wood-based energy accrue to the forest products industry, rural areas, and national governments. For the forest products industry, it would create an additional source of stable revenue from the sale of energy. It would also provide incentives for the industry to diversify into the production of multiple products. This shift increases the value-added to the raw timber and reduces the risk inherent in the forest products industry from business and building cycles. The income and employment multipliers of industrial activities based on forest resources would be increased through higher industry revenues. Use of wood-based energy would reduce the costs of local and national energy, and it would improve the reliability of the energy supply in rural areas by replacing the use of fossil fuel as a source of power. Wood-based energy would promote rural development by creating new jobs and income streams in rural areas. The costs of local disposal of wood residues—and its consequent pollution—by the forest products industry would be mitigated. Finally, a more sustainable forest products industry would create a higher, more stable source of government revenue.

Since the late 1970s, wood processing industries in developed countries have turned rapidly to energy self-sufficiency and to the sale of excess power to local electric grids as the costs of fossil fuels have risen sharply. Most companies realized the benefits that we have noted above (Klass 1988). However, in developing countries, the production of wood-based energy is rare because many sawmills do not require heat or steam energy. High costs for small-scale energy production, subsidized low-cost power in rural areas, biased governmental policies against the private generation of power, and lack of technical information—all discourage the production of energy from wood in developing countries.

Measuring Energy Potential

Quantifying the potential for energy from wood residues in developing countries provides a rough estimate of the possible opportunities for adding to existing sources of energy in rural areas. The estimate is a function of the total amount of wood residues produced by each processing mill, the combustion and efficiency of the energy system, and adjustment factors, such as economic and ecological limits (Gowen 1985). In most cases, unfortunately, the adjustment factors still must be collected. Consequently, the estimates for the potential of wood-based energy in this chapter are the highest possible and probably are not sustainable given the current harvesting practices of the industry.

The average amount of waste material available to produce energy from each of the forest products industries ranges from 20% to 55% of the unprocessed log, depending on the type of processing, available markets for wood by-products, and efficiency of processing. On a percentage basis, veneer and sawmills have the greatest potential of wood residues available, and the pulp and paper industry have the lowest (figure 2.5). Sawmills that cut large logs for quality lumber can typically have the highest percentage of waste by-products, from 50% to 60%. Some of the wood by-products are saleable to pulp mills, if chips are utilized, and some to fiberboard and particleboard mills or compressed fuel-log plants. In the United States, perhaps 30% of the original unprocessed log is saleable to other forest products industries, so only about 20% is potentially available for the production of energy. Wood processing plants, such as veneer and plywood mills, produce, on average, lower total amounts of wood residues than sawmills, but variations can range from nearly 0% to 65% of their volume of log inputs, depending on the efficiencies of the plants.

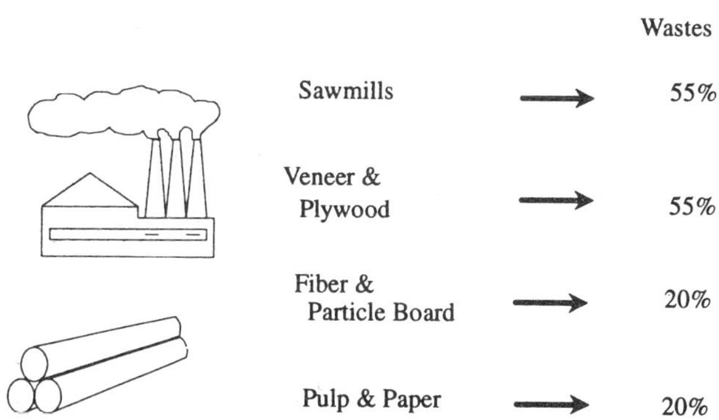

Figure 2.5. Average Wood Wastes from Forest Products Industries (Percent unprocessed log).

The total volume of available wood residues at a plant is also a function of the size of the plant. Sawmills may have a higher percentage of waste per processed log, but the small sizes of mills in many developing countries precludes the installation of economically viable energy systems. In contrast, pulp and paper plants, despite their low percentage of wood by-products, have high fiber through-put. Particleboard and fiberboard plants, along with pulp and paper mills, have an average of 20% of their raw timber input available to generate energy. In developed countries, the pulp and paper industry switched rapidly to energy self-sufficiency based on wood residues when fossil-fuel prices rose (Klass 1988).

Table 2.5 (Asia), table 2.6 (Latin America), and table 2.7 (Africa) present estimates of the energy potential from wood residues by country and global region. The conversion factors from wood products to wood residues assume 55% residues at sawmills and plywood and veneer factories and 20% residues at particleboard, fiberboard, and pulp factories. A factor of 0.9 converts m^3 to MT (MT = metric ton), and we assume 35% moisture content for an ODMT (ODMT = oven-dry metric ton), which has a heating value of 6,650 BTU/lb (BTU/lb = British thermal units per pound). A sawmill, or plywood or veneer factory, should have a 15% efficiency of production of wood-based electricity with a 1 MWH to 5 MWh system. A particleboard, fiberboard, or pulp factory should have a 20% efficiency with a 5 MWh to 20 MWh system. Our formula is:

MWh = [(x ODMT) (2205) (6650 BTU/lb) (efficiency coefficient) $(3413/\text{kWh})^{-1}$ $(1000)^{-1}$, where x = the amount of ODMT potentially available for production of energy.

To summarize energy potential by global region in MWh:

Potential Electric Power from Wood Residues in all Developing Countries Five-Year Average: 1985-89 (in 000 MWhs)

	Solid wood residue potential	Pulp residue potential	TOTAL
Asia	18,637	256	18,893
Latin America	6,967	432	7,399
Africa	2,191	28	2,219
TOTAL	27,795	716	28,511

Figure 2.6 (Forest Product Production), figure 2.7 (Wood Residue Potential), and figure 2.8 (Potential Electric Power from Wood Residues) summarize the data in graphic form and distinguish between A.I.D.-assisted and non-A.I.D. assisted countries. Asia, without question, has the best current opportunities, but the current potential is not sustainable. Latin America's opportunities are fragile, and Africa's are limited, but interesting. In the next section, we develop these points with more analysis to integrate the resource base with the potential for energy from wood residues.

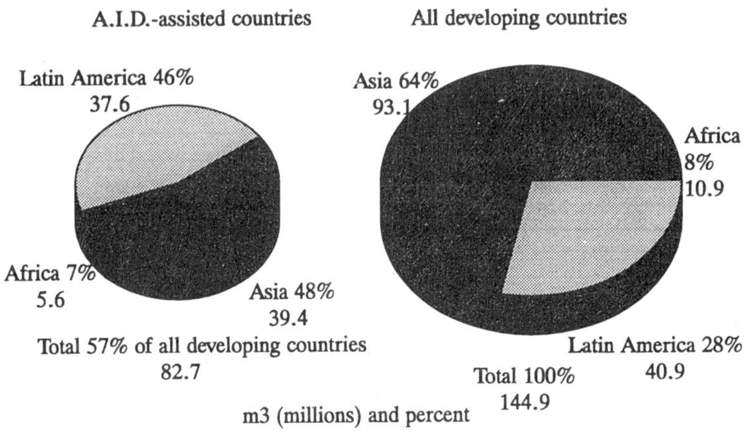

Figure 2.6. Forest Product Production* for Developing Countries
5-Year Average: 1985-1989.
* Including wood pulp.
Source: FAO (1991a).

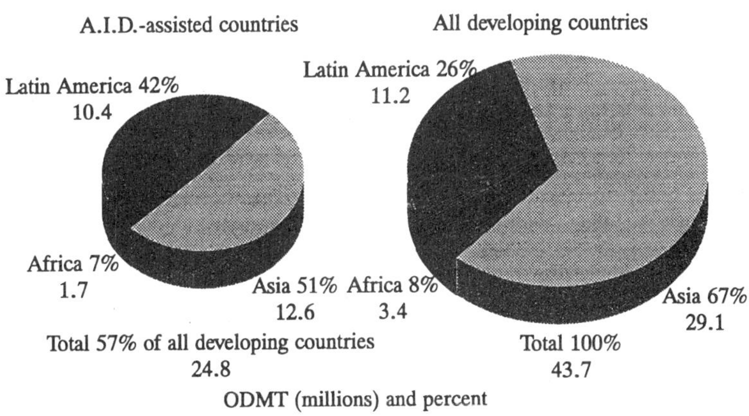

Figure 2.7. Wood Residue* for Developing Countries
5-Year Average: 1985-1989.
* Including wood pulp.
Source: FAO (1991a).

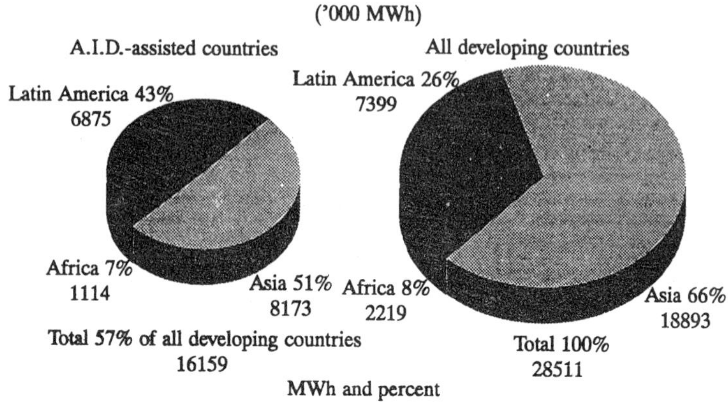

Figure 2.8. Potential Electric Power from Wood Residues* for Developing Countries 5-Year Average: 1985-1989.
* Including wood pulp.
Source: FAO (1991a).

REGIONAL ANALYSES OF FOREST USE, TIMBER PRODUCTION, AND ENERGY POTENTIAL

The following sections summarize the state of forests, timber production, and the resulting potential for wood-based energy on several continents. This review highlights the countries with the best opportunities, either because of their high potential or because they have specific industries with the appropriate conditions and policies favoring the production of energy from wood.

Asia: High Use and Future Scarcity

Forest Management

Since forest and forest-estate management had been practiced in Asia prior to independence from colonial rule, countries in Asia have more commercial forests under management than those in Africa and Latin America. At the end of 1980, about 18% of commercial forestland in Asia was managed, with India, Burma, and Malaysia having the most area (figure 2.9, table 2.1). In Africa, managed areas comprised only 1% of commercial forestland, mostly in Ghana, Uganda, and Morocco (figure 2.17, table 2.3). Even less commercial forestland in Latin America—0.1%—was managed, principally in Nicaragua and Cuba (figure 2.13, table 2.2).

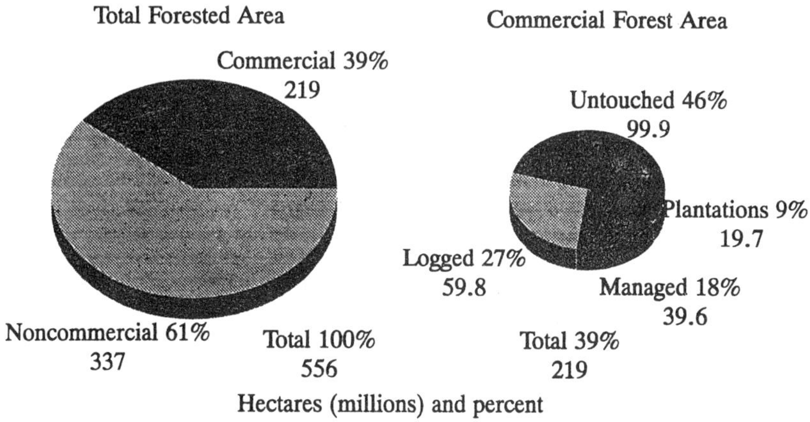

Figure 2.9. Forest Area at the end of 1980 for Developing Countries in Asia.
Source: FAO (1988).
Noncommercial includes parks, and unprotected, unproductive, open, shrub, and bamboo forest.

By the end of 1980, nearly 19% of mature Asian forests had been logged since 1930 (table 2.1). The high rates of deforestation in Asia, especially in Thailand, Malaysia, and the Philippines (figure 2.2), may have been caused by the pressure of increased human population, forest policies that promote rapid commercial development, low stumpage prices and royalties, and high demand for Asian timber. Economic and forest management policies that favor the logging industry have stimulated the rapid development and, as many believe, exploitative exportation of Asia's forests. Costs of timber production per hectare are lower in Asia than elsewhere in the world, partly because each hectare harvested has a higher yield of commercially valuable timber. For example, a logger finds from 25 m^3/ha to 50 m^3/ha of commercially valuable wood in Asian forests—some 10% to 40% of the total stand. In contrast, the average volume of commercially valuable wood is 5 m^3/ha in Africa and 10 m^3/ha in Latin America —only 2.5% to 5% of the total stand (estimates developed by Bender during 1989 consultancy).

Indonesia, with its holdings of nearly 35% of commercial forests, dominates Asia (figure 2.10). Over 75%, 160 million ha, of Indonesia's land mass was classified as forestland at the end of 1980 (FAO 1988), and Indonesia has some of the largest areas of valuable tropical hardwood in the world. At the end of 1980, only 34% of Indonesia's mature forests had been untouched, about 31% of its forests had been logged, and 35% remained unproductive or protected in parks since 1930 (table 2.1), but the pace of deforestation in Indonesia has increased. Behind Indonesia in deforestation from 1981 to 1985 rank Thailand, Malaysia, India, Burma, and the Philippines (figure 2.2).

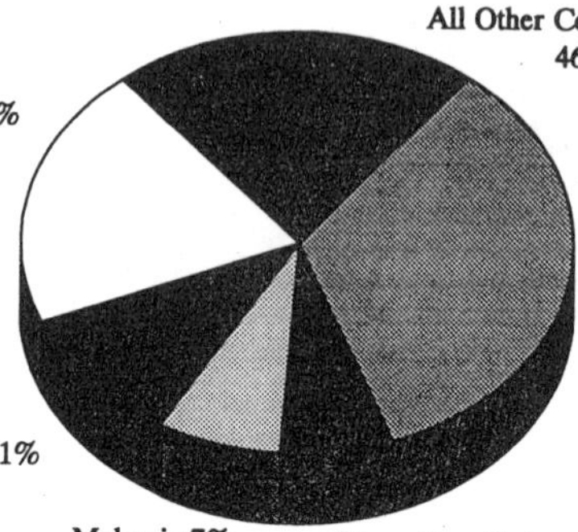

Figure 2.10. Countries with the most Commercial Forest Area in Asia.
Source: FAO (1988).
Commercial forest includes untouched, logged, managed, and plantation forest.

Timber and Production of Wood-Based Energy

Asia had the most developed forest products industry of all developing countries from 1985 to 1989 (figure 2.6). The region cut the highest volume of lumber (67%) and produced the largest amount of plywood (84%) and other wood-based panels such as veneer, particleboard, and fiberboard (50%), but it accounted for only 36% of the wood pulp produced in all developing countries (tables 2.5, 2.6, and 2.7). Given the high level of production of lumber and wood-based panels, this region offers the best opportunity for wood-based energy because these two industries produce such large amounts of currently unused wood residues (figures 2.7 and 2.8).

The 12 A.I.D.-assisted countries in Asia produced an average of over 39 million m^3 annually of wood products, including sawnwood, plywood, veneer, wood-based panels, and other products, from 1985 to 1989 (table 2.5). The dominant producers of wood products in the region, in order of total volume, were India, Indonesia, the Philippines, Thailand, and Burma. India and Indonesia, with almost 18 million m^3/yr and 16 million m^3/yr in lumber and wood-based panels, respectively, produced 10 times the volume of their closest competitors in Asia—the Philippines (1.6 million m^3/yr), Thailand (1.3 million m^3/yr) and Burma (500,000 m^3/yr).

Historically, the forest products industry in Asia has been regarded as an engine of economic growth, providing foreign exchange as well as meeting the internal demand for wood. The split between production for domestic use and for export varies widely among the countries. In India, for example, domestic consumption dominates. Approximately $3.5 billion/yr of sawn timber is sold internally, over 6,000 times the value that is obtained from exported sawnwood

(FAO 1991a). In contrast, Indonesia and the Philippines, which have far more forest resources per capita than India, use wood exports to bring in foreign exchange.

With the greatest amount of wood residues in the developing world, Asia's forest products industry has the highest potential for the production of wood-based energy and the most well-developed markets for industrial wood-based energy systems to provide rural electrification (figures 2.7 and 2.8). If we consider total potential for the production of wood residues and wood-based energy, India slightly exceeds Indonesia and the other A.I.D.-assisted countries with 5.98 million ODMT of wood residues available to potentially produce nearly 3.9 million MWh of electricity (figures 2.11 and 2.12). Most of this potential for wood residues comes from sawmills (97% of the total production of wood residues) that are scattered throughout the country, but this lack of geographical consolidation poses problems for making wood residues a financially attractive energy feedstock. However, given the vast supply of resources and the industrial diversity in India, many mills may be interested in using their wood residues for energy.

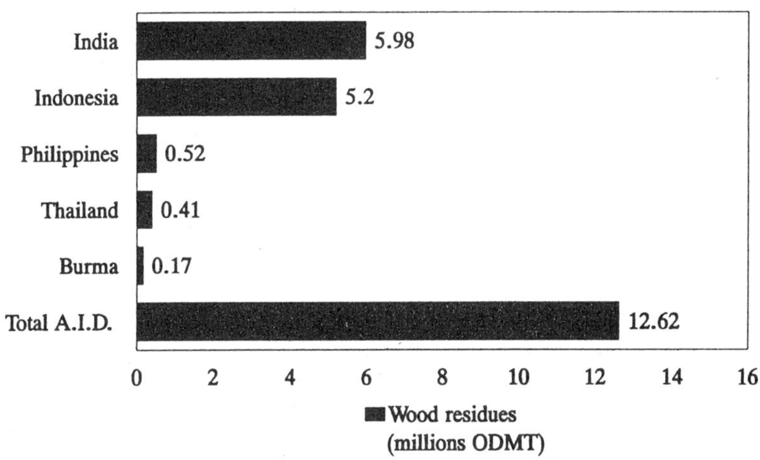

Figure 2.11. Wood Residues* in Top 5 A.I.D.-assisted Countries in Asia: 1985-89.
* Including wood pulp.
Source: FAO (1991a).

Indonesia, with almost 3.4 million MWh potentially available (figure 2.12), may also possess important opportunities to develop wood-based energy because of the scale and distribution of its wood processing plants. There are a large number of plywood plants in Indonesia, and off-grid costs of electricity are high in rural areas (Palmedo et al. 1987). Weingart and Jezek in this volume (1994) identified over 100 plywood factories with sufficient amounts of wood residues available for internal thermal-electrical production in Kalimantan. The market for small generating systems that use wood residues, on average 3 MW, could offer both indigenous and foreign suppliers of equipment a total of $141 million in investment opportunities. Seven Indonesian plywood factories generated electricity to be used in their factories from their wood residues in 1987. Malaysia's MechMar Energy has built three wood-fired power plants that use residues from their wood processing plants for internal energy consumption (Winrock International 1991).

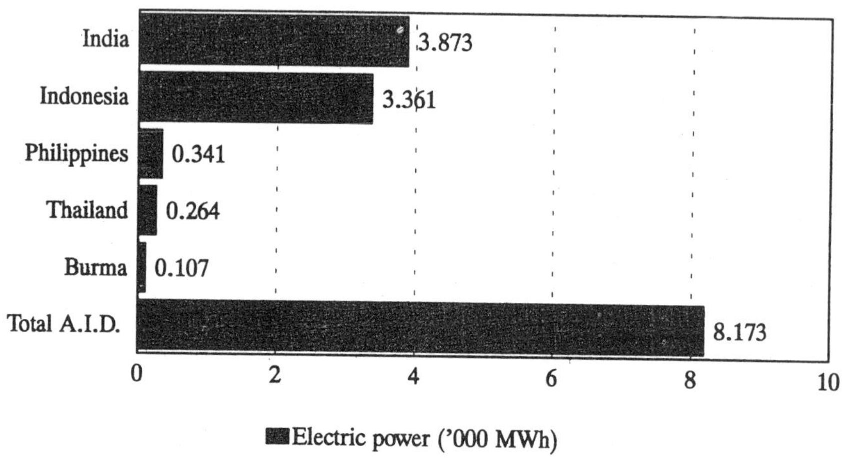

Figure 2.12. Potential Electric Power from Wood Residues* in Top 5 A.I.D.-assisted countries in Asia: 1985-89.
* Including wood pulp.
Source: FAO (1991a)

The Philippines, with a total potential to produce over 340,000 MWh of wood-based energy (figure 2.12), has new incentives for using wood residues because recent legislation was enacted for the production and use of private electrical power in the country. The new laws should encourage larger sawmills and plywood factories to consider the use of systems based on the combustion of wood. The five dendrothermal plants built in the 1980s mix wood from plantations with other, supplementary fuels. The Philippine government is establishing 10 large preserves to maintain remnants of their tropical forests as a response to rapid deforestation.

Several factors work for and against Thailand's use of its wood residues. Almost 88% of its total potential for wood-based energy—264,000 MWh (figure 2.12)—comes from sawmills, but most are small units. Consequently, high costs affect the economic potential for the country. Thailand's Maharaja Company returned to the production of energy from wood residues for its sawmills and associated industries in the late 1980s as the cost of rural electricity rose (Winrock International 1991). Although other Asian countries do not match the top five in total potential, several forest concessions, as well as chip and plywood mills in other Asian countries, such as Fiji and Papua New Guinea, possess the total volume of wood residues as well as the steam-electric needs to produce wood-based energy.

The importance of plantation forestry for future timber production is best realized in Asia (Sutton 1994, this volume). China had the largest area of plantations in the developing world at the end of 1980 (FAO 1988)—almost 13 million ha. Malaysia excels at the development of commercial plantations. Experiences in these countries, when coupled with energy production—as in Malaysia—demonstrate how production of commercial timber can be sustainable and intensively productive.

Latin America: The Fragile Potential

Forest Management

This region holds the largest undeveloped forest base in the developing world. Fragile soils and easily degradable forest ecosystems, especially in the Amazon Basin, limit the potential for these forest resources to be assets for development. The Amazon Basin is a region where the intricate relationships among forest dwellers, timber and nontimber uses, and global biochemical cycles make the area highly vulnerable to exploitation and to the permanent loss of productivity (Repetto and Gillis 1988, World Bank 1991). As a result, there is an ensuing political controversy regarding the use of natural forests by the forest products industry.

Latin America had 688 million ha of natural, mature forests at the end of 1980, an area which far exceeded the combined total forestland of Asia and Africa, of about 535 million hectares (table 2.2). The potential in Latin America lies in its untouched forests. These forests, 87% of the commercial forestland in Latin America at the end of 1980 (figure 2.13, table 2.2), typically have a lower market value than those in Asia. The remainder of forests in Latin America was primarily in its parks and protected lands.

Amazonia, which dominates Latin America's forest base, is the largest tropical forest area in the world. The Amazon Basin covers over 2 million mi^2, over half the size of the continental United States, and contains 31% of the world's tropical forests. It has a high percentage of the world's biological diversity (Myers 1988). The Amazon remains an elusive jewel for developers and ecologists. Despite its vast potential, the area has seen few commercial enterprises based on timber, agriculture, or grazing that have succeeded financially (World Bank 1991). Many scientists propose minimal timber development, with a focus on fruits, nuts, and other extractive products.

Following Brazil, Peru, Colombia, Mexico, and Venezuela are the next top-ranking countries in area of commercial forests (figure 2.14, table 2.2). Extremely rapid rates of deforestation have occurred in several Latin American countries, particularly in Nicaragua, Ecuador, and Colombia (figure 2.2). Costa Rica has also seen its mature forests disappear at the alarming rate of 1.8%/yr

(WRI 1991). Honduras, Guatemala, Bolivia, Peru, and Ecuador have experienced rapid changes in land use. Continued population pressures will threaten their forests in the future. If forests and the forest products industry are to survive in Latin America, new rules in land ownership and use, as well as economic incentives and forest management techniques, must be enacted.

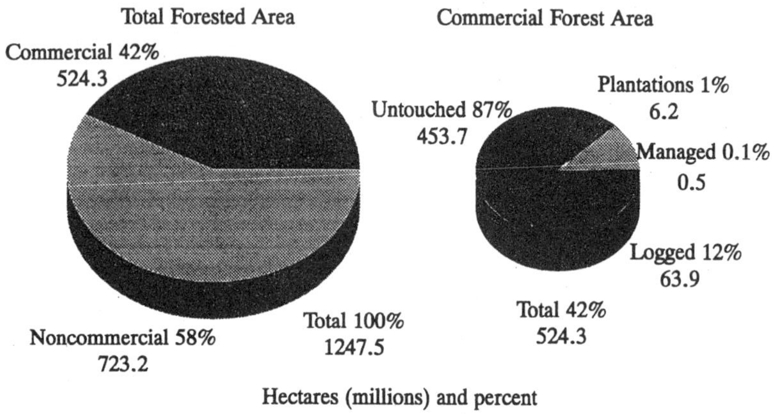

Figure 2.13. Forest Area at the end of 1980 for Developing Countries in Latin America.
Source: FAO (1988).
Noncommercial includes parks, and protected, unproductive, open, shrub, fallow, and bamboo forest.

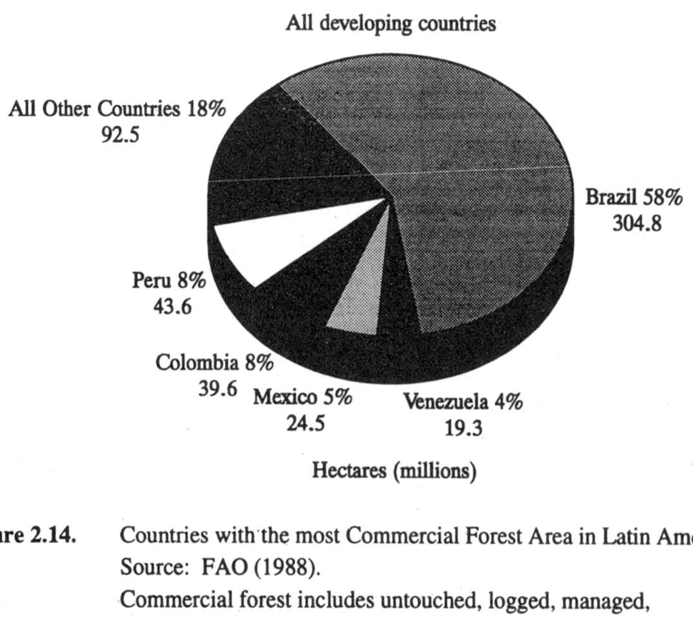

Figure 2.14. Countries with the most Commercial Forest Area in Latin America.
Source: FAO (1988).
Commercial forest includes untouched, logged, managed, and plantation forests.

Plantations are already playing a crucial role in supplying growing populations with wood for domestic and commercial use. Brazil was promoting reforestation efforts in Latin America by the end of 1980 (figure 2.3) with its program for the production of wood-based energy at pulp and paper facilities. In Ecuador, the plantations of eucalyptus and pine that grow on the cordilleras are supplying 25% of the wood needed by industry. Such efforts offer both economic and ecological advantages. In some areas, wood from plantations is cheaper and closer to the marketplace, where the alternative is to haul wood from the tropical forests of the Amazon or the Pacific coast. Use of wood from plantations also relieves the stress on the fragile ecosystems of the tropical forest, which are severely damaged from excessive timber harvesting.

Timber and the Production of Energy from Wood

Latin America holds the key to sustainable forest management and to the production of wood-based energy for many reasons. First, it is the world's greatest reservoir of both forests and diversity of species. Second, its forest products industry is under pressure to change by becoming more efficient and by shifting its reliance on timber from natural forests to alternative, fast-growing species. Third, as elsewhere, its wood residues are underutilized and are pollutants to local areas. Latin America, then, has the option of either maintaining the *status quo* of poor land-management practices or of implementing sustainable multiple-use patterns for resources that may include the production of wood-based energy where it is appropriate.

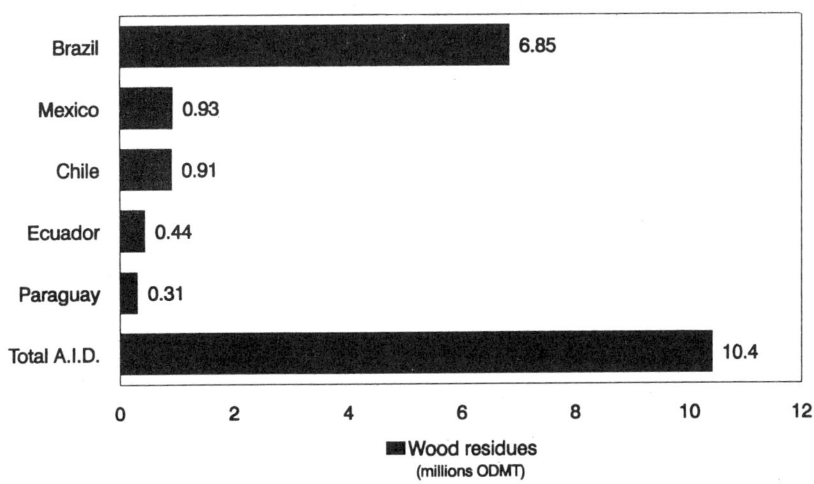

Figure 2.15. Wood Residues* in Top 5 A.I.D.-assisted Countries in Latin America: 1985-89.
* Including wood pulp.
Source: FAO (1991a).

In the production of wood products, the 19 A.I.D.-assisted countries in Latin America had an average annual production that exceeded 38 million m^3 between 1985 and 1989 (table 2.6). The volume of wood products, including wood pulp, from these countries varies from 14,000 m^3 in Haiti to almost 25 million m^3 in Brazil. The top five producers of wood residues that are assisted by USAID in Latin America are Brazil, Mexico, Chile, Ecuador, and Paraguay (figure 2.15). Among all countries in this region, Argentina is also among the top producers of wood products.

The potential for wood-based energy in Latin America represents 26% of the total potential among developing countries, with Brazil dominating the scene (figures 2.8 and 2.16). In this region, the top five A.I.D.-assisted countries have approximately 6.2 million MWh of potential electricity that could be produced from the wood residues of the forest products industry (figure 2.16). We find over 73%—4.5 million MWh—of this potential energy in Brazil, where wood residues from sawmills dominate feedstock sources. Mexico, Chile, Ecuador, and Paraguay hold the next higher-ranking potentials for the production of wood-based energy, from higher to lower, of A.I.D.-assisted countries. Using FAO (1991a) production estimates as the basis of potential power from residues, the range for these countries goes from Mexico, with 621,000 MWh, down to Haiti, with 3,000 MWh (table 2.6). Chile and Ecuador have 606,000 MWh and 282,000 MWh potentials, respectively, still about 20% of Brazil's potential (figure 2.16).

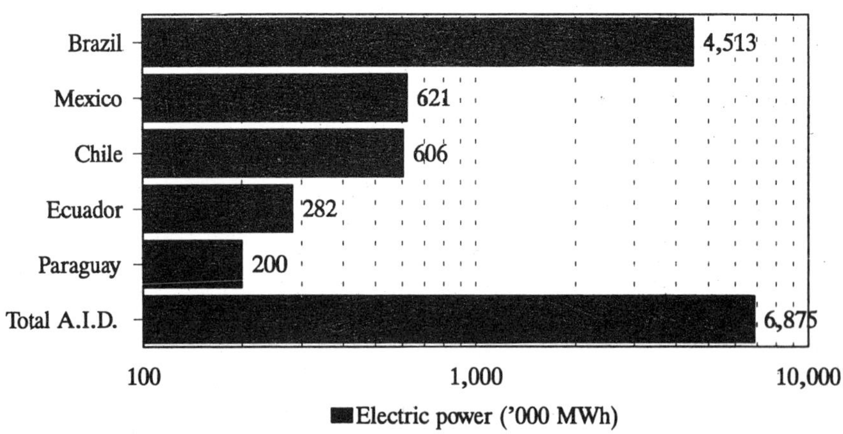

Figure 2.16. Potential Electric Power from Wood Residues* in Top 5 A.I.D.-assisted Countries in Latin America: 1985-89.
* Including wood pulp.
Source: FAO (1991a).

In Ecuador, the ENDESA plywood, veneer, and blockboard factories near Quito use their wood residues for the production of steam, but they buy electricity from the grid (Winrock International 1991). A recent study, sponsored by USAID, of the sawmill industry in Honduras showed a significant potential contribution to the national energy supply that could supplement the use of fossil fuel by sawmills. This use of wood for energy would be sustainable because it would rely on yellow pine in secondary growth areas (Wimberly et al. 1993). However, sale to the grid would be limited by the negotiation of fair and reliable prices for energy between the utility company and independent energy producers.

Africa: Limited and Concentrated Options

Forest Management

The wide range of forest ecosystems on the African continent, from the dense tropical forests in the Congo to the arid deserts, make any generalizations about this region misleading. In figure 2.17, we see that much of the potential for the continent, 71% of the commercial forestland, was untouched at the end of 1980. Almost half of the commercial forests can be found in Zaire (figure 2.18), which has the largest undisturbed timber potential of 80.1 million ha. Zaire also has almost 26 million ha in the protected classification, which is the largest area in Africa with some level of constraints on harvesting and land clearing (table 2.7).

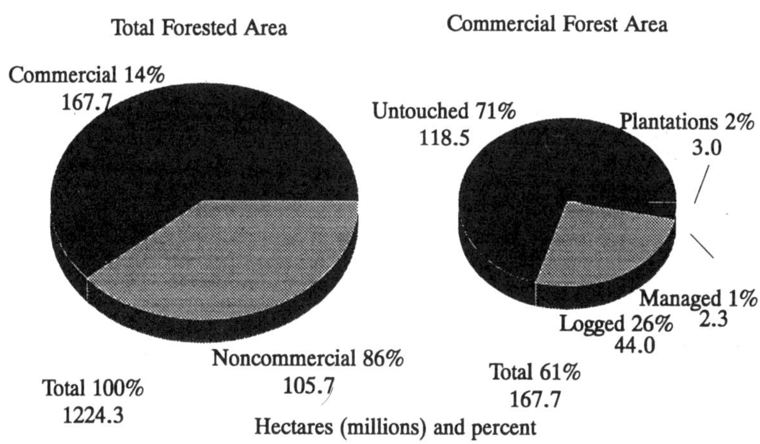

Figure 2.17. Forest Area at the end of 1980 for Developing Countries in Africa.
Source: FAO (1988).
Noncommercial includes parks, and protected, unproductive, open, shrub, fallow, and bamboo forest.

The annual rates of deforestation in particular African countries have been higher than those in other regions: the Côte d'Ivoire had a rate of 5.2%; Malawi, 3.5%; and Nigeria, 2.7% from 1981 to 1985 (figure 2.2). Most likely, this deforestation reflects rapid population-growth rates, poor definition of property rights, and little enforcement of either public or private forest boundaries.

Plantations offer Africa the best hope of future sustainability. At the end of 1980, the leaders in plantation forestry among A.I.D.-assisted countries (table 2.3) were Morocco (322,000 ha), Madagascar (266,000 ha), the Sudan (189,000 ha), and Kenya (181,000 ha). Many plantations are classified as nonindustrial, but their timber will be used for wood products if it grows to commercial size. In Burkina Faso (12,000 ha) and Gambia (1,000 ha), where plantations are the primary source of wood, small cottage industries that manufacture forest products have come into existence with the support of USAID programs. Wood from plantations will be the future source of timber raw materials for industrial products in most of Africa. This future supply will present new options and challenges to those in the forest products industry, particularly in the area of secondary products, such as wood-based energy.

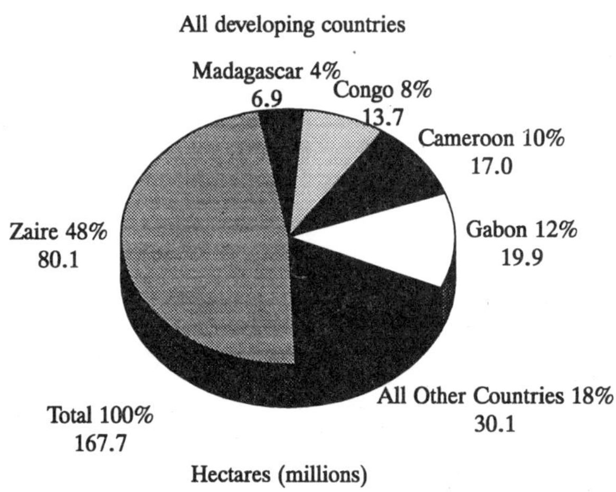

Figure 2.18. Countries with the most Commercial Forest Area in Africa. Source: FAO (1988). Commercial forest includes untouched, logged, managed, and plantation forests.

Timber and the Production of Wood-based Energy

In Africa, total production of timber and wood-based energy falls far below that in other parts of the world. Several key African countries, however, have wood processing plants or plantation forests. Both favor the production of biomass energy from wood. In Africa, there were 29 A.I.D.-assisted countries that produced an annual average of over 5.6 million m^3 of wood products between 1985 and 1989 (figure 2.6, table 2.7). Within these countries, there is considerable variation of production: from only 1,000 m^3 in Burkina Faso, Chad, and the Gambia to 0.7 million m^3 in the Cameroon, and 1.8 million m^3 in

Egypt. Nigeria, Egypt, the Côte d'Ivoire, the Cameroon, the Central African Republic, and Ghana have the highest volume of forest products and revenues, excluding pulp production. If we include the production of pulp, we must recognize Swaziland (322,000 m^3) and Morocco (312,000 m^3) with their large pulp mills (table 2.7).

Like Latin America, domestic consumption of wood in Africa accounts for most of the use of wood in these countries, 75% of the total volume in the Cameroon and 99% in Kenya during 1988 (FAO 1991a). Exports of sawnwood and wood-based panels represent exceptions to this pattern of wood use in the Cameroon, Swaziland, and Zaire, with pulp exports an additional exception for Swaziland. Again, reasons for this use pattern are that industries are poorly developed and undiversified because there is a lack of wood, especially quality wood, in these countries.

Many countries in Africa have experienced drastic declines in the production of forest products (Poore et al. 1989). For example, Nigeria's forests were rapidly depleted when it exported timber in the mid-1900s (Bushbacher 1990). The 29 A.I.D.-assisted countries, when compared with all African countries, produced 39% of total sawnwood and railroad ties, 74% of total wood-based panels, and 93% of total wood pulp on the African continent (FAO 1991a). Countries with the most area of commercial forests include Zaire, Gabon, the Cameroon, the Congo, and Madagascar (figure 2.18, table 2.3).

With respect to total worldwide production, Africa has the lowest potential of any region for the production of wood residues and wood-based energy. Africa has only 2.2 million MWh, or 8% of the total wood-based energy potential, for all developing countries (figures 2.7 and 2.8). However, since rural electrification is quite costly and highly unreliable in many African countries, there are some very specific, and possibly important, future markets for this type of energy.

Egypt has the largest potential among A.I.D.-assisted countries for wood residues and wood-based energy, approximately 0.6 million ODMT and 383,000 MWh (figures 2.19 and 2.20, table 2.7). Behind Egypt, for A.I.D.-assisted countries in Africa, is the Cameroon, with 158,000 MWh from its large sawmills. Ghana ranks next, with almost 104,000 MWh of potential energy because it has many small sawmills and a few plywood and veneer mills. In Ghana, most of the mills are located on the grid and, thus, have access to cheap power, but there is only one off-grid mill that has an operational cogeneration system (UNDP/World Bank 1988). Liberia and Madagascar follow in potential for producing wood-based energy. Swaziland has nearly 43,000 MWh of energy potential with its pulp and paper industry, the largest in Africa. Several of Swaziland's mills burn wood residues to generate steam for their dry kilns, but most buy cheap imported power from South Africa. In Madagascar, sawmills are relatively large, but only a few mills are large enough to produce sufficient amounts of wood residues for moderately sized wood-based energy systems in Liberia.

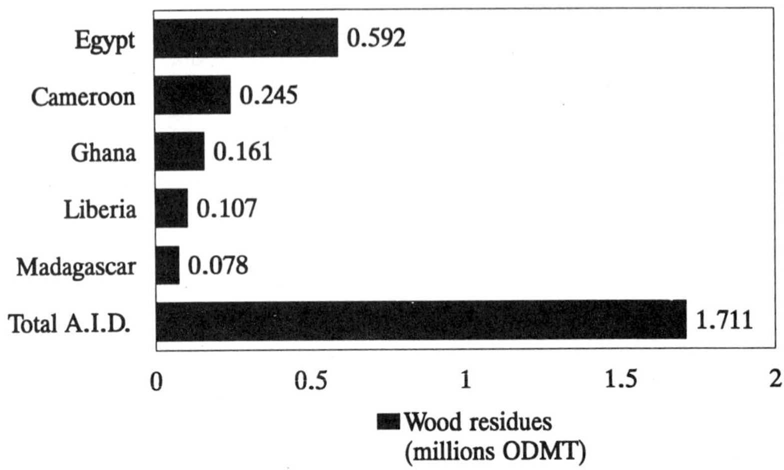

Figure 2.19. Wood Residues* in Top 5 A.I.D.-assisted Countries in Africa: 1985-89.
* Including wood pulp.
Source: FAO (1991a).

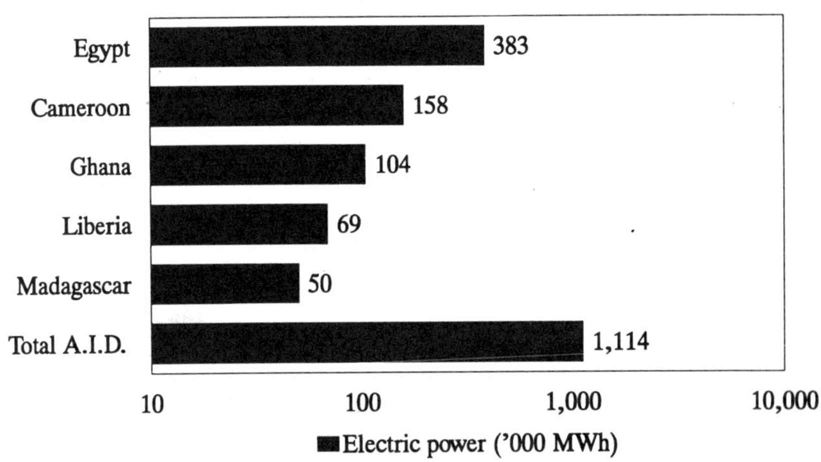

Figure 2.20. Potential Electric Power from Wood Residues* in Top 5 A.I.D.-assisted Countries in Africa: 1985-89.
* Including wood pulp.
Source: FAO (1991a).

Zaire and a few other nations still have significant areas of mature forests that might be harvested on a sustainable basis, but most African nations face severe timber and food shortages over the foreseeable future. Consequently, in relation to other regions, there is a very low potential for the forest products industry to contribute to the production of energy in Africa. Despite this overall gloomy picture, however, some most interesting opportunities exist for the sustainable management of natural forests and timber plantations in several countries. In some cases, ownership by poor people, either as individuals or as tribal groups, appears worth exploring.

SUMMARY OF WOOD-ENERGY POTENTIAL AND RURAL ENERGY NEEDS

Asia has the largest volume-flow through its wood markets and, as a consequence, has the largest potential of all developing regions to produce energy from wood residues (figures 2.7 and 2.8). It has the largest source of wood residues and potential for wood-based energy production, roughly 66%, 18.9 million MWh, of the total for all developing countries. In contrast, wood residues in Latin America comprise 26%, 7.4 million MWh, of the total potential, and Africa represents only 8%, or 2.2 million MWh (figure 2.8). These estimates are based on current industrial production of wood products and have not been adjusted for country-specific ecological and economic constraints. Wood-fired energy systems are important providers of energy in several countries—e.g., Malaysia, Thailand, Indonesia, the Philippines, Ecuador, and Swaziland—but such cases are the exceptions. Typically, subsidized rural power, small-scale wood-processing operations, and the inability to sell excess power to a grid delay the production of energy from wood residues in many countries.

The potential production of wood-based energy may be shifting as we look to future opportunities. For example, the total potential in Asia may be contracting, while the opportunity in Latin America could be expanding. The rapid depletion of Asia's forests suggests that future sources for energy in rural areas will have to come from plantation forestry. Asian nations are now at or past their peak for the production of timber products and wood-based energy (Sutton 1994, this volume). Fast-growing plantations on steeper, marginal agricultural sites and harvested forestlands are the key to Asia's long-term potential. In many cases, plantations can be established on degraded lands with the direct involvement and ownership by resource-poor people. In the future, plantations will offer an environmentally responsible source of energy production for these countries. Recent changes in policies concerning electricity, policies that now allow private generation of electrical power for sale and that remove subsidies for fossil fuel, will encourage reconsideration of investments in timber products, plantations for dendrothermal-energy production, and wood-based energy plantations in several countries—India, Indonesia, Malaysia, the Philippines, and Thailand.

With its vast untapped potential, it is tempting to view Latin America as our next frontier for the utilization of the forest. Land conversion—primarily from expansion of agriculture, with an annual rate of 6%—accounts for most of the rapid depletion of tropical forests in Latin America during the past few decades.

However, most foresters and ecologists agree that some possibilities exist for expansion of the harvest, mainly on secondary growth areas. Many forests are on ecologically fragile sites that cannot be used economically for the production of commercial timber if any reasonable standards for soil conservation or sustainability apply.

The serious degradation of forests and soils that is associated with land conversion in Latin America has created many opportunities for secondary growth forests and for rehabilitation, including medium- to fast-growing timber plantations, that might be transferred to resource-poor families. As in Asia, it will be necessary to identify a mix of fast-growing species for the industry from which energy as well as production of timber and wood products can be sustained. Recent opportunities for the production of energy, identified by such organizations as USAID and UNDP/World Bank, for Costa Rica and Honduras need to be expanded and explored further.

Overall, Africa's timber production probably peaked during the latter stages of the colonial period. Select countries, such as Egypt, the Cameroon, and Swaziland, may have the appropriate mix of wood-residue production and sizes of plants that encourage economically viable options. Although many mills exist that could produce electrical power, the availability of inexpensive power prevents the wide-spread commercialization of wood-based energy in most countries. Changes in policies and institutions are needed for the production of wood-based energy to make inroads into most of these highly productive countries.

It is critical that the prices for energy in rural areas accurately reflect the real social-opportunity costs to these areas and to the specific country. Currently, rural energy is cross-subsidized for social welfare. Yet, there are highly distorted perceptions of the supplies of energy that are created by these subsidies, and the real costs of diesel systems in rural areas are prohibitive (Schramm and Warford 1989). Wood-based energy might often be far cheaper and more reliable if the true costs were internalized. While wood-based energy may be a fraction of the total production of electricity in selected countries, it may be a critical source of power in rural areas and for some industries. Where discontinuous grids in India, Indonesia, and the Philippines make costs of energy quite high in rural areas, wood-based energy may fill critical niches at the edge of a grid. Adding electrical energy can be the catalyst for rural industrialization and improvements of the quality of life. In Nigeria, Swaziland, and Ecuador, for example, weak rural infrastructure could be strengthened by wood-based energy to attract new, small-scale industries.

SUMMARY AND RESEARCH NEEDS

The potential for wood-based energy is large in many developing nations in the tropics. But this potential is predicated on the design and implementation of schemes for sustainable forest management. Production levels of the forest industry and rates of harvest, as well as the flows of wood and fiber residues, are obviously correlated. The flows of wood residues can be used to produce steam or electricity, rather than dumped or burned as is now often the case. Biomass energy from wood holds a promise as a critical source of rural electrical power, and in several developing countries, it is already supplementing or replacing fossil fuel as a source of energy (Winrock International 1991).

As small-scale wood-based power plants have become more efficient, the costs of energy have fallen at wood processing facilities, making costs quite competitive with energy from fossil fuel (Weingart and Jezek 1994, Wimberly et al. 1994, both in this volume). The present barriers to the adoption of wood-based energy systems are typically not technical—they are economic, ecological, and political. In many developing countries, legislation is needed to encourage cogeneration and the purchase of excess power at reasonable prices by the local utility company.

Wood-based biomass energy can be a positive force, if it leads to higher expected values for timber in the future. However, expectations of higher values must be in a context where there are rational expectations of receiving future benefits and where the value-added is sufficient to offset opportunity-costs of alternative land uses. If future benefits are not secure, the incentives are to harvest now and not to invest in forests for the future. This leads to the "tragedy of the commons." The very act of not harvesting timber is in itself the critical investment required for future timber growth and harvests. In most cases, enabling legislation for cogeneration by the forest products industry might be coupled with revisions of crucial elements of forestry policy, especially those defining the ownership rights of land and timber for public and private organizations and the incentives that would encourage forest conservation rather than its depletion.

The institutional infrastructure required to permit cogeneration, similar to the institutional needs to implement sustainable forest management in the tropics, is complex. However, the growing scarcity of timber can be a positive force to impel nations toward sustainable systems if price-increases stimulate a demand for reserves of future harvests. This desire to reserve timber for the future effectively delays current harvests and encourages investment in management. For future-price expectations to have such a positive impact, however, timber and forest lands must be "owned," in the sense that decision-makers—either individuals or communities—have security of receiving future benefits. Without the security of such ownership rights, higher prices will increase current harvest levels and conversion of forestland to livestock, crops, and other uses.

Property or tenure rights in developing countries need not follow a western form (Fortman and Bruce 1989, Peluso 1992, Padoch 1994, this volume). These rights may accrue to farmers, industrial firms, public forest services, or a wide variety of nongovernmental and nonprofit entities that represent rural poor people. The lack of tenure rights of most rural people and marginal farmers, the instability inherent in concessionaire rights, and the inability to enforce boundary and timber-trespass laws on many public forests discourage forest conservation.

These forces may be compounded by population pressures and institutional incentives to exploit timber resources or convert forests to other land uses.

Converting any of these possibilities into realities, however, requires considerable investment in field research, education, and action. Needs for research that can assist the implementation of sustainable wood-based energy systems include:

- Selecting and targeting the best opportunities for conducting field activities in Asia, Latin America, and Africa. The criterion for selection should be sustainable timber sources for the wood processing firms. Development of case studies in developing countries would provide insight into the realistic options for integrating the production of wood-based energy with rural development
- Focusing on designing and monitoring acceptable ecological and institutional methods of sustainable forest management and comparing the advantages and disadvantages of natural-forest and mixed-species plantation management for supporting wood-based energy systems
- Comparing tenurial systems in various regions and under different schemes of management that favor the production of sustainable wood-based energy production
- Establishing a network among the environmental, wood processing, forest management communities, and rural utilities to integrate the necessary scientific, economic, and social information into the design and operation of wood-based energy systems

To succeed, this effort will require an on-going synthesis of information about rural electrification, forestry, and agricultural and other natural resources. Then we may integrate our efforts and field projects to improve the livelihoods and futures of rural people with economic and demographic data.

Literature Cited

Ascher, William, and Robert Healy. 1990. *Natural resource policy-making in developing countries.* Durham: Duke University Press.

Bender, Ozzie. 1994. Experiences with energy from wood residues in the western United States. In *Forest resources and wood-based biomass energy as rural development assets.* (W.R. Bentley and M.M. Gowen, eds.). New Delhi: Oxford & IBH Publishing Company Pvt., Ltd.

Bushbacher, Robert J. 1990. Ecological analysis of natural forest management in the humid tropics. In: *Race to save the tropics: ecology and economics for a sustainable future.* (Robert Goodland, ed.) Washington, D.C.: Island Press.

Ciriacy-Wantrup, S. von. 1963. (Rev. ed.) *Resource conservation: economics and policies.* Berkeley: University of California, Division of Agricultural Studies, Agricultural Experiment Station.

FAO. 1988. An interim report of the state of forest resources in developing countries. Forest Resources Division. Forestry Department. Rome: FAO. FO:MISC/88/7.

FAO. 1990. An interim report of forest resources assessment 1990 project. Committee on Forestry, Tenth Session. Rome: FAO. COFO-90(a). Mimeo paper.

FAO. 1991a. *Yearbook of Forest Products 1978-1989.* FAO Statistics Series No. 24 and No. 97. Rome: FAO.

FAO. 1991b. The sample survey design. FAO forest resources assessment 1990 project. Rome: FAO. Tables. Mimeo paper.

FAO. 1991c. Second interim report on the state of tropical forests. Forest Resources Assessment 1990 Project. Rome: FAO. Data presented at World Forestry Congress, Paris, France, September, 1991.

Ferrero, G.L., G. Grassi, and H.E. Williams. 1987. *Biomass energy: from harvesting to storage.* Essex: Elsevier Applied Science Publishers Ltd.

Fortman, Louise and John Bruce. 1989. *Whose trees? Proprietary dimensions of agroforestry.* Boulder: Westview Press.

Greeley, W. 1950. *Forest policy.* New York: McGraw Hill Publishing Company.

Goodland, Robert, ed. 1990. *Race to save the tropics: ecology and economics for a sustainable future.* Washington, D.C.: Island Press.

Gowen, Marcia M. 1985. *Renewable energy assessments: an energy planner's manual.* Honolulu: East-West Center.

Gowen, Marcia M. 1989. Understanding economic development: a resource planner's dilemma. In: *Proceedings from Hardwood forest products opportunities: creating and expanding business.* (S.B. Jones and J.A. Stanturf, eds.). State College: Penn State University Press.

Gupta, Turuth. 1983. *Economics of minor forest products in India.* New Delhi: Oxford & IBH Publishing Company Pvt., Ltd.

Jones, Donald W. 1994. Energy use are fuel substitution—lessons learned and applications to developing countries. In: *Forest resources and wood-based biomass energy as rural development assets.* (W.R. Bentley and M.M. Gowen, eds.). New Delhi: Oxford & IBH Publishing Company Pvt., Ltd.

Klass, Donald L. 1988. The U.S. biofuels industry. Paper presented at The International Renewable Energy Conference, Honolulu (September 18-24, 1988).

Laarman, Jan C. and Roger A. Sedjo. 1991. *Global forests: issues for six billion people.* New York: McGraw Hill.

Lai, C. and A. Kahn. 1992. Participatory forestry in Bangladesh: first steps towards sustainability. In: *Agroforestry in South Asia: problems and applied research perspectives.* (W.R. Bentley, P.K. Khosla, and K. Seckler, eds.). New Delhi: Oxford & IBH Publishing Company Pvt., Ltd.

Leonard, H. Jeffrey. 1987. *Natural resources and economic development in Central America: a regional environmental profile.* New Brunswick: International Institute for environmental Development.

Levine, G., W.R. Bentley, B. Brockbank, and B.P. Ghildyal. 1986. Problems and solutions: lessons from rural development projects. Ford Foundation Discussion Paper No. 20.

Miller, Sandra, C.W. Shinn, and W.R. Bentley. 1994. *Rural resource management: problem solving for the long-run.* Ames: Iowa State University Press.

Myers, Norman. 1988. *Natural resource systems and human exploitation systems: physiobiotics and ecological linkages.* Washington, D.C.: The World Bank.

Padoch, Christine. 1994. The woodlands of Tae. In: *Forest resources and wood-based biomass energy rural development assets.* (W.R. Bentley and M.M. Gowen, eds.). New Delhi: Oxford & IBH Publishing Company Pvt. Ltd.

Palmedo, Philip F., T. Carroll Owen, Robert Nathans. 1987. Land use and energy utilization: a final report. Upton, New York: Policy Analysis Division, Nature Center for Analysis of Energy Systems, Brookhaven National Laboratory. Springfield, Virginia.

Peluso, Nancy. 1992. *Rich forests, poor people: resource control and resistance in Java.* Berkeley: University of California Press.

Peters, Charles M., Alwyn H. Gentry, and Robert O. Mendelsohn. 1989. Valuation of an Amazon rainforest. *Nature* 339–655.

Poore, Duncan, Peter Burgess. John Palmer, Simon Rietbergen, and Timothy Synnot. 1989. *No timber without trees: sustainability in the tropics.* London: Earthscan Publications Ltd.

Repetto, Robert, and Malcolm Gillis. 1988. *The forest for the trees? Government policies and the misuse of forest resources.* Washington, D.C.: World Resources Institute.

Saterson, Kathryn A. 1990. Integration of biological conservation with development policy: the roles of ecological analysis. In: *Race to save the tropics: ecology and economics for a sustainable future.* (Robert Goodland, ed.). Washington, D.C.: Island Press.

Schramm, Gunter, and Jeremy J. Warford. (eds.). 1989. *Environmental management and economic development.* Baltimore: Johns Hopkins University Press for the World Bank.

Singh, K.D. 1991. FAO tropical forest resources monitoring programme. Rome: FAO. Appendix. Mimeo paper.

Sutton, W.R.J. 1994. Increasing global timber supply to meet the world's growing demand for wood. In: *Forest resources and wood-based biomass energy as assets for rural development.* (W.R. Bentley and M.M. Gowen, eds.). New Delhi: Oxford & IBH Publishing Company Pvt., Ltd.

UNDP/World Bank. 1988. Ghana: sawmill residues utilization study. (2 vols.). Energy sector management assistance program. Activity Completion Report No. 074B/87.

Weingart, Jerome M., and Peter A. Jezek. 1994. A prefeasibility assessment of the potential of wood waste power systems for the Indonesian wood products industry. In: *Forest resources and wood-based biomass energy as rural development assets.* (W.R. Bentley and M.M. Gowen, eds.). New Delhi: Oxford & IBH Publishing Company Pvt., Ltd.

Wells, M. and K. Brandon. 1992. *People and parks: linking protected area management with local communities.* Washington, D.C: International Bank for Reconstruction and Development.

Wimberly, James, C.B. Briscoe, Kelton Grubbs, Gregory Morris. 1994. Energy from sawmill wastes in Honduras: a case study. In: *Forest resources and wood-based biomass energy as rural development assets.* (W.R. Bentley and M.M. Gowen, eds.). New Delhi: Oxford & IBH Publishing Company Pvt., Ltd.

Winrock International. 1991. Industrial energy and electric power from wood residues. Biomass energy systems report. Biomass energy systems and technology project. Washington, D.C.

World Bank. 1991. The forest sector. A World Bank Policy Paper. Washington, D.C: International Bank for Reconstruction and Development.

WRI (World Resources Institute). 1992. *The 1992 Environmental Almanac*. Boston: Houghton Mifflin Company.

Part Two

Some Alternative Experiences with Wood-Based Biomass Energy

INTRODUCTION

Wood energy has fueled people's basic needs since the beginning of human development. Only in the past one hundred years have modern technologies been able to vastly increase the quality and quantity of the energy delivered from other sources. This section presents case studies from India, Honduras, Indonesia and the United States on a broad spectrum of modern wood energy use and its opportunities. Overarching themes in all country analyses are that sustainable wood energy management is feasible and integral to rural people's quality of life and economies. Opportunities to better utilize all the wastes from the forest products industry and agroforestry projects in many cases were overlooked in an effort to "modernize" rural villages and regions. Similarly, integrating sustainable, local forest management into rural development strategies often is ignored. Over and over again, these analyses identify distorted forestry and energy policies that fail to capture the environmental externalities of all fuels as critical barriers. Wood energy, as a consequence, cannot fill its proper niche in the energy profiles of many countries.

The Honduran, Indonesian, and United States case studies demonstrate that often no technical reasons prevent more efficient use of all forest products. Large volumes of unused wood residues are available. The forest products industry in most developing countries could utilize these resources if economic incentives existed. Wood energy use at large wood-fired industrial facilities, as discussed in the U.S. case studies, have their problems, but also present valuable learning experiences to developing countries. In Honduras and Indonesia, medium scale projects (3-20 MW) would be viable given institutional changes. Energy pricing policies, forestry concessions, and power generation laws often restrain uses that would demonstrate wood energy is sustainable economically and environmentally. Wood-based energy is economically viable in many countries if the appropriate environmental and energy policies are passed and enforced.

CHAPTER 3

ENERGY USE AND FUEL SUBSTITUTION: LESSONS LEARNED AND APPLICATIONS TO DEVELOPING COUNTRIES

D. W. Jones

MYTHS ABOUT ENERGY AND DEVELOPMENT

Myths often shape attempts at rational thinking, and the intersection of economic history and economic development is particularly subject to myth-shrouded analysis. It is often difficult to identify exactly what one's own myths are and consequently to understand how they shape our analysis. However, my study of the modernization of energy as part of planning economic development makes me suspect that how we believe that energy was developed in industrialized countries during the past two centuries shapes the way we think about modernization of the energy supply and its use in the contemporary Third World. Bruton (1985) has suggested that people in developing countries frequently look to experiences of the developed countries as models for imitation, often without accurate knowledge of what the experiences of developed countries actually were. I would like to develop Bruton's suggestion further—that at least in the case of energy development, the developed countries may not have a clear image of their past either. While the historical reconstructions of energy use during the previous century are impressive in some developed countries, I am struck by the gaps in our knowledge that are often filled in by beliefs and assumptions. Planners in the developing world would be better served by more solid knowledge of the similarities and differences of their present circumstances with those of industrialized countries when they transformed their energy systems.

Table 3.1 compares the characteristics of energy use in developing and developed countries. The most obvious differences are the relative importance of household energy-use and the much greater importance of biomass energy in developing countries.

Use of energy per unit of output and per capita in the United States from 1850 to 1950 (figures 3.1 and 3.2) and similar diagrams for Britain and Italy (figure 3.3) suggest apocalyptic projections for developing countries. The best news is that energy use per unit of output eventually declines. The bad news is that it may first increase roughly threefold over a period of about 50 years.

Table 3.1. Comparisons of Energy Use Between Developed and Developing Countries

Characteristics of Developing Countries	Characteristics of Developed Countries
large household consumption shares—35%-75% or more	much smaller household consumption shares—15%-7%
small portions of national energy supplied by electricity—1%-8%	modest shares of national energy supplied by electricity—12%-20%, with about 25%-35% of energy consumed by households
relatively large shares of electricity consumed by households—25%-60%	consumption by the transportation sector—20%-35%
large shares of total energy supplied from wood or other biomass fuels—30%-90%	consumption by industry—30%-40%
households consuming most of the available biomass energy—50%-99%	negligible consumption of biomass fuels
industrial energy consumption—25%-50% of nonbiomass energy	

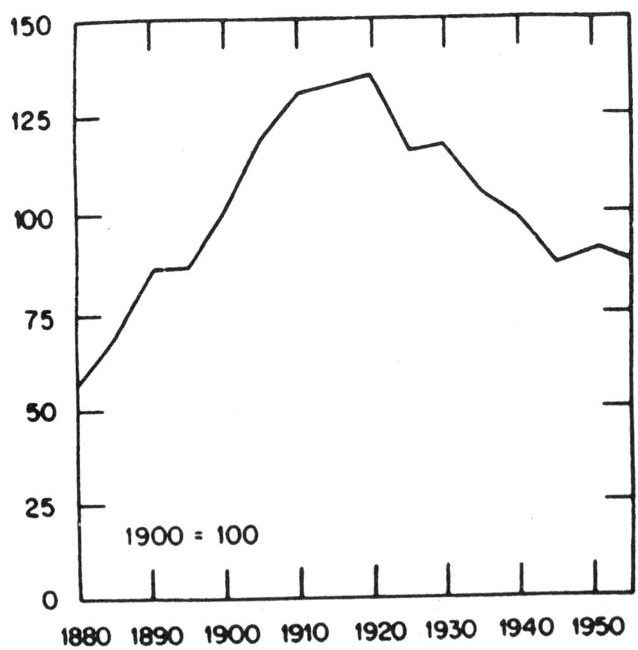

Figure 3.1. Energy Consumption per Unit of GNP, Excluding Wood, United States, 1880-1950.
Source: Schurr & Netschert (1960).

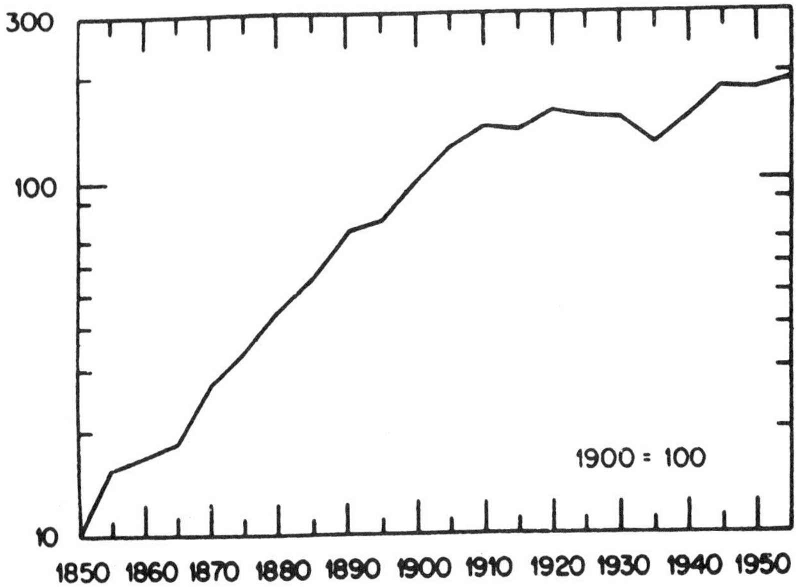

Figure 3.2. Energy Consumption per Capita, Excluding Wood, United States, 1880-1950.
Source: Schurr & Netschert (1960).

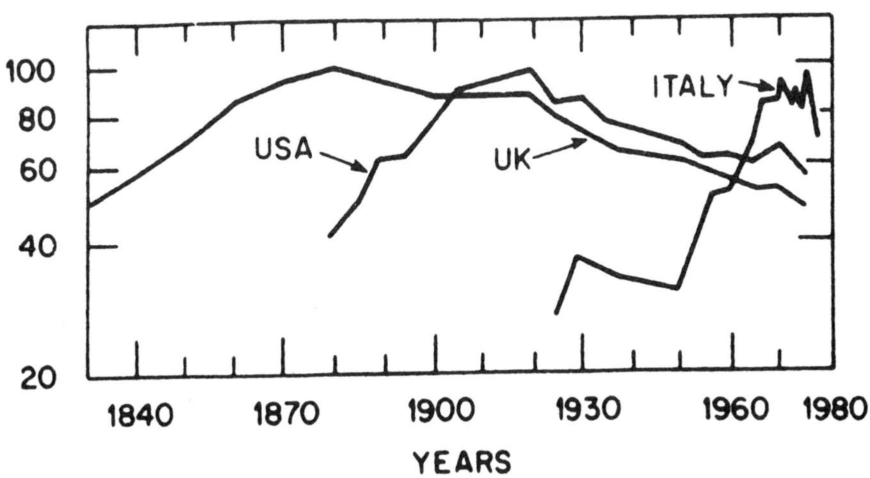

Figure 3.3. Energy-Output Trends, Excluding Traditional Energy Sources in the United States, the United Kingdom and Italy, 1830-1975.
Source: Proops (1984).

The really bad news is that overall consumption per capita rises continually, and consequently, energy consumption *per capita* might rise by as much as fifteenfold over a 50-year period. These diagrams include only modern energy. The use of wood has received less attention. Figures 3.4 and

3.5 show that total consumption of energy per unit of output rises by less than 20% before beginning a substantial decline and that energy consumption per capita rises a little less than twofold.

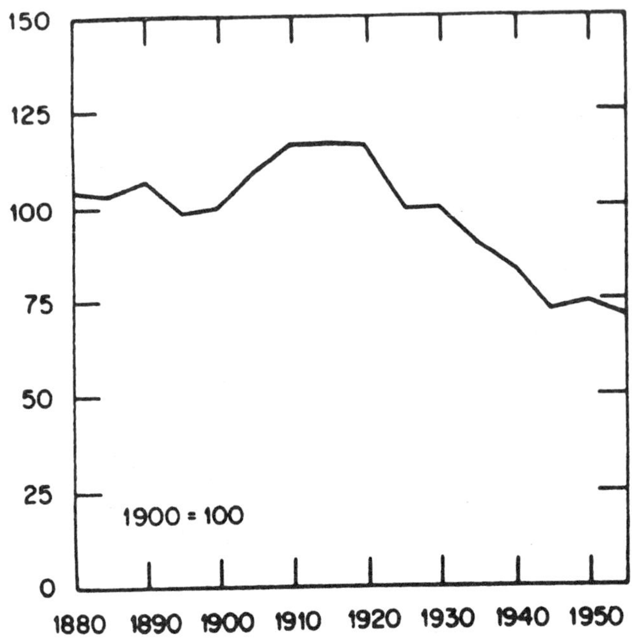

Figure 3.4. Energy Consumption per Unit of GNP, Including Wood, United States, 1880-1950.
Source: Schurr & Netschert (1960).

The household sector, rather than other sectors that contribute directly to GNP, uses most of the wood. Schurr (1983) contends that including wood fuels in measuring the production of total energy distorts the meaning of energy intensity per unit of economic production. While the substitution of modern sources of energy for wood fuels in developing countries dampens the overall net increase in total energy consumption, the purchase of modern energy, often by importation, poses a major problem for national solvency. Cases can be made on both sides of Schurr's argument, but the 50-year outlook for energy demands in developing countries is sobering. The problem of importing modern fuel is already upon many developing countries. Nevertheless, we may derive useful lessons from how patterns of energy use evolved in developed countries and from the contrasts and similarities between this evolution and the experience of many present-day developing countries.

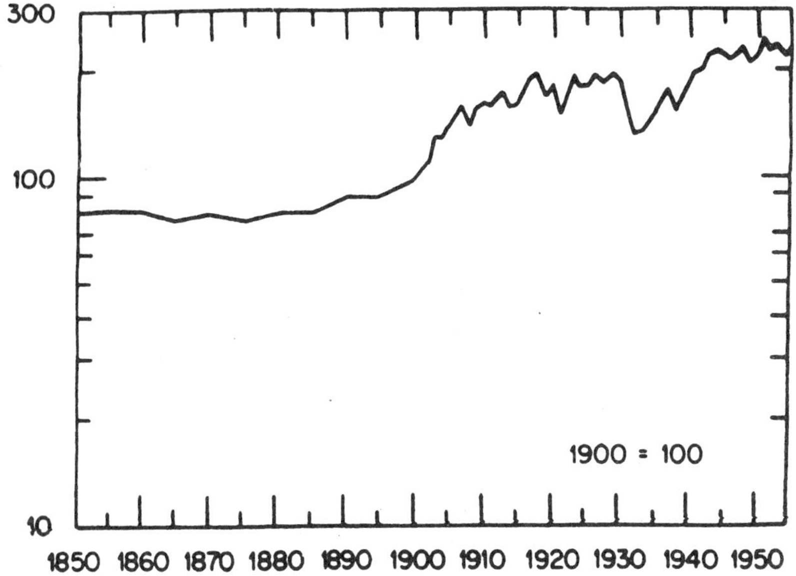

Figure 3.5. Energy Consumption per Capita, Including Wood, United States, 1880-1950.
Source: Schurr & Netschert (1960).

ENERGY USE IN THE UNITED STATES: 1800-1982

Sectoral Trends in Energy Consumption

Detailed energy balances are available for the United States from 1970 to the present (OECD 1984). Allocation records of primary fuel were published from 1947. Two important, but little-explored secondary sources consider energy use in the United States before 1947 (Putnam 1953; Dewhurst et al. 1955). Combined with the monumental work by Schurr and Netschert (1960), the Putnam and Dewhurst works provide a crosscheck.

Table 3.2 presents a distribution of primary energy-use between the household and the commercial sectors and the industrial and transportation sectors. Estimates are given for every 10 years between 1800 and 1970. In addition, Schurr (1983) offers a point estimate for the distribution of energy between households on the one hand and industry and transportation on the other in 1880. That estimate falls inside the Putnam-Dewhurst spread. Shares of energy use in industry and transportation have been estimated by assigning the sum of Putnam's estimated process-heat and work-energy shares to industry and transportation. According to Putnam (1953), 97% of residential and commercial energy-use went to comfort heat and 3% to work, and all process heat went to industry in 1947. For the years 1850 to 1950, Dewhurst's estimates of the share of energy used for work have been considered in conjunction with Putnam's estimates to provide interval estimates for the industry and transportation share. In the absence of better information, Putnam's estimate of the share from process heat is added to Dewhurst's estimate of the share from work-energy to arrive at a total share for industry and transportation.

Table 3.2. Sectoral Energy Shares, United States, 1800-1982

Year	Household and Commercial	Industry and Transportation	Year	Household and Commercial	Industry and Transportation
1800	0.95	0.05	1900	.42-.59	.41-.58
1810	0.93	0.07	1910	.42-.52	.48-.58
1820	0.91	0.09	1920	.45-.47	.53-.55
1830	0.89	0.11	1930	.40-.60	.40-.60
1840	0.86	0.14	1940	.37-.52	.48-.63
1850	.84-.87	.13-.16	1950	0.39	0.61
1860	.76-.77	.23-.24	1960	0.34	0.66
1870	.76-.79	.21-.24	1970	0.35	0.65
1880	.64-.68 (.67)[a]	.32-.36 (.33)[a]	1982	.42[b]	0.58
1890	.63-.65	.35-.37			

[a] Schurr's (1983) point estimate.
[b] Includes miscellaneous uses.
Sources: Putnam (1953), pp. 89, 416; Dewhurst (1955), pp. 1114-1115; Schurr (1983), p. 213.

The shares exhibit a continual shift from the household and commercial sectors toward industry and transportation. Three periods demonstrate particularly large shifts:

- the 1850-60 decade when the industry and transportation share was 50% or more
- the 1870-80 decade
- the 30-year period from 1890 to 1920 (probably most in the single decade 1890-1900)

The period from 1890 to 1920 was one of both industrial expansion and major innovation. Electrification was introduced on a large scale and remained the primary power-source for manufacturing. High electricity rates in the early years of electrification precluded most residential use (Rybczynski 1986). Urban electric trains quickly replaced horse trams in the mid-1890s, although the truck did not make intracity horse traffic obsolete until the second decade of the twentieth century. Truck registrations increased from 700 in 1904, to

10,000 in 1910, and to 1.1 million in 1920 (U.S. Department of Commerce 1975). These innovations increased demand for fuels from both the residential and commercial sectors. Production of iron and steel continued to expand rapidly from 1890 to 1910. Railroad freight ton-miles nearly doubled between 1890 and 1900. Passenger miles doubled between 1900 and 1910.

The relatively wide intervals of energy-use shares in 1930 and 1940 suggest that industry and transportation shares may have declined during those decades. The estimates indicate that the industry and transportation shares have leveled since the 1960s. As figure 3.6 shows, Putnam's estimated work-energy shares climb steadily, while Dewhurst's estimates, which were higher than Putnam's in the earlier decades, peak in 1920 and then gradually decline.

The trends in work shares from 1870 to 1950 are of some interest (table 3.3). The hydroelectric work-share, which should parallel the work share for all electricity, declines steadily from its appearance in 1890. This decline reflects increasing use of electricity for lighting, probably in the commercial and residential sectors. Simultaneously, the work share of petroleum rises. In 1880, about 85% of crude was refined as kerosene, which was used primarily as an illuminant. Production of fuel oil expanded around 1905 and maintained a steady share of about 50% until 1930.

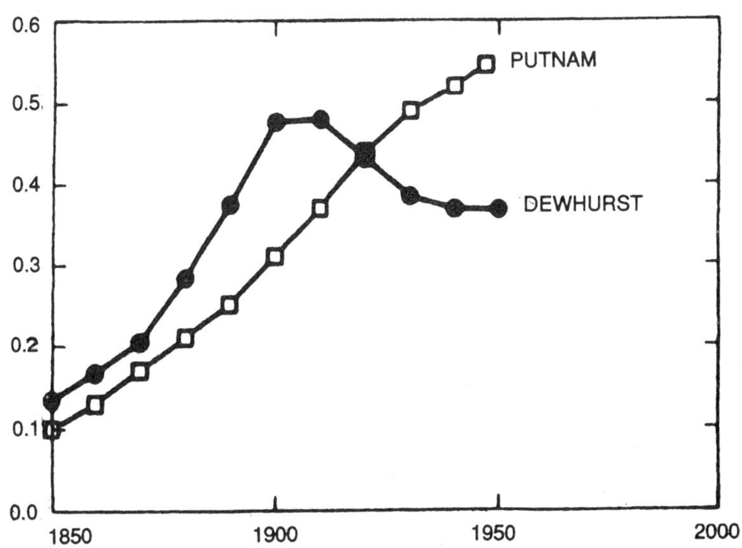

Figure 3.6. Putnam's and Dewhurst's Estimates of the Fraction of Energy Used for Work, United States, 1850-1950.
Source: Putnam (1953); Dewhurst (1955).

Table 3.3. Sources of Primary Energy, United States, 1870-1950

Energy Source	1870 Share by source	1870 Fraction of energy source devoted to work	1880 Share by source	1880 Fraction of energy source devoted to work	1890 Share by source	1890 Fraction of energy source devoted to work
Wind	0.00075	1	0.0006	1	0.00042	1
Water						
Direct drive	0.00231	1	0.00189	1	0.00124	1
Hydroelectric	0	1	0	0	0.00014	0.71053
Coal	0.26368	0.65501	0.40452	0.65136	0.57009	0.64023
Petroleum	0.00606	0	0.03003	0	0.0373	0
Natural gas	0	0	0.0015	0	0.0377	0
Fuelwood	0.72721	0.04004	0.56145	0.02996	0.35311	0.02
Total	1		1		1	

Energy Source	1900 Share by source	1900 Fraction of energy source devoted to work	1910 Share by source	1910 Fraction of energy source devoted to work	1920 Share by source	1920 Fraction of energy source devoted to work
Wind	0.00032	1	0.00019	1	0.00013	1
Water						
Direct drive	0.00087	1	0.00055	1	0.00058	1
Hydroelectric	0.00065	0.68945	0.00132	0.6399	0.00351	0.58008
Coal	0.72072	0.64702	0.78238	0.58043	0.73141	0.49936
Petroleum	0.03848	0.14982	0.07338	0.26991	0.14781	0.39004
Natural gas	0.02676	0.01959	0.03364	0.10005	0.04051	0.14504
Fuelwood	0.21219	0.013	0.10854	0.00994	0.07605	0.00699
Total	1		1		1	

Energy Source	1930 Share by source	1930 Fraction of energy source devoted to work	1940 Share by source	1940 Fraction of energy source devoted to work	1950 Share by source	1950 Fraction of energy source devoted to work
Wind	0.0001	1	0.00004	1	0.00001	1
Water						
Direct drive	0.00019	1	0.0001	1	0.00006	1
Hydroelectric	0.00557	0.51999	0.0074	0.48346	0.01088	0.42523
Coal	0.59847	0.40457	0.509	0.32855	0.38785	0.23822
Petroleum	0.23941	0.50998	0.31122	0.55002	0.36902	0.58001
Natural gas	0.09209	0.18999	0.11414	0.235	0.20301	0.27999
Fuelwood	0.06417	0.00504	0.0581	0.00306	0.02917	0.00189
Total	1		1		1	

Source. Dewhurst, Appendix 25-3, Tables J & K, pp. 1114-1115.

The gasoline share of petroleum began to expand from around 10% of refined crude in 1910, to around 20% by 1920, and to around 40% by 1930—a level which was maintained through 1955 (Schurr and Netschert 1960). Estimates for petroleum work-shares double between 1900 and 1910 and then increase by 50% between 1910 and 1920. Between 1920 and 1930, petroleum work-shares increased again by nearly one-third with the widespread introduction of the automobile. Dewhurst's estimated work share for petroleum in 1950 is 58%. According to Putnam, the transportation sector allocated 99% of its energy for work and consumed 55% of petroleum products in 1950 (Morrison and Readling 1968).

The work share of coal remained steady at 63% to 65% from 1870 to 1910, and then declined by nearly two-thirds between 1910 and 1950. This decline corresponds with the increased use of coal-fired central heating in commercial and residential buildings. Competition for railroad freight by petroleum-fueled trucks, conversion to the diesel locomotive, and the relative decline of the railroad industry also contributed to the diminished use of coal. Transportation claimed 20.8% of coal use in 1947, 14.3% in 1950, and only 3.7% in 1955.

Dewhurst claims that natural gas was used almost exclusively for heating and lighting prior to 1900 (Dewhurst 1955). However, Schurr and Netschert show that industry claimed about 70% of natural gas in 1906, and the residential and commercial sectors claimed 30% to 40% until 1920. Residential and commercial uses of natural gas fell to an approximate level of 25% after 1930 (Schurr and Netschert 1960). The two sets of figures are not necessarily in conflict since industrial uses cannot be assigned to either work or heating. There is no present basis for questioning Dewhurst's 28% work-share estimate for natural gas in 1950.

Use of Wood in the United States

Data on wood use are the most unreliable of those for any energy source, but clear trends are evident. Dewhurst estimates the shares of energy from wood in the United States at 90% in 1850 and 83% in 1860. Table 3.3 reports Dewhurst's estimates of the energy share from wood every 10 years from 1870 through 1950. The wood share declined at an increasing rate between 1870 and 1910—from 73% to 11%. We can deduce that most wood was used for space heating. The work share of wood was 5% in 1850 and 1860, 4% in 1870, and fell steadily in subsequent years.

For 1880, Schurr and Netschert (1960) and Schurr (1983) show some disaggregation of wood use in the United States that is tabulated in table 3.4. Their overall shares of energy from wood agree with Dewhurst's estimate for 1880. Relying on Sargent (1879), Schurr and Netschert (1960) estimate that the industrial share of wood use was between 3.5% and 5% in 1879-80, with railroads claiming 29% and manufacturing, 50% of industrial wood. Schurr's (1983) estimates imply that wood supplied 9% of energy use for industry and transportation and 85% of household energy-use in 1880. Overall, households consumed two-thirds and industry and transportation, one-third of the national supply of energy. Both sets of figures are comparable to those found for present-day low-income developing countries.

How we regard wood-based energy has major consequences for what we believe happens to the use of aggregate energy as per-capita income increases. If wood is excluded from the measure of aggregate energy, energy use per unit of output increases threefold and energy use per capita surges about fifteenfold between 1880 and 1910. If wood is included, however, the estimated increases drop to 20% and twofold.

Schurr (1983) argues that an energy-output measure that excludes wood is more appropriate because wood was not used extensively in production. Wood

Table 3.4. Fuel Shares in the United States, 1880

Household's total energy share	67%
Wood's share of household energy use	85%
Industrial and transportation total energy share	33%
Wood's share of industrial energy use	9%
Percent of total energy from wood	56%
Percent of wood used by households	95%
Percent of wood used by industry	5%
Percent of total energy from mineral fuels	44%
Household share of mineral fuels energy	25%
Industrial/transportation mineral fuels energy	75%

Source: Calculated from Schurr (1983), p. 213, fn. 4.

was, in fact, predominantly used for space heating in the household and commercial sectors. Coal, not wood, was the most important industrial fuel before 1880. On the other hand, omission of wood-based energy from energy-output measures introduces an asymmetry because coal, natural gas, and petroleum products that replaced wood in the residential and commercial sectors.

Currently, the residential and commercial sectors consume nearly 40% of energy used in the United States. Excluding the traditionally used energy sources of those sectors, but including the modern fuels that replaced them, introduces a large element of confusion and error. In addition, the income generated in supplying wood fuel is included in the measure of income, but the energy supplied is not. The magnitude of this error, however, is relatively small.

Use of Household Fuel

In the United States, heating was the major demand for household energy before the introduction of electrical appliances. Heating and cooking, however, are difficult to separate over a good part of the year. Lighting was produced with candles and whale oil between 1800 and 1860, with whale oil and sperm candles becoming less significant as fuel sources after 1830. Kerosene, first from asphalt and later from petroleum, was a cheaper lighting fuel. It entered the domestic market in the 1860s and was employed into the early twentieth century. Gas provided some urban domestic lighting from the

1830s through the 1890s. After a long dormant period as a domestic fuel, gas emerged as the major heating source for households after WWII (Schurr and Netschert 1960; Morrison 1982). Household work was performed with animate power until the introduction of some electrical appliances in the first decades of the twentieth century.

Few data exist on the distribution of household energy-shares for heating, cooking, and lighting in the nineteenth century. Even though wood was the dominant fuel for domestic heating and cooking, there is little record of its use. Cole (1970) attributes this paucity of records on fuelwood to the pervasiveness of wood use and the informality of its supply systems. Most urban wholesale-price indices from 1800 through 1860 include anthracite and bituminous coal, but not wood. Cincinnati is the only exception, with a 0.5% weight for cordage in a wholesale-price index for the period from 1824 to 1846, otherwise wood is not mentioned between 1816 and 1860 (Cole 1938). Lindstrom (1978) documents the importance of fuelwood in Philadelphia—in 1851, 35% of the coastal cargoes entering the city were fuelwood. In New York, the fuel and lighting weight, excluding fuelwood, in the wholesale-price index was 8.1% for 1860 (Cole 1938). However, Cole separately reports the sale of nearly 4 million cords of wood in New York State in 1864—"undoubtedly in considerable measure to Manhattan."

The entry of coal into the domestic fuel market for New York around 1800 (Cole 1938); Philadelphia in 1826-27 (Schurr and Netschert 1960); and Cincinnati in the 1820s, 1830s, and again around 1855 (Berry 1943) is documented by mostly anecdotal evidence. Berry notes, "In 1833, Pomeroy coal entered into domestic use in Ohio River towns. The operators of the mines tackled the transportation problem with towboats and barges, and approached housewives with a 'specific price to families' in an effort to induce them to change their cooking and heating habits." Table 3.5 reports the percentage of families using various fuels for domestic heating in 1880, 1908, and 1940.

The transportation cost of wood compared to coal encouraged increasing use of coal as urbanization concentrated demand for fuel into small areas. Lindstrom (1978) reports market ranges around Philadelphia in 1830 for cordwood—6 miles by wagon and 29 miles by canal—and for coal—19 miles by wagon and 175 miles by canal. Despite its lower transportation costs, charcoal remained almost exclusively an industrial fuel in the United States. In addition, the expense of urban land discouraged the stockpiling of large quantities of wood that would supply the same amount of energy as a smaller quantity of coal. The capital expense of manufacturing and installing coal-fired furnaces prevented the widespread use of coal as a heating fuel long after its technology had been developed. Residential furnaces became available in the 1850s, but only 1% of houses in the United States had central heating in 1920. By 1940, 42% of houses had central heating, and by 1970, 78% (table 3.6).

Table 3.5. Families Using Particular Domestic Heating Fuels in the United States, 1880-1940

	Total U.S.				Urban		Farm	
	Wood	Coal	Oil	Gas	Wood	Other	Wood	Other
1880	65%	35%	0	0	16	84	98	2
1908	36	63	1	0	9	91	95	5
1940	23	55	11	11	6	93	67	33

Source: Lebergott (1976), p. 276, Table 14.

Table 3.6. Penetration of Central Heating in U.S. Households, 1900-1970 (percent of households)

	1900	1910	1920	1940	1950	1960	1970
Total	<1	<1	1	42	50	66	78
Urban	<1	<1	2	58	63	76	79

Source: Lebergott (1976), p. 278, Table 15.

Compared to heating and cooking, lighting in the 19th century was a minor component of energy demand. Candles and nonpetroleum oil were the major lighting sources before 1860. At that time, kerosene entered the household lighting market, and despite some gas lighting, it dominated the market until 1910. Between 1910 and the mid-1920s, gas lighting prevailed, and it remained an important source of domestic lighting until 1940. Electricity lit 15% of American homes in 1910 and steadily replaced gas lighting thereafter. With increased effectiveness of lighting and decreased cost, lighting's share of household energy use in the 1980s was greater than in the beginning of the twentieth century. Lighting accounted for 19% of energy consumption by the commercial sector in 1979 (Samuels 1985b).

Use of Electricity

Initially, electricity was an energy source for industry in the United States. Figure 3.7, which includes industrial autogeneration, shows the initial low levels of electricity consumption by the residential sector and its slow increase to the current 30% level. Initial electricity rates for residential consumers were three to four times those for industrial consumers (table 3.7). The share of household energy supplied by electricity has been low in the United States (table 3.8). The relative rate of consumption for the residential sector fell more precipitously than the share of residential-electricity consumption, which rose from 1907 to 1970.

Table 3.7. Electricity Prices (cents/kWh), United States, 1902-1970[a]

	Residential[b]	Large Light & Power		Residential[b]	Large Light & Power
1902	16.20		1940	3.84	2.06
1907	10.50	2.70	1945	3.41	1.73
1912	9.10		1950	2.88	1.81
1917	7.52	2.10	1955	2.65	1.67
1920	7.45	2.83[c]	1960	2.47	1.69
1925	7.30	2.71[d]	1965	2.25	1.59
1930	6.03	2.66	1970	2.10	1.59
1935	5.01	2.46			

[a] Information insufficient to determine whether prices are current or constant dollar.

[b] Average of four average block rates.

[c] 1922.

[d] 1926.

Source: U.S. Dept. of the Census, Historical Statistics of the United States (1975), p. 827.

Table 3.8. Electricity Share of U.S. Household Energy Consumption, 1900-1980

	1900	1910	1920	1930	1940	1950	1960	1970	1980
Electricity (direct)	<1	<1	<1	1	2	4	9	15	25

Source: Morrison (1982), p. 219, Table 11.1.

Until 1930, lighting was the principal demand for household electricity. Table 3.9 shows the increasing use of major and minor household electrical appliances between 1900 and 1970. By the mid-1930s, the use of electrical appliances, and even electrical lighting, was still an upper-income phenomenon in urban areas (Rose and Clark 1979, see also Cowan 1976 and Rose 1984). Electric vacuum cleaners and washing machines were initial household appliances. Electric refrigerators were not generally used until 1940, and electric ranges were relatively rare even in 1950. Gas was a more common cooking fuel until the mid-1970s. In fact, the number of households using electricity for cooking surpassed those using gas only in 1974. By 1980, 52% of American households cooked with electricity, compared to 40% with gas (Samuels 1985b). Electric heating has been a more recent development. Houses heated primarily by electricity represented only 10% of new houses in 1950, but the numbers grew rapidly in the early 1960s and reached nearly 50% of new houses by 1980 (Samuels 1985b).

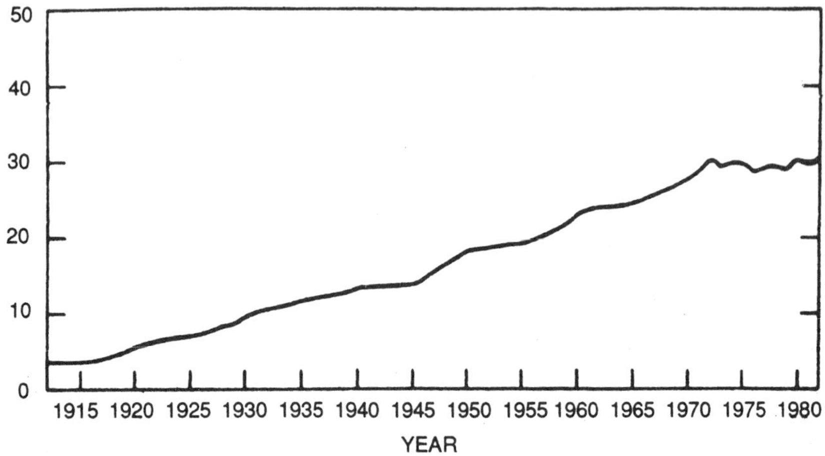

Figure 3.7. Residential Share of United States Electricity Consumption, 1912-1982.
Source: U.S. Department of Commerce (1975).

Developments in Transportation

Mechanization of transportation occurred first in ships and railroads. Coal was used in shipping on the Ohio River as early as the 1820s, although "wood...remained the staple fuel on the river at least as late as 1834..." (Berry 1943). Locomotives shifted from wood to coal in the early 1870s. Local and intraurban transportation, however, relied entirely on animal power until the late 1880s. In 1890, steam-powered street railways in the United States carried 287 million passengers, while horse-powered conveyances carried 1.23 billion (McKay 1976). In the early 1890s, urban electric-railways prevailed over horse-tram omnibus systems and clumsier, more expensive steam systems (McKay 1976). Intraurban freight traffic still moved by horse-powered wagon until the wide diffusion of the truck in the early 1920s.

Farm animals and draft animals for transportation never contributed more than 0.64% of total national energy between 1850 and 1920, when horsepower was widely used for local and intraurban transportation (table 3.10). The share of animal-based energy began to decline sharply from 1900 to 1910 (table 3.10), but the use of farm animals continued from 1920 through 1940. Animals were gradually replaced by the more energy-intensive, mechanized technologies for transportation. In addition, the passengers and freight carried by these new systems reflected a relatively high-income society. They were able to offer a high value-added relative to the capital and energy used.

Table 3.9. Electric Appliance Penetration, United States, 1900-1970 (percent of households owning)[a]

	1900	1910	1920	1930/32	1940/41	1950/53	1960	1970
Washing machine	0	0	8	24	50	70	73	70
Refrigerator (mechanical)	0	0	<1	8	44-51	80-84		99
Range (electric)	0	0	1[b]	4	12	21		56
Dishwater	0	0	0	1		3		26
Vacuum cleaner	0	0	11[b]	30	41	54		92
Iron	0	0	0	65		82		100
Toaster	0	0	0	27		65		93
Radio	0	0	0	61		94		100
Television	0	0	0	0		43		99
% of dwellings with electric service		16[c]	35	68	79	94	99+	99+

[a] Lebergott's figures are for percents of U.S. households while Dewhurst's are for percents of wired households. An attempt has been made to adjust Dewhurst's figures to percents of U.S. households.
[b] 1925.
[c] 1912.

USE OF ENERGY IN THE UNITED KINGDOM: COAL AND THE HOUSEHOLD SECTOR IN THE NINETEENTH CENTURY

"Coal was king" in the United Kingdom from an early date. It was used as a domestic fuel because of rising prices for wood following severe deforestation. A successful reforestation effort was completed by the last quarter of the seventeenth century, so that charcoal did not replace coal (coke) in the iron industry until the early eighteenth century. Charcoal was still the preferred fuel for iron forging until the spread of Cort's puddling process in the early 1790s (Ashton 1962, Hyde 1979). Pawson (1979) notes that steam power was not generally employed in the United Kingdom by 1800 and that "for far longer than convention has admitted, the water wheel, windmill, horse-gin and manually driven machine persisted." He cites a 2.5 to 1 ratio of hand looms to power looms as late as 1830. Coal first replaced other fuels and sources of power in the iron industry, then in the steel and copper industry, and later in the textile industry. In 1812, the value of textile output was three times that in the iron, steel, and copper industries. It is widely believed that coal supplied nearly 100% of the energy for domestic use in the nineteenth century. The following discussion explores the empirical grounds for this belief.

Figure 3.3 shows that the United Kingdom doubled its energy-output ratio between 1830 and 1880, and then it gradually declined. That trend in energy intensity—called "the development effect" in energy use—has been identified for a number of countries, including the United States (see figure 3.1). We expect that developing countries will exhibit this trend as well. However, I would like to cast reasonable doubt on the history of energy use that is indicated by the data for the United Kingdom in figure 3.3.

Table 3.10. Animal Energy Shares, United States, 1850-1940 (percent)

	1850	1860	1870	1880	1890
Percent of total primary energy supplied by animals evaluated by horsepower-hour contributions	0.57	0.61	0.53	0.55	0.51
	1900	1910	1920	1930	1940
	0.45	0.28	0.18	0.11	0.07
	1850	1860	1870	1880	1890
Percent of total primary energy supplied by animals evaluated by fodder requirements	0.6	0.64	0.56	0.58	0.53
	1900	1910	1920	1930	1940
	0.47	0.29	0.19	0.12	0.07

Source: Dewhurst (1955), pp. 1114-1115, Tables J and K. Fodder adjustments made with Revelle's (1976) estimates for Indian bullocks.

Humphrey and Stanislaw (1979) omitted energy from wood, wind, and water in their estimates of energy consumption in figure 3.8. However, their estimates of energy supplied by wind and water in 1760 represent an extremely wide range of possibilities—from 5% to 164% of coal output. They did not attempt to quantify later consumption of fuel other than coal and particularly warned their readers of their exclusion of Irish peat as a fuel. The crucial question is whether consideration of only coal in the nineteenth century seriously underestimates energy consumption. Industrialization occurred much earlier in the United Kingdom than in the United States, so it may be safe to assume that noncoal sources of industrial energy-consumption were negligible. The household sector is the principal place where we may find evidence of using fuel other than coal in significant quantities.

Deane and Cole (1962) estimate that the domestic sector used 33.2% of coal produced in the United Kingdom in 1840, 18.7% in 1869, and 20.6% in 1887. Using the appropriate census data, these shares imply annual per-capita consumption rates of coal—0.574 tons, 0.684 tons, and 0.616 tons for these three years. The Flux-Jevons estimate of per-capita coal consumption in 1869 which is not independent of the later Deane and Cole estimate for that year, since both were based on the 1871 Coal Commission report—was 0.672 tons (Jevons 1903). That amount for 1869 was prepared by Flux in his third edition of reworking Jevon's 1865—66 estimate of domestic coal consumption. Jevons had allowed one ton per person, and Flux dropped the estimate to 12 cwts (cwt = hundredweight, a unit of weight in the British Imperial System equal to 112 lbs.) on the basis of the 1871 Coal Commission report.

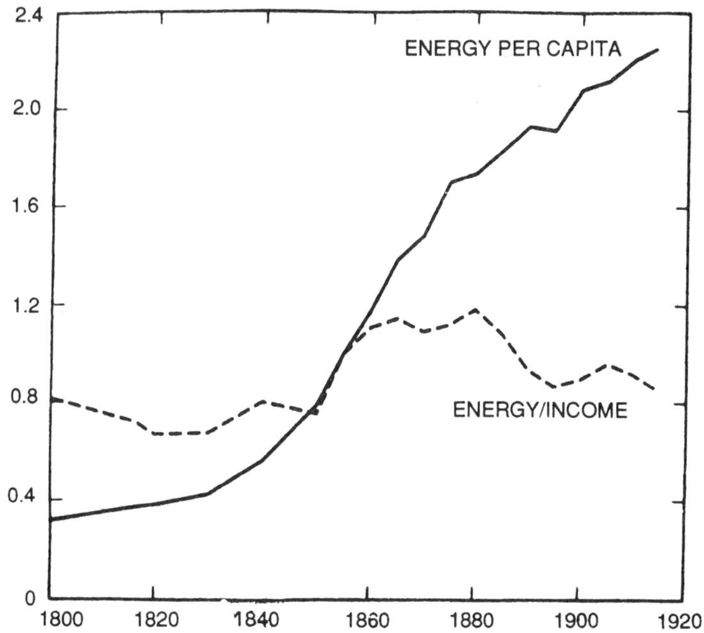

Figure 3.8. Energy Per Capita and Energy/Income Ratio, United Kingdom, 1800-1914.
Source: Humphrey and Stanislaw (1979), *Energy Policy 7:29–42.*

These estimates of per-capita consumption are substantially below comparable estimates for the United States for 1880. Lebergott (1976) estimates urban consumption of coal in the United States at seven tons per household per year. An average family of three persons—the model size for a sample of Rhode Island families in 1875, when one-third of families had six or more persons (Pryor 1972)—yields an annual per-capita consumption of 2.33 tons of coal. An independent estimate for 1880, based on Putnam's estimate of total energy consumption (1953) and Schurr's (1983) estimate of the household share of energy consumption, yields 3.112 tons of coal-equivalent per capita per year. Coal prices in the United States and the United Kingdom were roughly similar. The British retail prices were possibly a bit lower, but American wood prices were very cheap, which is reflected in the higher estimate of consumption based on Putnam's estimate of total energy that includes wood. To reduce the per-capita consumption in the United States to the 1869 rate in the United Kingdom, the average size of the American family would have had to have been over 10 persons per household. It is possible that British heating practices in the nineteenth century kept indoor winter temperatures cooler than their counterparts in the United States. However, it is difficult to believe that different heating practices could account for a 3.4-fold difference in fuel consumption or that lower wood prices in the United

States coupled with different heating practices could account for a 4.6-fold difference.

The most plausible reconciliation of the estimates for consumption in the United Kingdom with that in the United States is that the British household sector extensively used peat and other fuels, rather than coal, in the mid- and late nineteenth century. Gilboy (1934) cites the use of wood, peat, and cow dung as fuels in the United Kingdom in the late eighteenth and early nineteenth centuries. Cole and Postgate (1939) note that dung was still regarded as a marketable commodity in Lancashire in the early 1840s, although its use as fuel or fertilizer is not specified. Further direct evidence on the use of fuel in the household sector has not been found for the United Kingdom in the nineteenth century.

The share of energy consumption for the household sector—33.2% cited by Deane and Cole for 1840—seems low for the mid-nineteenth century, but it is not entirely implausible for a country as industrialized as the United Kingdom at that time. However, the 43% drop in the household share to 18.7% in 29 years is less plausible, despite the fact that the smaller share implies a slight increase in per-capita household consumption. The residential share in the United Kingdom was 27% in 1982. Eighteen percent would be a low household-sector share in a developed country in the 1980s, and middle-income developing countries have household-sector shares that range from 30% to 45% of national energy consumption. It is possible that rapidly increasing consumption of coal for railroads could account for the drop, but Deane and Cole's (1962) data by sectoral distribution do not identify railroads. In 1841, 67% of the population in the United Kingdom was still rural. It is possible that biomass fuel contributed half of the energy supply for household consumption, if we assume that urban households used more fuel and that rural households used some coal. It may be possible that by 1869, an increased demand for coal by both railroad and steam navigation induced substitution of biomass fuels for coal in households. In 1871, 45% of the population was still rural, so some small supplementation of coal with wood for urban domestic use could mean that biomass fuels were still contributing 50% of energy for the household sector in 1869. These estimates that 50% of fuels used in households were biomass in 1840 and 1869 are conservative, but they increase the share of national energy consumption from the household sector and reduce the steepness of the increase in the energy-income ratio in figure 3.3. This is the aim of the exercise.

Figure 3.8—which presents energy per-capita and energy-income ratios for the United Kingdom (England, Wales, Scotland, and Ireland) from 1800 to 1914—offers another hint of exclusion of biomass fuel from estimates of energy consumption in the United Kingdom. The energy-income ratio has been calculated independently of Humphrey and Stanislaw's (1979) estimate. Figure 3.8 should be examined with comparable figures for the United States in figures 3.1, 3.2, 3.4, and 3.5. The data for the United Kingdom show a 7.17-fold increase in energy consumption per capita between 1800 and 1914, but only a 1.78-fold increase between the lowest and highest values of the energy-income ratio (E/Y). The rise in per-capita income in the United Kingdom, relative to the rise in E/Y, looks much more like figures 3.1 and 3.2 for the United States, which omit wood use, than it does like figures 3.4 and 3.5, which include wood. In fact, working backwards in time in figure 3.8, energy consumption per capita in the United Kingdom would look much like the pattern for the United States in figure 3.5, if we allow for probable price differences for wood and if the solid line flattened out around 1850 at an index of around 0.75 and continued back to 1800 largely unchanged. That would imply about a 50% share of wood or other biomass fuel in aggregate energy consumption in the United Kingdom during the first half of the nineteenth century. This result agrees with my previous suggestions.

PROFILES OF ENERGY USE IN DEVELOPED AND DEVELOPING COUNTRIES

This section compares some aggregate characteristics of energy use in developing countries with those in developed countries. The information reviewed in the previous two sections on past energy use in the United States and the United Kingdom adds perspective to the present snapshot comparisons of developed and developing countries. The purpose of this comparison is to gain insights into current and prospective energy problems in developing countries.

Demands for Energy by Sector

Three major sectoral classifications of users—residential, industrial, and transportation—account for 70% to 80% of national energy demand. In earlier periods, the residential sector of industrialized countries consumed as much as 80% or more of national energy supplies. The economic history of energy use recounts the shift of energy consumption from households to industry and transportation. Currently, developing countries face the task of shifting their energy consumption from households to industry and transportation, just as industrialized countries have done over the past 100 to 150 years. However, the technologies involved in the competition among sectors have changed, and some major end-use characteristics of demand differ.

Use of Residential Energy

Table 3.11 shows residential-energy shares of national energy consumption for 12 developing and 11 developed countries. The household sector has the largest claim on energy supplies in developing countries with lower incomes—e.g., Burma, Sudan, Togo—and still consumes around one-third in mid-level developing countries—e.g., Ecuador, Peru, Thailand. The household sector,

Table 3.11. Residential Energy Shares

A. Developing Countries	Year	Percent of Total Energy	Percent of Biomass Energy	Traditional Fuel Share of Household Sector Energy
Liberia	1983	69.7	3.0	97.9
Ecuador	1984	32.8	84.8	57.4
Solomon Islands	1981	40.7	56.2	94.3
Fiji	1982	15.9	17.9	64.0
India	1976	36.0	94.7	79.8
Pakistan	1984/85	48.3	88.7	81.7
Burma	1982/83	86.4	100.0	99.3
Sudan	1987	77.8	93.0	98.4
Peru[a]	1970	41.3	80.1	78.3
Peru[a]	1975	38.4	78.9	73.6
Peru[a]	1981	39.2	85.9	69.3
Bolivia[a]	1981	46.7	83.4	74.4
Togo[b]	1982	71.3	99.7	96.4
Thailand	1983	32.8	75.4	85.9
1980s Average		51.1	80.4	83.5

[a] Includes the commercial sector.
[b] Includes artisanal activity.

B. Developed Countries	Percent of Total Energy 1970	Percent of Total Energy 1982
United States	22.0	18.7
Australia	7.3	12.4
Austria	32.3	32.0
Belgium	31.4	26.4
Britain	26.0	26.9
France	27.9	27.5
Italy	26.4	17.6
Japan	16.8	22.7
Netherlands	30.6	27.5
New Zealand	23.0	16.2
West Germany	32.8	24.1
Average	25.1	22.9

Sources: UNDP/World Bank, Issues and Options in the Energy Sector (various dates); Pakistan National Energy Balance, 1985/85, OECD Energy Balances, 1970/82 (1984).

which accounts for over 75% of energy used in the economy, is the major user of traditional, biomass fuels in every country, except Fiji and the Solomon Islands. Additionally, traditionally used fuels supply over one-half of the household sector's energy consumption. This proportion compares roughly with the situation of the United States in 1880 (table 3.4), except that modern transportation is a much higher claimant of energy in developing nations.

In 1880, most household energy in the United States went to heating, which is a minor use in most developing countries, where household fuel generally goes to cooking. For example, in 1978, 83% of wood and 91% of charcoal used in Kenya went for cooking, and the remainder, for heating. Kenya's altitude makes its demands for heating higher than those in many developing countries, although India and Pakistan have mountainous areas with high heating demands as well. The United States made over a fivefold improvement in energy efficiency of household heating systems between 1880 and 1950 because of major technological changes.

Improving cooking efficiency is of paramount importance today in developing countries. The end-use efficiency of air-dried wood in cooking is 5% to 10%, and that of charcoal is 15% to 25%, excluding the three to sixfold ratio of wood to charcoal. Charcoal is an urban household fuel in an urbanizing Third World. Households in the United States shifted directly from fuelwood to coal without a transition of using charcoal, but household sectors in most developing countries will not eliminate that step. Kerosene is the modern form of energy that is most likely to replace charcoal for cooking as income increases. Kerosene has an efficiency of 30% to 40%. LPG (liquefied petroleum gas), with an efficiency of 45% to 60%, is available is some developing countries. Assuming the net technical efficiency of fuelwood and charcoal to be roughly the same—say 7.5%—the improvements in energy efficiency alone that are possible by shifting to kerosene and LPG range from 4.6- to sevenfold. Demands for energy in the household sector may increase markedly before appliances and fuel substitution decrease demands. Major shifts to modern gas and electric cooking in developing countries are not likely in the near future.

An issue often raised about fuels for cooking is the effect of fuel substitution on culinary habits and preferences. Certainly, American culinary habits have changed substantially over the past hundred years with changes in cooking technologies, food prices, and food tastes. Whether food habits in other societies will change substantially or whether the culinary change is peculiarly American is an open question.

Household-energy shares in developed countries are currently less than half those in developing countries. They have continued to fall between 1970 and 1982. There are smaller shares for household energy in developed countries despite a greater range of energy-using activities in households than is common in households of developing countries. Biomass fuels are used more extensively in households of less-developed countries because of their low relative prices. Greater end-use efficiency from modern fuels accounts substantially for the myriad energy-using activities, with a small share of national energy use, in the household sector of developed countries. Jones and Younger (1988) estimate that substitution of kerosene for fuelwood and charcoal in Sudanese urban areas technically could reduce the end-use energy from 49% to only 7.7% of national energy consumption, although there are a number of economic and political barriers that obstruct a shift of that magnitude.

Use of Industrial Energy

Table 3.12 shows energy consumption shares from industry for the same group of developing and developed countries. Industry in the developing countries claims about 19% of national energy supplies, compared to roughly 37% for the developed countries. However, the industrial sectors of developing countries claim a much larger share of modern energy, an average of 30%. This fact may be more relevant as a basis of comparison. The residential sector is a dominant user of biomass fuels in developing countries, and modern industry will rely increasingly on modern energy as development continues. In the developed countries, shares of energy consumption are very close to the industrial contribution to GNP. The same is true for developing countries. Compared with developed countries, industrial activity in developing countries is relatively energy-intensive.

However, we must be careful when we interpret the ratio of energy shares to GNP shares. Since agriculture is a relatively much larger contributor to GNP, the share-ratio in developing countries is biased upwards relative to that in developed countries. Agriculture in developing countries uses almost no energy. In addition, several other factors contribute to this high ratio in developing countries—particularly, industry, poor operational efficiency and subsidized energy prices relative to output prices.

Another problem with a simple interpretation of the ratio of energy consumption shares to GNP shares (table 3.12) is that a single industry can cause a large shift in the ratio in smaller countries. For example, Liberia appears to have a remarkably energy-efficient industrial sector, but mining, which is excluded from both GNP and energy consumption in table 3.12, contributed 12.5% of GNP in 1981 and used 34% of Liberia's modern energy in 1983. In Fiji, the much smaller percent of nonbiomass energy relative to the percent of total energy used in industry reflects the industrial mix of the country. Sugar refining, which is fired mostly from bagasse and a small amount of wood, is a major industry that uses 84% of the industrial energy supply (East-West Center 1982a). Agriculture is even more dominant in the Solomon Islands where agricultural processing is included in primary industry accounts (East-West Center 1982b). The ratios for the larger developing countries—India, Pakistan, and Thailand—are less subject to influence by single industries.

Use of Energy for Transportation

Table 3.13 shows shares of national energy consumption from transportation for the same developing and developed countries. The transportation sector uses more modern energy than the rest of the economy. Its shares of total energy supplies for developing countries are close to or lower than those in developed countries. However, transportation claims much higher shares of modern energy in developing countries and much higher shares of energy than developed countries during their earlier transportation revolutions.

Table 3.12. Industrial Energy Shares

A. Developing Countries	Year	Percent of Total Energy	Percent of Non-Biomass Energy	Share of manufacturing, construction & electricity, gas & water in GNP, 1981
Liberia	1983	1.0	3.0	11.4
Ecuador	1984	17.3	18.9	20.0
Solomon Islands	1981	2.9	1.1	
Fiji	1982	58.7[a]	25.3	21.5[b]
India	1976	39.3	53.4	23.9
Pakistan	1984/85	29.9	47.7	24.5
Burma	1982/83	4.9	34.5	11.4
Sudan	1981	5.8	31.5	13.8
Peru	1970	14.3	21.4	28.3
Peru	1975	17.6	24.4	30.7
Peru	1981	18.2	23.4	29.4
Bolivia	1981	19.6	28.0	20.2
Togo	1982	16.7	53.7	15.5
Thailand	1983	24.2	29.1	26.8
1980's Average		19.2	30.3	21.3

[a] Includes the commercial energy use.
[b] 1908.

B. Developed Countries	Year	Percent of Total Energy	Share of manufacturing, construction & electricity, gas & water, in GNP 1981
United States	1960	33.3	35.8
United States	1965	32.6	36.1
United States	1970	32.9	33.1
United States	1975	30.6	30.7
United States	1982	28.0	30.1
Australia	1982	39.1	29.9[a]
Austria	1982	32.6	38.6
Belgium	1982	41.5	36.2[b]
Britain	1982	31.8	27.6
France	1982	34.8	33.9[c]
Italy	1982	37.9	41.4[c]
Japan	1982	49.4	42.3[c]
New Zealand	1982	37.8	30.6
West Germany	1982	35.8	47.0[b]
1982 Average		36.9	35.8

[a] 1979.
[b] 1980.
[c] 1981.

Sources: UNDP/World Bank, Issues and Options in the Energy Sector (various dates); Pakistan National Energy Balance, 1984/85; OECD, Energy Balances, 1970/82 (1984); East-West Center, Energy Mission Report: Solomon Islands (1982).

Table 3.13. Transportation Energy Shares

A.

Developing Countries	Year	Percent of Total Energy	Percent of Non-Biomass Energy	Transportation (& communication) shares of GNP (at factor cost), 1981
Liberia	1983	16.1	51.0	7.8
Ecuador	1984	38.8	49.9	6.8
Solomon Islands	1981	8.2	25.8	
Fiji	1982	24.1	55.9	9.9
India	1976	16.8	24.1	6.8
Pakistan	1984/85	11.9	21.4	7.7
Burma	1982/83	7.6	53.5	5.6
Sudan	1981	11.0	61.9	10.5
Peru	1970	21.8	36.6	5.4
Peru	1975	24.8	38.7	6.5
Peru	1981	25.3	37.0	6.8
Bolivia	1981	30.7	52.6	11.4
Togo	1981	11.1	35.8	7.2
Thailand	1983	32.9	52.5	6.5
1980's Average		18.7	42.7	6.7

B.

Developed Countries	Year	Percent of total energy (= % non-biomass energy)	Transportation (& communication) shares of GNP (at factor cost, 1981)
United States	1960	27.4	6.6
United States	1965	27.6	6.4
United States	1970	28.2	6.4
United States	1975	33.2	6.2
United States	1982	33.8	6.8
Australia	1982	35.9	8.1[a]
Austria	1982	24.5	7.3
Belgium	1982	18.6	6.9[b]
Britain	1982	24.9	7.7
France	1982	25.8	6.3[b]
Italy	1982	25.3	7.0
Japan	1982	20.1	6.3[b]
New Zealand	1982	33.4	8.3[b]
West Germany	1982	21.9	6.3
Average (1982)		26.4	7.1

[a] 1978.
[b] 1980.
Sources: UNDP/World Bank, Issues and Options in the Energy Sector (various dates); Pakistan National Energy Balance, 1984/85; OECD, Energy Balances, 1970/82 (1984); World Bank, (1983).

Table 3.2 shows estimates of industry and transportation shares in the United States at 53% to 55% in 1920, when the initial effect of the automobile and truck was felt on energy demands. Transportation shares in GNP are nearly the same in developing and developed countries, but this sector uses nearly double the share of modern energy in developing countries. The share of petroleum products used by the transportation sector are even higher in developing countries.

Further, the extensive latent demand for freight presents another impressive aspect of the transportation sector. For example, in India in 1976, 14% of total rural energy was derived from bullocks—16% in transportation and 84% in agriculture (Revelle 1976). In comparison, in 1850, only 0.64% of the total national energy supply of the United States was contributed by animals. Rural transportation and agriculture powered by animals in India represents an enormous potential addition to the demand for modern fuel when those sectors become mechanized. In Nairobi, in 1976, 30% of trips taken by middle-income residents and 65% to 80% taken by lower-income people were on foot or bicycle in 1976 (McGranahan et al. 1979). These urban pedestrians also represent substantial additions to demands for transportation fuel over the next generation.

The transportation sectors of most developing countries are carrying relatively low value-added cargoes and passengers with relatively high-cost capital equipment and fuels. Both equipment and fuels for transportation trade in international markets at international prices. Despite motorized passenger transportation with high people loads, the result is a high energy cost to provide a service with a relatively low final value. There is little substitutability for the short-term between capital and fuel, and substitutions between labor and capital—e.g., reduced maintenance—cause more fuel use per unit of service.

Fuel and most equipment entail foreign-exchange costs. Therefore, the way they are used must yield foreign-exchange earnings within a fairly short time to avoid increasing indebtedness that will contribute to macroeconomic problems. Intersectoral linkages are often weak in developing countries. Even in many developing countries that are more economically advanced, policies for industrial protection have inhibited the development of intersectoral linkages. In addition, imports are absorbed by a domestic production sector and may generate indirect export earnings only slowly and haltingly (Chenery et al. 1986). Many present-day developed countries either entered the modern age of transportation with more sophisticated structures for interindustry linkage, which enabled them to convert imports into foreign exchange or sources of domestic fuel for modern transportation, or they were able to rely to a large extent on domestic fuel sources for modern transportation, or both.

Few options for intermediate technology exist in modern transportation between head loads and bullock carts, on the one hand, and modern motor vehicles, on the other. However, subsidization of some or all motor fuels exacerbates many problems for developing nations. Non-highway transportation—rail and water—is not highly developed. According to a sample of developing countries from 1976 to 1982, an average of 85% of energy for transportation was for road transport, while a sample in 1976 of developed countries showed an average of 83%. India has a much more extensively developed rail system than most other countries. In fact, developed and developing countries have similar structures of energy use for the four major modes of transportation—air, road, rail, and internal water. Yet, this classification is misleading because it disguises other modes of transportation, such as pipelines for liquid fuel, which are not major options for most developing countries.

Industrialized countries developed rail and inland-water transport systems for freight before the advent of trucks, and these systems were appropriate for income levels at that time. In urban passenger transportation, the electric intraurban streetcar and interurban train were an intermediate technology. Both were widely used in industrialized countries before the shift to motor vehicles and petroleum-fueled transport. These more fuel-intensive, mechanized transportation systems were introduced in more economically advanced circumstances.

The developing countries have modeled their initial transport systems on those of economically advanced countries. Except for rail systems in many British colonies, they have usually skipped the intermediate development stage. But what were the options of the developing countries? First, both rail and inland-water transport require large investments in infrastructure, which are probably greater than those required for motorized road transport. Second, rail systems require fairly sophisticated management inputs, which are quite scarce in most developing countries. The high degree of divisibility and low management requirements of motorized road transportation avoid these two problems. The early emphasis on roads in developing countries may well be a rational response to their limited resources and has been encouraged to some extent by government policies such as subsidies for motor fuel. Additionally, the inefficiencies of existing railroad systems in developing countries—in circumstances in which a simple comparison of rail and road tariffs would appear to favor rail transport—encourage shipping even bulk loads by road rather than rail (Mwase 1987).

The pattern and evolution of energy shares for transportation in India supports this explanation of "the transportation energy imbalance" in developing countries. In 1976, 54% of India's energy for transportation was used by the rail system, and 41% by road transport. The rail systems of other developing countries—Bolivia, Togo, Nigeria, Brazil, Indonesia, Mexico, and Egypt—used an average of around 2% of energy for rail systems and 75% to 90% for road transport between 1976 and 1982. India's transportation consumed a correspondingly low 24% of the country's modern energy in 1976, compared with an average of 43% for 11 other developing countries in the early 1980s (table 3.13). In table 3.13, Pakistan shows a similarly low transportation share in national energy consumption, between 1976 and 1984, and its rail share of transportation energy, though smaller than India's, is larger than that of most other developing countries, in the range of 14% to 18%. Since 1976, India's road transportation has increased its share of energy consumption in the transportation sector at the expense of the rail system, but the overall share of national, modern energy consumption from the transportation sector had risen to only about 26% by 1982. Between the mid-1970s and the early 1980s, the quantity of energy used in the rail transport sector in India decreased by about 16%. The share of rail energy has fallen relative to that of road transport. A plausible hypothesis is that while road transport has boomed in India, shares of freight and passenger volume have not shifted as heavily toward road transport as have those of energy consumption. Consequently, the overall transportation share in national energy consumption has not risen.

As I described earlier, the industrialized countries had the opportunity to develop their transportation systems gradually over 60 to 80 years—from rail and steamboat systems to automobiles and trucks. Mechanization of industry preceded most of the revolution in petroleum-based transportation and increased the value-added of cargoes. Developing countries have not had this luxury. Indeed, their industrial mechanization lags behind the mechanization of their transportation systems.

Consumption of Electricity

Table 3.14 compares important, characteristic patterns of electricity consumption in developing and developed countries. Electricity comprises a smaller share of total energy supply in developing countries, whether measured as a proportion of total energy or only modern energy. At the same time, the residential share of aggregate electricity consumption is somewhat larger in developing countries (see table 3.14, panel A). Developing countries have household shares of electricity consumption that are much higher than those of the United States (figure 3.7) or the United Kingdom (table 3.15) in comparable periods when their electricity systems were being developed. The residential shares cited in panel A of table 3.14 are recorded data. They do not reflect the substantial loads from unreported residential hookups which are currently reported as part of "losses." The small consumers typical of residential loads are more expensive to service, and their bills may be harder to collect than bills from large industrial consumers. Additionally, residential loads may contribute more to peaks than do industrial loads, which can be spaced more easily. In developing countries with a low percentage of residential loads—such as Egypt with about an 18% residential-load share in 1975 (DOE 1979)—these loads often contribute heavily to peak demands, and thus degrade the overall system far more than their low load-shares would suggest.

Low prices for residential electricity encourage use. Industrial users are discouraged from relying on grid-supplied electricity in many developing countries because of its unreliability, a reflection of the financial problems that utilities have. Subsidized fuel costs for thermal generation or capital and maintenance costs for hydrogeneration are not usually covered by the collected revenues, so the utility becomes a burden on the national treasury or the international donor community. Low-income developing countries can ill-afford the low value-added activities provided by 25% to 50% of their electricity generation. Lighting is the most prevalent domestic use. Televisions are remarkably common, even in rural areas. Domestic refrigerators, irons, and room air conditioners are also popular household appliances in homes with somewhat higher incomes.

Table 3.14. Characteristics of Electricity Consumption

A.	Developing Countries	Year	Electricity as % of Total Energy Consumption*	Electricity as % Modern Energy Consumption	Residential Share of Electricity Consumption*
	Liberia	1983	8.5	27.1	35
	Liberia	1986	2.4	8.8	61.9
	Ecuador	1984	6	7.8	65.9
	Solomon Islands	1981	1.3	4.2	32.1
	Fiji	1982	8.3	19.2	20.5
	India	1976	6.8	10.9	8.7
	Pakistan	1984/85	7.0	12.5	45.1
	Burma	1982/83	1.1	7.5	30.1
	Sudan	1981	1.0	5.5	38.3
	Peru	1970	5.6	9.3	28.8
	Peru	1975	6.5	10.2	31.9
	Peru	1981	8.0	11.7	33.0
	Bolivia	1981	5.8	9.9	44.4
	Togo	1981	3.5	11.4	21.3
	Thailand	1983	8.3	13.2	24.7*
	1980's Average		5.3	11.8	6.7

* Excludes autogeneration; uses 1983 figures for Liberia.

B.	Developed Countries	Electricity as % of Total Energy Consumption 1982	Residential Share of Electricity Consumption 1982
	Australia	15.1	36.2
	Austria	15.8	27.1
	Belgium	11.9	33.2
	Canada	18.5	29.6
	France	14.4	30.1
	Italy	13.9	25.4
	Japan	19.1	24.0
	Netherlands	10.1	27.9
	New Zealand	24.5	40.8
	Norway	41.7	31.8
	Sweden	24.7	27.0
	Switzerland	19.2	25.9
	United Kingdom	14.4	36.9
	United States	13.7	36.2
	West Germany	15.0	27.7
	1982 Average	16.4[a]	29.9[b]

[a] Excluding Norway; 18.1 with Norway.

[b] Excluding Netherlands; 30.7 with Netherlands.

Sources: UNDP/World Bank, Issues and Options in the Energy Sector (various dates); Pakistan National Energy Balance, 1984/85; OECD, Energy Balances, 1970/82 (1984); GOL/DOE (1987).

The share of electricity consumption for the developing countries in table 3.14 exclude autogeneration, which is a large nonresidential component of energy—perhaps of substantial magnitude—in many countries. Higher-income residences, which are connected to the national grid, often have oil-fired back-up power systems which are used during periods of grid failure. In 1983, capacity for autogeneration in the Sudan was estimated as equivalent to one-half of the National Electric Corporation's supply from the grid and isolated systems (UNDP/World Bank, July 1983). In 1981, autogeneration by iron-mining companies in Liberia amounted to 114% of the supply of electricity generated by the grid—Liberia Electricity Corporation, or LEC (Samuels 1985a). Independent research estimated that the electricity autogenerated by iron-mining companies in Liberia in 1986 was 283% of LEC-supplied consumption—excluding transmission losses of LEC (GOL/DOE 1987). Generation losses from the mining companies in 1986 were 68%, comparable to generation losses of 69.4% for LEC. The commercial and agricultural/forestry sectors had slightly higher generation losses of 76.2% and 72.0%, although relative economic efficiency cannot be inferred without information on generation costs. Thus, the additional alleged pressure on supplies of fuel oil and diesel from autogeneration is only marginally higher than the pressure on petroleum that supplies the national grid for the same load. This marginal difference in Liberia probably characterizes small-scale autogeneration in other developing countries as well, and it may evaporate for large, industrial autogenerators.

In general, the addition of autogenerated electricity to national energy supplies would reduce the residential share of total electricity consumption for most developing countries. However, the residential share of publicly supplied electricity is a good inverse indicator of the financial strength of the public electricity system. In the early years of electrification in the United States, industrial autogeneration was also common. In 1919, 43% of electric-motor capacity in manufacturing was autogenerated (Devine 1983; also see Woolf 1987), and by 1940, 21.5% of electricity in the United States was still autogenerated by industry. Autogeneration fell below 1% in the United States in 1961. Figure 3.7 shows shares of residential consumption in total production of electricity in the United States, including industrial autogeneration. Excluding industrial autogeneration in figure 3.7 to convert shares of residential consumption to shares of utility-produced electricity would raise those shares by as much as 42% in 1920, 25% in 1940, and 10% in 1960. However, these larger residential-consumption shares of grid-produced electricity in the United States are still smaller than those in most present-day developing countries.

Shares of electricity consumption by households in developed countries have only recently reached their current levels, although they have been subsided in some Western European countries since the early 1970s. Figure 3.7 shows the long, slow growth in the share of residential electricity in the United States, and table 3.15 shows a similar trend in the United Kingdom from 1895 to 1982. In the first decade of electricity supply in the United Kingdom, commercial and public lighting were the principal loads, but industry and railroads very rapidly dominated. In 1920, the first year for which separate data for domestic loads exist, the domestic share was 8%, compared to 80% for industry and railroads. The residential share in the United Kingdom increased more rapidly than its counterpart in the United States. By 1950, it had reached levels that were achieved in the United States in 1980. In 1974, residential shares in the United Kingdom peaked at 41%, a high figure by current standards of developed countries. The ratio of residential to industrial prices for electricity in the United Kingdom fell from 3.4 in 1921 to 1.5 in 1948, similar to the decline in relative rates in the United States during the same period (Hannah 1979). The patterns in both the United States and United Kingdom are consistent with initial installation of transmission and distribution systems and their subsequent expansions, which were financed first by industrial users. When these systems were mostly in place and paid for, costs of additional residential hookups were much lower and limited to kWh charges and some portion of additional costs for capacity. The current sectoral structure of electricity loads in developing countries looks as if a different strategy of financing transmission and distribution costs has been followed.

Table 3.15. U.K. Electricity Consumption Shares, 1895-1982 (percents)

Year	Domestic	Industrial and Traction	Year	Domestic	Industrial and Traction
1895	N.A.	<1.0	1935	22.0	60.7
1897	N.A.	1.7	1939	26.7	58.2
1898	N.A.	4.7	1940	25.7	61.9
1899	N.A.	5.0	1945	28.1	60.3
1900	N.A.	815.7	1946	33.5	54.6
1905	N.A.	45.4	1948	33.5	52.9
1910	N.A.	62.1	1970	37.7	42.2
1913	N.A.	64.8	1974	40.9	39.5
1920	8.0	80.0	1975	39.7	39.1
1925	11.2	74.8	1980	36.8	38.6
1930	16.7	67.2	1982	36.9	36.6

N.A.: Domestic consumption cannot be disaggregated from commercial consumption and public lighting.

Sources: 1895-1948, Hannah, Electricity Before Nationalization (1979), pp. 426-429, Table A.1. 1970-1982, OECD, Energy Balances, 1970/82 (1984), pp. 366-378.

The high residential shares of electricity consumption in developing countries are, in large part, artifacts of deliberate subsidization policies and should not be expected to grow—despite the growth in Peru's residential share between 1970 and 1981, as shown in table 3.14—even with increased use of appliances that is associated with growth in real income. On the other hand, industrial shares will not grow rapidly with unreliable grid supplies. Either pricing reforms in developing countries will cut household demands or retard their growth, offering the possibility of a more reliable supply of electricity to industry, or major supply systems will experience failures of an enormous magnitude, one that would signal opportunities to reorganize the system. Only significant activities that earn foreign exchange—at resource rather than financial costs—can offer financial viability to large-scale electricity supply. These earnings may be achieved both directly through productive activities and indirectly through productively financed household demands.

As with the transportation imbalance, the high share of residential electricity consumption in developing countries has another side. Electrification was initially supplied for industrial or agricultural application in a sound financial manner, but authorities were forced to yield to demands for residential hookup to avoid sabotage to the facilities. *Fairness* includes access to electricity, but consumers are less concerned about the financial soundness of the electrical system. In addition, availability of electrical household appliances in developed countries grew concurrently with the extension of household electrification. Householders in developing countries face a fully developed array of such devices even before their residences are electrified and before they can afford to pay the full cost of owning and operating them. These electrical devices to provide comfort undoubtedly affect the popular concept of *fairness* for residential consumption of electricity, which in turn affects the political limits of policy for rational pricing.

ENERGY IN DEVELOPING COUNTRIES: PROBLEMS AND PROSPECTS

The sectoral structure of national energy consumption in developing countries looks similar to that of the United States in the late nineteenth century. The household sector consumes most of the national energy supply, usually generated by burning biomass fuels. The industrial sector is a relatively minor claimant, reflecting the small size of industry in most developing economies. However, this sector uses proportionately more energy than its counterpart in developed countries. The sectoral balance between industry and agriculture—rather than various inefficiencies of energy use that are attributable to price distortions and operation and maintenance—probably causes the apparently disproportionate intensity of energy use in developing countries. The combination of industries in most developing countries probably has an effect in the opposite direction.

The major structural imbalance appears to be in the transportation sector. Mechanized transportation is energy-intensive relative to its share in national income, which is magnified strongly in developing countries because there is little or no intermediate technology between domestically priced animal and internationally priced motor power. Cargoes are low-valued relative to the equipment and fuels that provide the transport service, and there are few effective substitutes available that save foreign exchange. The revolutions in transportation of the nineteenth and early-twentieth centuries in the developed countries evolved over nearly a century and followed or paralleled the mechanization of industry. The developing countries in the last third of the twentieth century have not had this advantage of gradual change: major competitors of mechanized transportation cannot compete in an integrated international economy, and industrial mechanization has usually followed that of transportation. Reform for pricing of motor fuel or development of some intermediate technologies of transportation might alleviate the situation. However, solutions to the economic imbalance in sectoral demands for energy may ultimately require that the industrial sectors in economically less-advanced countries grow into their transportation technologies. Presently, their transportation systems often employ new technologies highly inefficiently.

There is also a serious imbalance between residential and industrial sectors in the structure of electricity consumption. Residential loads in developed countries grew slowly from shares of less than 10% to 30% to 40% of national consumption over periods of several decades as the ratio of residential-to-industrial prices for electricity fell from as high as seven-to-one to near parity. Prices for residential electricity in developing countries are typically subsidized. Although the social reasons for this policy have their roots in equity or fairness, the result is weak supply systems that offer unreliable electricity to industries which are reluctant to depend on the national grid. Residential shares probably cannot grow much beyond their present levels, but industrial shares—and hence financial solvency for utilities—will not increase until reliability improves substantially.

The large household shares are probably less menacing than they appear, except for their potential to cause deforestation. In the developed countries, heating dominated demand for household energy, but cooking dominates in developing countries. Substitution of modern, petroleum-based fuels—with relatively simple electrical appliances—for traditionally used fuelwood and charcoal may provide end-use energy services that are equal to half of current national demands for energy with less than 20% of the energy that was formerly generated. These substitutions entail problems, but the technical possibility exists if the economic costs can be successfully addressed.

I have discussed energy problems in developing countries in terms of shares, but questions of total amounts of energy are important as well. There are "development effects" on energy use where use per unit of national income may increase threefold. Energy use per capita may increase fifteenfold over a 50-year period before gradually leveling off. These effects have been much less apparent in developed countries than has been reported. However, the good news is grim enough. Continued urbanization over two generations, with no increases in per-capita income or in the extent of industrialization, could increase per-capita energy demands by one-third to one-half (Jones 1989) for nearly two billion of the world's low-income population.

Comparison of energization in developed countries over the past century with prospects for developing countries is limited because the technological options are vastly different. Industry and agriculture were mechanized relatively slowly in developed countries, but those sectors in developing countries can now switch directly from fuelwood and animal power to highly sophisticated electrical machinery and equipment that is powered by petroleum-based fuels. The aggregate improvements in energy efficiency, which occurred during 150 years in developed countries, can be achieved in a much shorter time in developing countries.

The rapid change that is possible for developing countries may exacerbate structural imbalances if mechanization proceeds at different paces in different sectors and may, in turn, foster instability in overall development. The relative overdevelopment of the transportation sector in developing countries exemplifies this possibility for unpropitious imbalance. In particular, problems of foreign debt for many small African countries may illustrate such attendant problems of instability.

However, we should recall the social upheavals of the industrial revolution in developed countries. The transition to modern energy in these countries was not smooth. Clearly, their energy transformation will not be reproduced exactly because there are differences in technological settings. The transferable lessons may be the circumstantial differences as much as events to be copied.

Literature Cited

Ashton, T.S. 1962. *The industrial revolution, 1760-1830.* Rev. Ed. London: Oxford University Press.

Berry, Thomas S. 1943. *Western prices before 1861: a study of the Cincinnati market.* Cambridge: Harvard University Press.

Chenery, Hollis, Sherman Robinson, and Moshe Syrquin. 1986. *Industrialization and growth: a comparative study.* New York: Oxford University Press.

Bruton, Henry J. 1985. The search for a development economics. *World Development* 13:1099–1124.

Cole, Arthur H. 1970. The mystery of fuel wood marketing in the United States. *Business History Review* 44:339–59.

Cole, Arthur H. 1938. *Wholesale commodity prices in the United States, 1700-1861.* Cambridge: Harvard University Press.

Cole, G.D.H., and Raymond Postgate. 1939. *The British common people, 1746-1938.* New York: Knopf.

Connell, K.N. 1950. *The population of Ireland 1750-1845.* Oxford: Oxford University Press.

Cowan, Ruth Schwartz. 1976. The "industrial revolution" in the home: household technology and social change in the 20th Century. *Technology and Culture* 17:1–23.

Deane, Phyllis, and W.A. Cole. 1962. *British economic growth, 1688-1959.* 2nd Ed. Cambridge: Cambridge University Press.

Devine, Warren D., Jr. 1983. From shafts to wires: historical perspective on electrification. *Journal of Economic History* 43:347–72.

Dewhurst, J. Frederic and Associates. 1955. *America's needs and resources: a new survey.* New York: Twentieth Century Fund.

East-West Center, Pacific Islands Development Program and Resource Systems Institute. *Energy mission report: Fiji.* Honolulu: East-West Center, 1982a.

East-West Center, Pacific Islands Development Program and Resource Systems Institute. *Energy mission report: Solomon Islands.* Honolulu: East-West Center, 1982b.

Gilboy, Elizabeth W. 1934. *Wages in eighteenth-century England.* Cambridge: Harvard University Press.

Government of Liberia, Department of Energy (GOL/DOE). 1987. *1986 Liberia energy balance.* Monrovia: Department of Energy, June. Mimeo paper.

Hannah, Leslie. 1979. *Electricity before nationalization: a study of the development of the electricity supply industry in Britain to 1948.* Baltimore: Johns Hopkins University Press.

Humphrey, William S., and Joe Stanislaw. 1979. Economic growth and energy consumption in the UK, 1700-1975. *Energy Policy* 7:29–42.

Hyde, Charles K. 1979. *Technological change and the British iron industry, 1700-1870.* Princeton: Princeton University Press.

Jevons, W. Stanley. 1903. *The coal question.* Third Ed., revised. A.W. Flux (ed.). New York: Macmillan. Reprinted by Kelley Reprints, 1965.

Jones, Donald W. 1989. Urbanization and energy use in economic development. *The Energy Journal.* 10:29–44.

Jones, D.W. and D.R. Younger. 1988. A comprehensive energy policy for Sudan. Oak Ridge: Oak Ridge National Laboratory. Unpublished paper.

Lebergott, Stanley. 1976. *The American economy: income, wealth, and want.* Princeton: Princeton University Press.

Lindstrom, Diane. 1978. *Economic development in the Philadelphia region, 1810-1850.* New York: Columbia University Press.

McGranahan, G., S. Chubb, R. Nathans, and O. Mbeche. 1979. Patterns of urban household energy use in developing countries: the case of Nairobi. W. Averell Harriman College of Urban Policy Sciences, State University of New York, Stony Brook. Mimeo paper.

McKay, John P. 1976. *Tramways and trolleys: the rise of urban mass transport in Europe.* Princeton: Princeton University Press.

Morrison, Bonnie Maas. 1982. Household Energy Consumption, 1900-1980: a quantitative history. In: *Energy and transport; historical perspectives in policy issues* (George H. Daniels and Mark H. Rose, eds.). Beverly Hills: Sage. pp 201-33.

Morrison, Warren E., and Charles L. Readling. 1968. *An energy model for the United States, featuring energy balances for the years 1947 to 1965 and projections and forecasts to the years 1890 and 2000.* U.S. Department of the Interior, Bureau of Mines, Information Circular 8384. Washington, D.C.: U.S. Government Printing Office.

Mwase, N. 1987. Railway pricing in developing countries: a comparative analysis. *Journal of Transport Economics and Policy* 21:189–217.

OECD. 1984. *Energy balances of OECD countries, 1970/1982.* Paris: OECD.

Pawson, Eric. 1979. *The early industrial revolution: Britain in the eighteenth century.* New York: Harper & Row.

Proops, John L.R. 1984. Modeling the energy-output ratio. *Energy Economics* 7:47–51.

Pryor, E.T., Jr. Rhode Island family structure: 1875 and 1960. In: *Household and family in past time.* (P. Laslett and R. Wall, eds.). Cambridge: Cambridge University Press. pp 571-89.

Putnam, Palmer C. 1953. *Energy in the future.* Princeton: Van Nostrand.

Revelle, Roger. 1976. Energy use in rural India. *Science* 192:969–75.

Rose, Mark H. 1984. Urban environments and technological innovation: energy choices in Denver and Kansas City, 1900-1940. *Technology and Culture* 25:503–29.

Rose, Mark H. and John G. Clark. 1979. Light, heat, and power: energy choices in Kansas City, Wichita, and Denver, 1900-1935. *Journal of Urban History* 5:340–64.

Rybczynski, Witold. 1986. *Home: a short history of an idea.* New York: Viking.

Samuels, Garland. 1985a. *Liberian energy consumption and sectoral distribution for 1981.* ORNL/TM-9424. Oak Ridge: Oak Ridge National Laboratory.

Samuels, Garland. 1985b. *The outlook for electricity supply and demand.* ORNL/TM-9469. Oak Ridge: Oak Ridge National Laboratory.

Sargent, C.S. 1879. The forests of the United States in their economic aspects. *Census of Manufactures*, Volume 9. Washington, D.C.: U.S. Government Printing Office.

Schurr, Sam H. 1983. Energy efficiency and economic efficiency: an historical perspective. In: *Energy, productivity, and economic growth.* (Sam H. Schurr, Sidney Sonenblum, and David O. Wood, eds.). Cambridge: Oelgeschlager, Gunn & Hain. pp 203-14.

Schurr, Sam H., and Bruce C. Netschert, with Vera F. Eliasberg, Joseph Lerner, and Hans H. Landsberg. 1960. *Energy in the American economy, 1850-1975: an economic study of its history and prospects.* Baltimore: Johns Hopkins University Press.

UNDP/World Bank. December 1985. *Ecuador: issues and options in the energy sector.* Washington, D.C.: UNDP/World Bank.

UNDP/World Bank. September 1985. *Thailand: issues and options in the energy sector.* Washington, D.C.: UNDP/World Bank.

UNDP/World Bank. June 1985. *Burma: issues and options in the energy sector.* Washington, D.C.: UNDP/World Bank.

UNDP/World Bank. June 1985. *Togo: issues and options in the energy sector.* Washington, D.C.: UNDP/World Bank.

UNDP/World Bank. December 1984. *Liberia: issues and option in the energy sector.* Washington, D.C.: UNDP/World Bank.

UNDP/World Bank. January 1984. *Peru: issues and options in the energy sector*. Washington, D.C.: UNDP/World Bank.

UNDP/World Bank. July 1983. *Sudan: issues and options in the energy sector*. Washington, D.C.: UNDP/World Bank.

UNDP/World Bank. June 1983. *Fiji: issues and options in the energy sector*. Washington, D.C.: UNDP/World Bank.

UNDP/World Bank. June 1983. *Solomon Islands: issues and options in the energy sector*. Washington, D.C.: UNDP/World Bank.

UNDP/World Bank. April 1983. *Bolivia: issues and options in the energy sector*. Washington, D.C.: UNDP/World Bank.

U.S. Agency for International Development/Pakistan. 1987. *Pakistan National Energy Balance, 1983/84*. Islamabad: USAID.

U.S. Department of Commerce, Bureau of the Census. 1965. *The statistical history of the United States from colonial times to the present*. Stamford: Fairfield.

U.S. Department of Commerce, Bureau of the Census. 1975. *Historical statistics of the United States, colonial times to 1970*. Part 2. Washington, D.C.: U.S. Government Printing Office.

U.S. Department of Energy (DOE), Assistant Secretary for International Affairs. 1979. *Joint Egypt/United States report on Egypt/United States cooperative energy assessment*, Volume 3, Annexes 2-5. Washington, D.C.: U.S. Government Printing Office, April.

Woolf, Arthur G. 1987. The residential adoption of electricity in early twentieth-century America. *Energy Journal* 8:19-30.

World Bank. 1983. *World Tables*. 3rd Ed. (Vols. 1 & 2). Baltimore: Johns Hopkins University Press.

CHAPTER 4

GENERATING SUSTAINABLE RURAL DEVELOPMENT OPTIONS AROUND WOOD-BASED BIOMASS ENERGIES: THE CASE OF LOCAL KNOWLEDGE SYSTEMS

Anil K. Gupta

INTRODUCTION

In ancient times, residence in the forest (*Vana Prastha*) was part of the duties a person was expected to complete before seeking final emancipation from the worldly cycles of life and death. Of the four stages one passes through—childhood, married life, stay in the forest and Sanyas—the stay in the forest is the most crucial for spiritual and intellectual growth.

The relationship with the forests was not entirely utilitarian. In parts of Africa, Latin America and Asia, people named different niches in the forests as abodes of various gods and goddesses. These beliefs along with sanctions in conservation of resources have come under stress with increasing scarcity of wood and related sources of energy. John Perlin (1989) provides a broad view of the role played by forests, wood in particular, in shaping the evolution of human civilization.

We are passing through a similar crisis at present. The prices of fire wood have dramatically risen in most Indian cities (CSE, 1985). The declining tree cover in urban regions has made the prospects of leaf and twig collection even more difficult[1]. For the very poor, survival is indeed difficult, given high inflation and declining budget support for public distribution of food grains. The current economic restructuring in India and many other developing countries is likely to transfer private costs to public accounts. These costs will show up in the national accounts after a lag, and include generalized decline in productivity and increasing turbulence in the ecological environments. And yet the rural poor are not the culprit. An earlier study (Gupta 1984) had noted that the tree density on cultivated lands was positively correlated with the land size in irrigated villages and inversely correlated in dry villages[2].

1 In Bangladesh, the leaf collection by the poor people in the cities is a very bouyant activity given much larger tree cover in some of the localities.

2 Such a relationship was identified, perhaps, for the first time in this study. There could be several reasons for this pattern but the important ones were: the poor people were dependent far

Rural development is a process of expanding decision making choices and extending the time frame (Gupta 1981). The first requires improvement in the access and abilities or skills. The second objective requires improvements in the institutional assurances. The higher the assurances—or the lower the risks—the longer the time frame. We have to understand how the different households adjust to the risks before appraising various technological and institutional choices presented. The developmental implications of forest and wood-based energies cannot be assessed in a partial framework. A multi-level and multi-market framework is necessary for this purpose.

The paper is organized in two parts. First, I present the conceptual framework for analyzing evolution and adaptation of household portfolios over space, sector, season and social exchange mechanisms, including access to resources and their conversion into investments given risk and institutions. Second, technological aspects of the local knowledge system are discussed drawing upon the insights from survey of innovations developed by farmers, pastoralists and artisans.

MULTILEVEL MULTI-INSTITUTIONAL APPROACH TO UNDERSTANDING HOUSEHOLD PORTFOLIOS

Portfolio as performance

When I plan my investments, I look at various alternatives in the context of personal needs, unforeseen circumstances, consumption preferences and sensitivity to my obligations to others. While economic evaluation is important, I do not base all my choices on economic criteria alone. Certain investments are made with other people in mind. Others are made just for fun or aesthetic needs. Still others are guided by life cycle perspective, age, and demographic stage. In some decisions, I consult a financial expert; in others I take advice of my spouse, children or parents; others I just gamble. The investment mix includes long term and short term, liquid and non-liquid, status-linked and status-indifferent, ancestral and acquired, coupled and uncoupled[3] or storable and non storable assets[4].

---

more on the livestock than those better off. They used (at least up until 1980) bullock for plowing the land, whether their own or leased, rather than tractors (the practice has been changing in Haryana). Skipping a spontaneously sprouted Prosopis Cineronia or any other tree is easier while plowing by bullock but it is almost impossible when tractors are used. The leaves of these trees are valuable source of fodder in the years of drought apart form other uses of these trees. The reason for slightly inverse pattern in irrigated villages is that the poor people have to intensify the cultivation on their small holdings. The affluent farmers have either private "waste land" or have more bund space to accommodate more trees (Gupta 1984a and b).

3 Coupling refers to such assets which are inter-linked and cannot be used alone. For instance a cattle shed and cattle are linked but not coupled because we cannot keep cattle without a cattle shed. However a bullock and a plough are coupled. We cannot use plough without bullock and bullock without plough or in some cases cart or thrashing roller or weeding hoe as the case may be. Thus decision of keeping a bullock may sometime guide acquiring another asset in the

I have previously argued (Gupta 1984, 1985 and 1990) that a household portfolio includes assets which may be maintained through the resources drawn from other private assets, open-access assets, or common property resources. Various property right regimes generate externalities of different kinds which may influence individual household choices. The conventions and customary rights to various resources may also influence the way households appraise their individual portfolios. The access to resources, the assurances of future returns, ability to convert access to investment influence are influenced by the attitudes shaped by the cultural traditions individual experiences and certainty and uncertainty about future outcomes.

Portfolios are assessed by assuming dependence in choices. If my neighboring farmer irrigates her field and water seeps through to my field, where I want to grow rainfed chickpea or gram, my choices are limited by my neighbor. If I cannot persuade my neighbor to change her irrigation pattern, my portfolio gets modified by my neighbor's choices. There could be other instances in the field of plant protection which are even more serious and generate externalities which modify individual household choices. With modification in one sub-system, say a land-related enterprise, choices in other subsystems are also influenced. Household choices, particularly by the deficit budget farmers cannot be appraised in credit, product or labor markets independent of the choices or constraints in other markets. (Gupta 1981, Bharadwaj 1974).

In addition to the cultural deference for others' feelings and individual predicaments, portfolio choices and outcomes may also be affected by religious and institutional preferences or limits. Farming system research, while taking note of linkages between crop livestock and tree sub-systems, has used a functional perspective and thus misses historical insights of the dynamics of portfolio design and evolution.

There is a view that sometimes portfolios have to be seen as a result of a performance (Richards 1989, Box 1989, Gupta 1990). There is a saying in India that traveling together may serve a greater purpose than just reaching

---

portfolio which may on its own economics be not justified. Since the decision to keep cattle is non negotiable for whatever reasons cultural or emotional, further decisions are then to be evaluated keeping the first decision as given. The price of inefficiency or sub-optimality may be paid on account of improved utility achieved from better utilization of bullock in such a case. There may be many other factors such as need for autonomy during crucial farming stages which may justify keeping bullock in the first place. Though enough work round the year may not exist for it. Such assets which are tied in use are called coupled in this paper.

4 The wood based biomass can be stored though different forms (logs, chips, filings, powder etc.) and thus will have varying costs, depending on distance traveled and consumers. Thus the local wood based developmental strategy may not only be more effective from a social point of view but also from an efficiency standpoint. Thus wood based energy production system have definite rural developmental implications in such areas. The only snag is that many of the same hill areas also have access to cheap hydro electrical energy. Thus only in cases where such alternatives do not exist, can one make case for wood based source of energy in a big way. Of course where due to lumbering and other activities such a biomass is indeed available, the alternative uses as soil amendment, fuel for house hold use vis-a-vis for commercial gassifiers or other energy generators etc., can be considered.

your destination. The students of bio dynamic agriculture learn that sowing according to the lunar calendar sometimes provides an effective way of synchronization to overcome the problem of pest population build up. Performing a ritual may thus be a way of achieving order in a process which may seem chaotic. Theory of Chaos teaches us that search for order sometimes has to proceed on a different plane than the one in which a particular act has been performed or perceived. I do not deny the irrationality of many rituals. And at the same time the rationality of some others cannot be denied just by definition.

Other dimensions of performance are simultaneity, coordination, deference for collective rewards and synchronization. The choice of technologies by the household can be understood better if we also look at portfolios as performance and then study which components have to be modified for improving the overall performance. Emphasizing a specific role may be justified until we understand the nuances of interactive performance. The outcome will be partial till we recognize the need for change in multi-role enactments.

Portfolio outcomes are interactions among space, season, sector and social exchange relations

Social scientists have a great preference for sequentialism and sectoralism. Contribution of space or ecological variables organized over space through niches was often unrecognized in the sectoral equilibrium oriented analysis of demand and supply constraints (Mabugunje 1979, Gupta 1981, 1984, Biswas and Biswas 1979, Richard 1985). Ancient conceptualization of agriculture and its relation with nature had always emphasized linking space, season, sector and social systems. This relationship was recognized in most ancient Indian and Chinese texts. Modern curricula are biased towards western models and neglect such traditional concepts and theories.

Several studies on farmers' adjustment to risk have shown a multi-market, multi-enterprise and multi-institutional approach to survival (Jodha 1975, 1979, Jodha and Mascarenhas 1983, Gupta 1981, 1984, 1988, 1990, and Ostrom, Picht and Feeny 1989). The multi-market approach refers to the farmers' attempt to adjust to risk through simultaneous operation in different factor and product markets. The factor markets include land, labor, capital and information. The product markets include crop, livestock, trees, and technologies of land and water use. The higher the risk in the environment, the greater the dependence between the decisions made in one resource market and those made in others. These links are important in developed regions but many imperfections in respective markets often can be offset through market mechanisms themselves over time and space. In high risk environments the cost at which these errors may be repaired may be far higher and thus more dependent on inter-market adjustments.

The multi-enterprise approach implies that farmers' adjustments to risks or evolution of portfolios cannot be understood by concentrating on any one enterprise such as crops, livestock, labor or trees etc.[5]. The 4-S model helps understand these linkages at the macro level (figure 4.1).

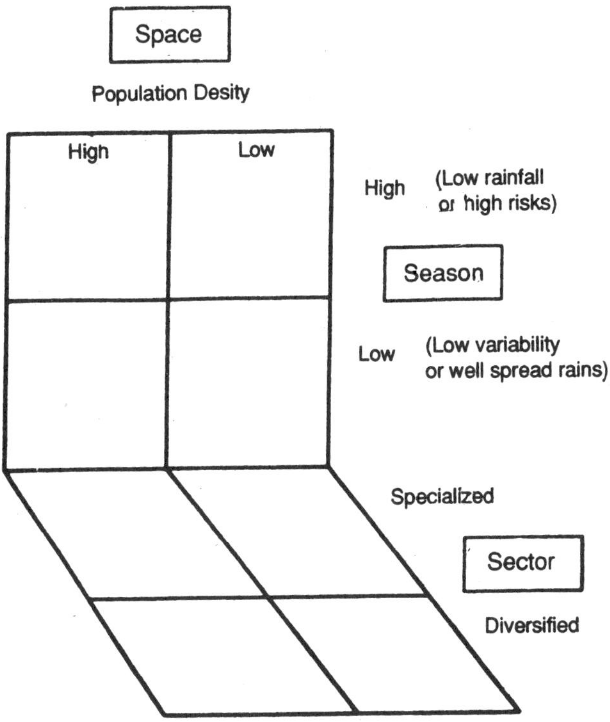

Figure 4.1. 4-S Model.

Each dimension can be dichotomized for the purposes of creating ideal types. The basic principle of logic that we use here is 'compare and contrast.' If we want to understand a phenomena it may be useful to compare and contrast the extreme values of its distribution.

For instance, 'space' can be dichotomized in terms of high or low population density. It could also be contrasted in terms of high land and low land, undulated and plain topography, higher slope and lower slope, etc.

5 The multi institutional perspective is helpful because various resources or enterprises, may be governed by various kinds of property right regimes together or separately. Livestock for instance may be managed by some households through biomass derived from only private land. In other cases it may be derived from private as well as common and/or open access lands. Thus various institutional arrangements, whether or not regulated by the state, market or both, further influence the choices at a micro-level. Any framework which ignores multi market, multi enterprise and multi institutional dimensions of household portfolios may generate only partial understanding of the survival logic of the people. The innovative technologies or institutional arrangements are a part of dealing with these complexities. Innovations of survival sometime may follow rules that are different from innovations for accumulation (Gupta 1988,1990).

Likewise 'sector' can be dichotomized as agriculture and industry, public and private, specialized and diversified, etc. 'Season' can also be contrasted into uni- or bi-modal rain fall regimes; arid and humid; low and high rainfall; etc.

Given any two parameters we can speculate about the third. For instance, in a region with low population density and high seasonality (low rainfall and high diurnal temperature variations in the arid plains and low diurnal temperature variations at the high altitudes) the sectoral characteristics may be highly diversified. Instead of a single crop farmers may prefer mixed or intercropping in several plots. Households may simultaneously pursue many activities such as crop, crafts, livestock etc., rather than being dependent on any particular activity. The social exchange relations in such regions will be quite different compared to the regions with high population density, low seasonality and specialized sectoral activities or diversification for accumulation rather than survival.

Some of the characteristic ways of social exchange relations may include the following: (a) the predominance of kinship and extended family networks over the nuclear family systems to avert risks (b) importance of non monetary exchanges and the informal mechanisms of pooling[6] of resources, (c) dominance of generalized reciprocities over the specific ones[7] and (d) choice of a longer timeframe to settle book of accounts[8].

---

6 Anthropologists have provided rich insights about the pooling mechanisms in various societies living in hill areas, arid regions or forests. Pooling of bullocks in Maharashtra for instance, is called irjik. As many as 10 to 12 pairs of bullocks can be seen plowing the land in a particular catchment area across the fields at a specific gradient. Since the moisture could recede faster near the ridge line, the plots along the contour towards the ridge have to be plowed first. It is possible that some people may contribute one pair of bullock though they may only have 1/2 an acre or even no land in that niche. While others may have much larger tract of land in this niche and yet contribute the similar pair of bullocks. The obligations for feeding the cattle and the ploughman are also worked out in many diverse ways. What is important to understand is that uncertainties over time and space may generate reciprocities which may be settled over a longer period of time and thus generate rationality of choice in the short term.

7 The generalized reciprocities refer to exchange of labor for, say, thatching the hut by plowing the land. It is very difficult to work out the equivalence between such unrelated activities. How critical thatching is before the rains only a poor family living in such hut can realize. Likewise the critical draft power in receding moisture conditions in light soil regions can be understood by someone who may miss the entire season in the event of failure of sowing a crop at the right time. The traditional economic theory can be of limited help because equivalence is not just the value of labor as assessed in the market place. Sometimes help provided in such a context may generate an I.O.U which may be redeemed much later. The specific reciprocities on the other hand refer to exchange of same goods or services. I have paid for your tea today you should pay for me tomorrow. Or I have given you five kilograms of wheat seeds and you return the same amount of the same crop later. The commercialized societies often would have dominance of specific reciprocities. Decision making with constrained resources cannot be analyzed without looking at these reciprocities.

8 Studies have shown that in the regions mentioned here, it takes far longer time periods, extending to sometimes several generations, for IOUs to be settled. A good or bad turn may provoke a gesture not necessarily on the same day, month or even year. Even the nature of factional leadership remains divided at village level for longer periods than at the state or national level, where loyalties can shift quickly without generating problems of legitimacy or social acceptance.

The communication system in these regions is far more metaphorical or analogical rather than digital. The strategies of technology transfer in forestry related on-farm or in-field research and extension systems would obviously have to be tailored to the typologies which can emanate from the simple matrix given above. One can make it more complex and generate richer insights but parsimony always has a price. I must acknowledge that the nature of institutions and market interventions can modify the initial conditions that may be predicted by the configuration of spatial, seasonal, sectoral and social variables. It might appear that some of the social relations are defined by the ecological variables in a deterministic manner. We have seen that a relationship between pastoral and cultivating communities in the Swiss Alps, (Netting 1972), in northern Pakistan (Buzdar 1988), Bhutan and other Himalayan mountain regions (Gupta and Ura 1990) have striking similarities, though specific parameters may vary due to cultural and religious differences. Over time however, formal institutional inroads and market developments do modify these strategies. Availability of a walkie talkie to a Swiss pastoralist does not require development or retention of specific whistling styles as observed in Andean mountains or in the Himalayan Mountain regions. However, the need for surviving collectively is felt in almost all such socioecological conditions.

Eco-Institutional Framework

Human choices in a given eco sociological configuration are circumscribed by the historical evolution of institutional structures. The institutions provide a framework of rules, sanctions and meanings which are commonly understood by groups of people within a boundary. Institutions provide assurance—formal and informal—to individuals and groups about various uncertainties faced over time and space. We deal with mainly two types of assurances—horizontal and vertical. The former includes the assurances that provide a guarantee about others' behavior in relation to one's own. Thus if I sowed my crop early will others also sow likewise? Or if I did not graze my animals on common land will others also cooperate? The vertical assurances refer to the future returns from present investments. If I plant trees on the common or private land will I be allowed to harvest them? Or if I apply organic fertilizer to a particular plot of land taken on lease will I be allowed to get the same plot next year also?

The assurances by themselves however are not sufficient. Assurances of better prices, better returns or collective behavior without access to the given resource, or the skill or ability to convert a resource into investment, are of little use.

The assurances help in generating cooperative behavior when we deal with common properties (Sen 1974, Rang 1984, Gupta 1985). In the case of private resources, assurances may stimulate demand for better access or technical skills or both. Likewise if we have an institution in which people have access to resources and also have assurances but do not have the skills or abilities, the investments would not follow.

All the three vectors of choice, that is, access, assurances and abilities must be synchronized to generate appropriate attitudes for change or maintenance of a resource use systems. Thus within a specific spatial, sectoral and seasonal configuration portfolios may vary within a given range because of changes in access, assurance and abilities.

As we note in figure 4.2, the access to natural resources, assurances from institutions, ability in terms of technology, and attitudes in terms of culture, collectively influence the household portfolios. Access to institutions, technology and cultural resources will also make a difference to the kind of assurances that are needed in resource, technology or cultural subsystems of life. Similarly, ability or skills to understand a resource, use institutions or interpret cultural norms will influence the choice and use of technology. This framework also helps in designing interventions and can be used as a tool to assess available information and generate choices. Given the complexity in the available system of access and the abilities of the people, we can anticipate the type of assurances that would generate or respond to the given attitudes. Attitudes here are both an outcome of historical experiences and are also inputs into the future choices.

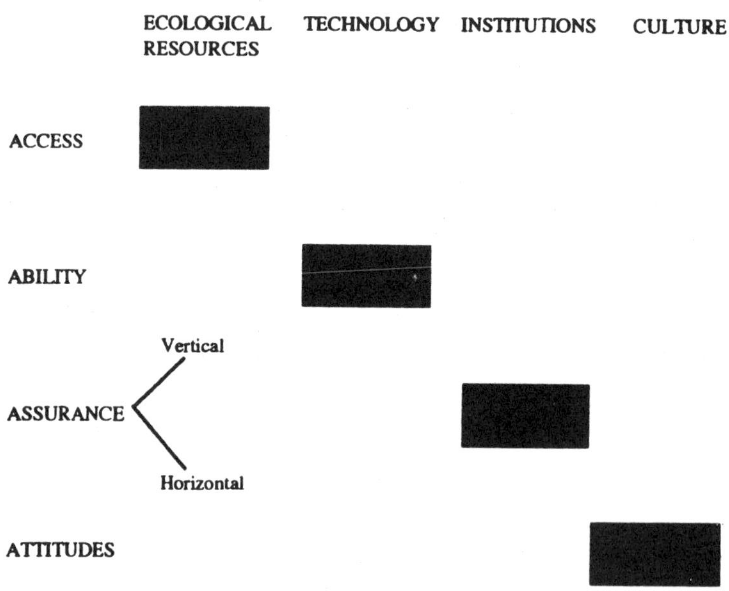

Figure 4.2. Eco-Institutional Model.

Articulation-Response Model

This model presents a relationship between needs—both felt and unfelt—and the response of the supply side, in this case, the institutions. Which needs are felt and which are not may be a function of historical experience, `learned-helplessness,' expectation of supply and sensitivity to the environment. Every time a new opportunity emerges, consumers may try to stake their claim for the relevant good or service. Once the needs have been felt they have to be voiced to generate pressure on the supply side. An isolated articulation is less likely to make supply side responsive then an aggregated articulation. A favorable response may encourage a household to articulate the same need more often.

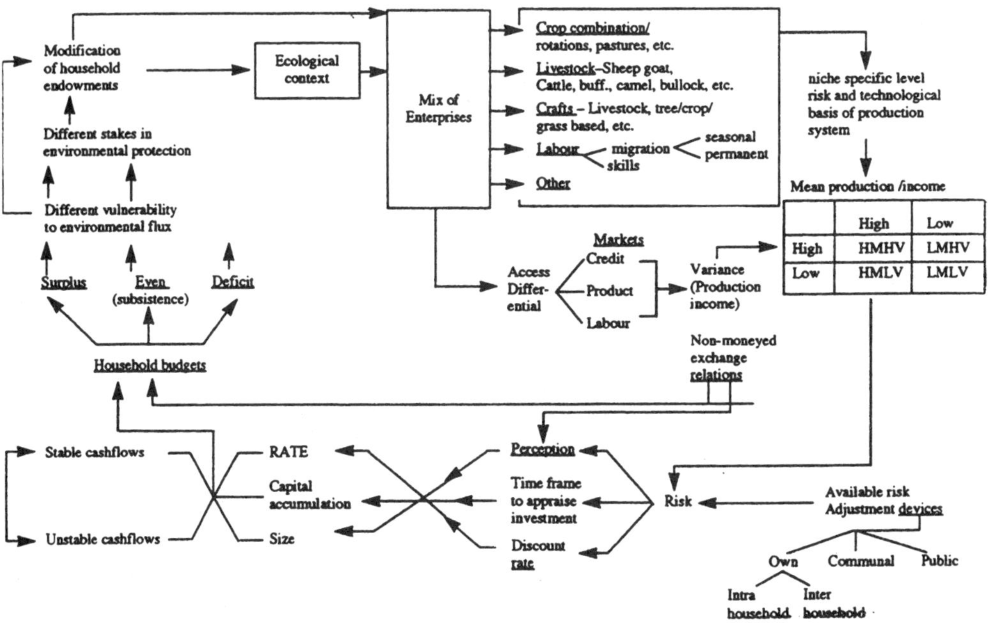

Figure 4.3. Eco-Sociological Paradigm for Analysing Household Economy in Dry Regions (Gupta 1985).

The portfolio of households and scientists thus gets modified by the institutional choices available in the given articulation-response framework. Social movements, particularly farmer's complaints, can result in quantum shifts in the capacity and willingness of the supply side to respond to various demands. At the same time we have to recognize a historical reality that farmer's needs are seldom noticed around the problems of growers of millets, pulses, oilseeds etc., and pastoralists tending small ruminants. It is the cash crop which often becomes the rallying point for agitations in the high growth regions. The implication is that their is a greater need for responding to unarticulated needs of disadvantaged farmers in high risk regions.

Within a given ecosociological and institutional context the articulation of needs modifies the portfolio evolution. Above elements can be interlinked in a dynamic ecosociological framework.

Ecosociological Paradigm

Assuming that: (1) Ecological conditions define the range of economic choices that can be sustained in a given region; and (2) The scale at which different enterprises are selected, is a function of the access to factor and product markets, kinship networks, public, private and common institutions, historical resource reserves, etc.

Ecological aspects of a geographic location need not be the major determinant of portfolio. The distant environment where the community has customary or traditional rights through migration or any other such means or where the market of local produce emerges must also to be taken into account.

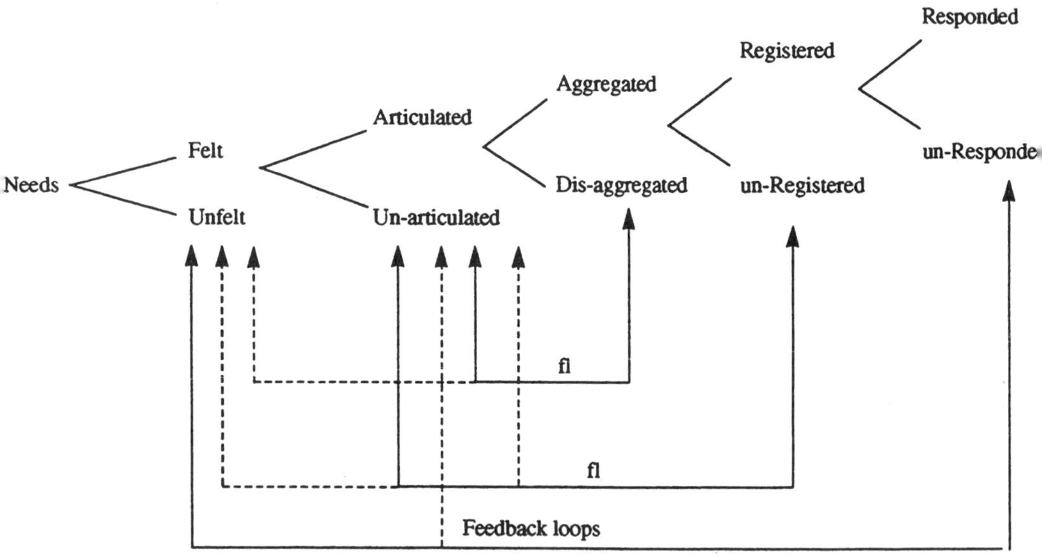

Figure 4.4. Articulation-Response Model.

Thus, once an enterprise or portfolio is selected, drawing upon resources from private, public, and common properties, the nature of risk inherent in these portfolios can be analyzed through a matrix of mean or average return and variance in returns. The high mean-low variance portfolios would obviously have different implications for individual and collective behavior than portfolios with low mean and high variance.

Given an initial portfolio and its mean-variance or risk-return characteristics households may respond to given risks in the environment through:

a) Household level risk adjustments;
b) Public and market risk reducing mechanisms; and
c) Communal and common property risk adjustments.

The household risk adjustments can be further analyzed at the intra-household level and inter-household level. The intra-household risk adjustments include measures which a household can take recourse to by negotiations within the households. For instance, asset disposal, migration, and reduction or modification of family consumption, etc. The inter-household risk adjustment strategies include tenancy, borrowing, labor contracts, group plowing, etc.

The public risk adjustment mechanisms imply availability of drought, or flood relief, public employment programmes etc. The market based risk adjustment option include forward trading, interlocking of factor and product markets, insurance cover, etc.

The communal risk adjustment strategies refer to the group based measures which require collective decision-making either for utilizing or preserving private or common property resources. The pooling of resources such as bullock or equipment is also part of communal risk adjustment strategies.

Once the range of risk adjustment options is known, the households may modify either their perceptions or responses or both by changing the discount rate or time frame used for appraising returns from each investment. Thus while discount rate captures the control a household has in a given resource market, the time frame may capture the certainty with which the household views a particular resource stream. In fact either of the two can be used to derive risk preference. The shorter the time frame in which households appraise their choices, the less likely it is for technology to be sustainable. It is obvious that not everybody's choices can be broadened at the same time and in the same proportion given the limitation of resources in a developing society. It is at this stage that an ecosociological framework has to become an ecopolitical framework. Constraining the choices of some while widening that of others is an institutional issue which is discussed elsewhere[9].

The uncertainty of an outcome may vary for different households depending upon (a) previous experience with a particular enterprise/crop including immediate past experience; (b) successive losses or gains and accumulated deficits or surpluses in the household cashflow; (c) future

9 Anil K Gupta. 1990. Politics of Articulation, Mediating Structures and Voluntarism : From 'Chauraha' to 'Chaupal', IIM, Ahmedabad, Working paper No. 894, September.

expectations of returns; and (d) relationship between other assets/enterprises and the proposed investment.

The cash flow of the households, due to a given portfolio modified by various risk adjustment options, may be in surplus, deficit or subsistence. In addition, the variability in these cash flows may be evened out over space, season, sector and social networks. The risks taken by different social groups in management of ecological systems would vary in each resource market.

The trick is to develop an analysis of how unequal risks taken by different groups in various resource systems or regimes generate a set of expectations which are equitable or appear equitable (given differences in cultural and social ways of perceiving returns) at the portfolio level of households. The fairness of these distributions cannot be estimated from the individual point of view only. The group level estimation of aggregated effects of individual portfolios may generate rules that modify the conditions for resource use, technology and institutions. Under extraordinary circumstances the cultural norms are also modified to accommodate ecological and sociological imperatives[10]

A deficit household budget influences choice differently than a subsistence or balanced budget or a surplus budget. It would be incorrect to club the deficit budget groups with the subsistence ones. Sustained deficit may shift the portfolio in favor of low mean-low risk assets and in some cases low mean-high risk provided the risk is not co-variant. In some cases low mean-high risk assets can be accommodated in the portfolio because much of the cost is transferred to the open access or common property resources. For example, sheep are seldom stall fed and are maintained by some of the poorest households.

At the aggregate level, shifts in the portfolios can be seen by differential growth rates of various species of crops, trees, livestock, etc. Public policy at the macro and micro level influences the portfolios through changes in the access modes, assurances (through various risk adjustment strategies) and abilities. The attitudes are also modified by the expectations of changes.

The changes in the individual stakes in various resource systems feed back into the ecological conditions. Once the ecological conditions are modified, changes in the enterprise mix become inevitable. This is not to underplay the

10 Aggarwal (1990) provides an interesting example of a village where the punishment for poaching in a common property was to offer grains to the birds standing barefoot under the sun. Such a logic cannot be analyzed in the classical tradition of institutional analysis. Such sanctions cannot be justified on economic ground at all. The reciprocities extend here to such claimants of resource who may not have any vote, that is, the birds. But in the process suffering in public by standing bare-foot in the sun generates a collective responsibility. It is recognized that the moral appeal may have a longer lasting effect compared to an economic tax or fine. The public display of the punishment may also generate guilt.

The cultural norms for individual and group behavior thus do modify the perception and response to the risks and resources. While crisis of fuelwood may generate tendency for poaching, the sanctions generated by the Institutions may safeguard, to some extent, the scarce and depletable common property resource. In southern Bhutan we came across a case where a group of villagers had put restrictions on bringing a male bull of exotic breed lest the local breed selected over centuries become contaminated (Gupta and Ura 1990). Even today many villages in South Asia follow similar practice.

importance of changes in the institutional conditions or the technology choices. However, a multi stage or a multi plane analysis requires that we do not mix assumptions necessary for analysis at one plane with the assumptions relevant at another plane.

The theoretical perspectives presented in this part provide relationships among various variables that are easily tested. At the same time specific hypothesis can be generated by changing the values of different parameters of each of the perspectives.

The frameworks presented above help in assigning the weights to various resource markets from which wood based biomass energy may be withdrawn. The property right relations in different resource markets are likely to favor different participants in a non-uniform manner. We need to understand the interaction between these rights and derivative access and preference. The choice of technology and market prices are not just influenced by local factors.

INDIGENOUS ECOLOGICAL AND TECHNOLOGICAL KNOWLEDGE SYSTEM AROUND FORESTS

A list of illustrative and innovative practices developed by the people dependent on the forests is presented in appendix one. Various recipes for veterinary medicine and other agricultural purposes have evolved around bark, leaves, twigs, roots etc. The purpose of listing these is to broaden the meaning of wood-based biomass energy. One should distinguish between direct and indirect sources of energy. If bark extract helps in keeping bullocks fit, then should the contribution of this specie's medicinal property be taken into account?

A study in Bhutan (Gupta and Ura 1991) provides details of different rights and the responsibilities people have towards forests. Bhutan is one of those rare countries which has decided to maintain highly forested area as a conscious policy.

Forest Institutions in Bhutan: blending culture and technology

Institutions evolve when conflicts arise in resource use. In general institutions arise more often than not in a higher risk environment. There are several positive and negative interactions between forest, farm, and livestock ecosystem.

Rights and responsibilities towards forests among cultivators and pastoralists include: cleaning of *Sokshing* (community forest); collection of firewood (*zashi*) on a particular day; prevention and control of forest fires; sustainable management of *Tsheri* lands (shifting cultivation practices); prevention of intrusion of cattle herds into reserve forests; protection of wildlife; use of trees (with prior permission) for domestic use; collection of shingles (flat thin pieces of wood cut for roofing) every few years to replace and repair roofs (the old shingles are piled in the field to be decomposed as manure or burned for preparing field for cultivating buck wheat); collection of leaves and pine needles for preparing bed for animal houses for eventual conversion into fertilizer. We provide only a few instances of institutional evolution next.

Forest Fires

It was noted in the National Assembly of Bhutan a few years ago that large-scale outbreaks of forest fires had become frequent. Forest guards appointed for forest protection were found inadequate. A suggestion was made that a person from the village be appointed as fire watcher to work under the village head. The fire watcher would not be required to contribute compulsory labor. The person found responsible for the fire would be imprisoned for three years. If the villagers could not locate the culprit or did not help in extinguishing the fire, they could be penalized.

There is no doubt about the damage fire causes to forests whether it is done to clear it for growing apple orchards or crops. However, doubt has been expressed about the potential damage since the 1860s and whether firing has been practiced in forests indiscriminately.

Another British officer writing on the forests of Kolloo observed in 1851 that the dry grass was fired to produce new tender shoots of grass and in the process killing myriads of young trees. He suggested that the villagers should have charge of the forests and be allowed to collect an *ad valorum* tax on wood cut there provided they did not fire the grass and prevented any trees being cut before maturity[11].

Present public policy assumes forest fires to be unmitigated disasters. Therefore, prevention measures have become part of the resource management strategy. There may be a case for scientific analysis of the conditions under which firing forests used for pastures might be treated as part of the management practices and at the same time other diverse forests might be protected from it.

People have a right as well as a responsibility to conserve forests and enhance their productivity. A country with 60% area under forest and with a clear policy of maintaining it at that level has to blend culture, technology and environment.

Reliance on indigenous knowledge systems thus is necessary. Some studies have been done to find out the scientific basis of traditional fodders. For instance, it was found that willow leaves (*Salix babylonica*) provide sheep with energy and protein requirements for body maintenance as well as growth. The leaves were so rich and palatable that they could sustain the animals without any additional supplement. There was no significant difference in metabolizable energy concentration of willow leaves in relation to Lucerne,

---

11 Edward Thornton, "Remarks on Tree Resources on Jelum Division", 12th July, 1851, No. 348 in Select Papers of the Agri-Horticulture Society of the Punjab from its commencement to 1862, Lahore Chronicle, 1868.

vetch, fodder maize and fodder oat[12]. It was recommended that willow leaves be considered the most suitable fodder for ruminant livestock in temperate regions of Bhutan.

It is unfortunate that the developmental programmes often ignore the accumulated wisdom of centuries of experimentation. In this case, out of a large number of possible leaves and grasses, only a few have been selected by the people in different regions. The emphasis on cultivated fodders should in fact be less than on tree fodders in fragile regions because tree ecology is much more sustainable than the cultivated crop ecology for higher altitudes as well as steep slopes. In sub-tropical Bhutan, leaves from fodder trees complement grazing. In sub-alpine and alpine regions, wheat is grown as green fodder for cattle and horses. The animals go through nutritional stress during the long winter in alpine regions. Buckwheat straw and even some crop are used as supplementary diet.

The gathering of leaves from a common grove for composting with animal manure or for animal bedding is a regular activity. Here again individuals can not strip the forage widely. Forage is collected from a common grove only on a publicly declared day so that all community members get their fair share. The forest is a source of many other forms of livelihood[13].

The conflicts between external advice for conservation-without-people and Bhutanese cultural ethos with conservation as a way of life are becoming stronger. Bureaucratic barriers have seldom proved helpful in insulating natural resources from human greed. The Indian and Nepalese experience is quite instructive in this regard. With the best technical skills, trained manpower and coercive authority available to the state, the degradation has continued to increase over time in both these countries. If a country can generate indigenous technologies and institutions capable of maintaining forests, water streams, biodiversity and ecological balance for so long, why should it be found wanting now.

---

12 P. Bajracharya, "Intake and Digestion and Nitrogen Retention in Sheep fed on Willow leaves (Salix Babylonica L.) and Rumen Degradation Characteristics of Willow Leaves," Food and Agriculture Organization of the United Nations, Thimbu, Bhutan, December, 1987.

13 Gathering of wild food is an important element of supplementing food shortages in some parts of Bhutan. In some districts of Bhutan such as Kheng, yields from the forest contribute substantially to the food basket. There are wild yams, wild avocados, wild taro, young frons, young buds of rattan, flowers of orchid and various kinds of mushrooms to name a few fruits and roots for the table. Wild yams, which achieve giant sizes—as long as four feet—are collected even as a normal part of food. Since it lies a meter or two below the surface, appropriate extractive tools are there for wild yams. Nowadays, the forest food gathering is a growing petty business for the farmers. The forest offers various raw materials like bamboo and rattan for baskets, needle yarns for making clothes, natural dyes and many other more obvious products. Also see, Tony Beck (1989) "Survival Strategies and Power Amongst the Poorest in a West Bengal Village" in I.D.S. Bulletin (April 1989), Vulnerability: How the Poor Cope, Vol. 20, No. 2, pp 23-32.

TAPPING LOCAL ECOLOGICAL KNOWLEDGE THROUGH COMPETITIVE STRATEGIES: BIO-DIVERSITY CONTEST

In a recently held biodiversity contest in Madurai, a shepherd boy who came first identified 116 species of trees and other plants along with their economic uses. The farmers who came first could identify 245 species. In the current context of social system, this student would have to unlearn all this knowledge because the society puts no premium on it. The question is: "can the sustainable system of wood based energy be designed in a truly participative manner without building upon the indigenous ecological knowledge systems?"

The wood based ecological and survival system needs proper conceptualization from demand and supply side so that implications of change in one sub-system for another can be worked out.

Darkoh (1991) has illustrated the wood product consumption scheme as given below:

CONSUMPTION OF WOOD PRODUCTS

Fuelwood and Charcoal	Wood Use		Charcoal
Agriculture			Gums, resins + Oils
House Building	Household Consumption	Industrial Consumption	Wood Poles
Joinery			Sawlogs
Weaving			Veneer Logs
Apiculture			Pulpwood
Other			Residue

Source: Darkoh (1990}.

Darkoh has argued that the disjunction between the political pronouncements and institutional responses has prevented sustainable alternatives for arid and semi-arid lands to emerge. Given the economics of fuel, fuelwood and charcoal are the cheapest fuels. The evidence in Kenya (and elsewhere) at the same time indicates that it is not the rural household energy consumption of wood which is responsible for large scale deforestation. It is the urban demand for charcoal in Kenya (and timber and fuelwood in other developing countries) which is responsible for the major degradation of resources. Darkoh also illustrates how the demand for woodfuel and charcoal has led to not only deforestation but also afforestation in the traditional reserved areas (sacred groves) such as Kaya forests. Despite this, the afforestation programmes are not able to make up for the trees cut. The story is not very different in other countries.

Wood Trade in Deserts

Pangare (1990) provides a very graphic account of wood trade in the Bikaner district of Rajasthan, part of the Thar desert in western India. About 30-100 carts come everyday to the city from adjoining villages, a total of approximately 650 quintals of wood per day. Most of the wood is Khejadi—*Prosopis cineraria* and Phog—*Calligonum polygonoides* even though harvesting of Prosopis is banned. The price of wood varies from two dollars to ten dollars per quintal depending upon demand and supply. The Phog wood is in great demand because it burns slowly, emits less smoke and therefore fetches a higher price. The wood is also used for charcoal by jewelers and other metal workers.

Phog and prosopis leaves and shoots are relished by livestock. The flowers and buds of Phog are consumed in the winter months with curd and also as a vegetable. Phog is a an excellent soil binder and helps in soil conservation and sand dune stabilization.

The non-wood uses of the forest are as important if not more so than wood-based uses in most tropical countries. In fact development of non-extractive, non wood-based uses (Myers 1990) may provide incentives for more efficient wood-based biomass energy utilization. It has been estimated that in Indonesia, non-wood products brought a foreign exchange worth of $200 million in 1982 up from $28 million in 1973 (Gillis 1986). The export earning from wood products increased from $583 million to $899 million in the same period. Ecological costs and benefits are not incorporated in these revenues. The implications of short-term revenue for the long-term loss of employment as well as income opportunities for local communities have to be drawn if biomass utilization has to be a means of rural development.

Approximately 20 to 30 years ago, efficient stoves using wood filings and powder as fuel were quite popular. These stoves provided heat in the winter and with a slight modification in design, were also used for cooking. Many of these options have disappeared with changes in relative prices of fossil fuel, coal and other agrowastes.

It is not inevitable that a wood-based economy necessarily lead to resource depletion. In mountain regions, most houses are built of wood and almost the entire economy revolves around the use of wood for cooking, making butter oil, churners, cheese presses, looms, farm implements, shingles for roofs, etc. Vibrant institutions exist in many places for sustainable resource management. The wood-based biomass energy has in fact been a very crucial input into evolution of social and cultural institutions.

requirements. The role of culture, religion and other collective social institutions in modifying individual needs has not been adequately appreciated. For example, it is customary that people go to the forest together on a particular day to collect shingles. There are several explanations for this practice.

a) While collecting wood on the steep slopes, if somebody falls down, there are people around to help in the emergency.
b) Everybody monitors everybody else's collection of wood.
c) Since collection of wood has to be done, the age, health, and condition of the tree is considered by all.
d) Some people are either too old, handicapped, weak or they require more help to manage, groups help in such cases and carry the extra burden.
e) There are sites which might have suffered some damage due to rain, landslides, or other catastrophe. The fact that such sites are observed together enables those involved to search for corrective action.
f) In addition to the utilitarian dimensions mentioned above, the group action is its own reward when there is music, fun and laughter around.

Thus, emphasis on only the economic aspect of a resource use would not provide sufficient information for building sustainable institutions that can help in managing resources. Development is possible only through creative institutions which constrain individual choices to some extent and yet provide scope for entrepreneurship.

SUMMARY

A comprehensive strategy for rural development requires recognition of local self-designing institutions. Institutions regulate individual behavior and provide an opportunity for making sense of complex and uncertain world. Sources of subsistence for any given resource, wood or a non wood, are derived from various resource regions governed by varying property right arrangements. Institutions may explicitly influence behavior for only certain resource regimes, they may generate incentives or discourage the taking of resources from private or public resource regimes. A sustainable strategy for resource management, therefore, requires recognition of the material and non-material, spiritual and physical, individual and collective, and utilitarian and non-utilitarian dimensions of portfolio management. Interventions by state or outside markets may focus on a specific resource for a specific purpose, i.e. energy production, but the implications for developing institutions which can be managed from within will have to be drawn for other resource requirements as well.

The extension of a timeframe, happens when institutional assurances are available regarding tenure or property rights. The assurances can be sustained through a collective decision-making process. This process is often costly. To sustain it one needs tangible incentives for the local population that satisfy cultural and social needs.

Without extending the timeframe, sustainable technologies may not become viable. The diversification of resource use is necessary if risk adjustment over space and season are to match the requirements of re-generation. Nature has few rules and we can violate them only at our own expense. Sustainable resource management is essential for local communities to evolve sustainable lifestyles. However, demand by the rural communities has to compete with demand of urban and affluent consumers. Societies must create a market for local resources such that extractive and non-sustainable resource use becomes commercially attractive. Rural development cannot be isolated from the options of urban development. Similarly, micro-level incentives for resource conservation are guided by macro-level policies acting as signals for trading future options vis-a-vis current ones.

I have tried to show how trade-offs can be appraised. Decisions about how the trade-offs are actually made will depend upon rights of the poor to pursue their cultural, social and economic goals, which may also be modified through evolving institutions. Investment in these institutions can become an efficient and effective way of both generating feedback on micro-policies and creating countervailing local power to correct the distortions introduced by powerful commercial, urban, interest groups.

This paper has covered a broad spectrum. The discussion on multi-institutional, multi-level framework to focus on the evolution of household portfolio of economic enterprises was intended to deepen our search for sustainable rural development alternatives.

It is hoped that while analyzing options relating to wood-based biomass energy systems and their role in rural development, consideration will be given to the institutional aspects. An issue which needs further exploration is whether certain kinds of energy systems require specific kinds of institutional arrangements. For instance in the case of fire wood (a frequently used energy source), higher uncertainty in the supply system may require greater individual reserves. Among the poor people, the ability to acquire and stock such a reserve is limited. In such a system, whether regular access to the source is more important than the ability to build inventories will depend upon specific ecological conditions and the system of assurances in the given socio-economic setting.

It is necessary to look at the political reality of a resource situation so that economic incentives are compatible with political incentives available to the decision makers. Dissemination of information about various alternative scenarios is important so that mobilization around sustainable alternatives becomes possible. The study of ecological movements around forest related conflicts raises the possibility of policy makers using the creative energy of these movements.

The study of indigenous ecological institutions and knowledge systems is vital for generating creative alternatives for institution building. It is obvious that the concept of energy need not be restricted in its direct sense. The indirect contribution through fodder (tree loppings), veterinary medicine (draft power), manure (soil fertility and production of grains and dry matter), etc., require a review so we may achieve a better understanding of resource relationships. The study of cultural and ethical aspects of local knowledge systems in this regard may help in anticipating the impact various technologies will have on the group exchange relations.

Finally, expansion of decision making choices and extension of time frame of the households requires interplay between access, assurances, abilities and attitudes of both the demand and the supply side factors. Rural development process can become a truly liberating alternative if we can link political, institutional, cultural and ecological knowledge systems.

Literature Cited

Aggarwal Arun. 1990. Personal Communication.

Bharadwaj K. 1974. *Production conditions in Indian agriculture.* Cambridge, U.K.

Biggs, S. 1981. *Linkage between formal and informal R and D ceres.* Rome.

Biswas, M.R. and A.K. Biswas. 1979. *Food, climate and man.* New York: John Wiley and Sons.

Buzdar, Nek. 1988. *Property rights, social organization, and resource management in northern Pakistan.* Environment and Policy Institute East West Center Honolulu, Hawaii, W.P. No. 5.

Chambers, Robert. 1983. *Rural development; putting the last first.* Longman, London.

Chambers, Robert, Arnold Pacey, and Lori Ann Thrupp. 1989. *Farmer first: farmer innovation and agricultural research* (edited). Intermediate Technology Publications, London.

Darkoh, M.B.K. 1990. Trends in natural resource use and prospects for sustainable resource management in Kenya's arid and semi-arid lands, *Land Degradation and Rehabilitation* (2)177–190.

Gupta, Anil K. 1981. A note on internal resource management in arid regions small farmers-credit constraints: a paradigm. *Agricultural Systems* (UK) 7(4):157–161.

Gupta, Anil K. 1981. *Viable projects for inviable farmers.* Presented at Symposium on Rural Development in South Asia, Amsterdam, April.

Gupta, Anil K. Why poor don't cooperate: lessons from traditional organizations with implication for modern organizations. In *Research Relationship, Politics and Practice of Social Research*, (Clare G. Wanger, ed). George Allen and Unwin, London. pp 107-127.

Gupta, Anil K. 1983. *Impoverishment in drought prone regions: a view from within.* (joint field study SDC/NABARD/IIM-A CMA, IIM Ahmedabad).

Gupta, Anil K. 1984. *Small farmer household economy in semi-arid regions.* CMA, IIM, Ahmedabad. Mimeo.

Gupta, Anil K. 1984. *Socio-ecology of land use planning in semi-arid regions*. September 1984, IIM Working Paper No. 525, p 33, paper presented at Commonwealth Workshop on Land Use Planning in Tropics, BARI and IIMA, January 28-February 2, Ahmedabad.

Gupta, Anil K. 1985. Socio-ecological paradigm to analyze the problem of poor in dry regions. *Eco-Development News* (Paris) 32:71–75.

Gupta, Anil K. 1985. *Sociology of stress: why do common property resource management projects fail?* Proceedings of The Conference On Common Property Resource Management April, BOSTID & National Research Council, Washington, D.C.

Gupta, Anil K. 1986. *Socio-ecology of grazing land management*. Published in Rangelands: A Resource Under Siege, proceedings of the Second International Rangeland Congress. Edited by P.J. Jose, P.W. Lynch and D.B. Williams, Australian Academy of Sciences, Canberra, (IIM Working Paper No. 524, 1984). pp 288-291.

Gupta, Anil K. 1986. *Socio ecology of stress: why do common property resource management projects fail?: a case study of sheep and pasture development project in Rajasthan, India.* Paper presented at conference on Management of CPR, organized by National Academy of Sciences, BOSTID, US, Annapolis, April 21-26, 1985; proceedings published by National Research Council, Washington, D.C. USA. pp 305-322.

Gupta, Anil K. 1987. *Musical chairs, snake and ladder and management of risks*. Seminar presented at University of Munich.

Gupta, Anil K. 1987. *Organizing the poor client responsive research system: can tail wag the dog?* Paper presented in a workshop on Farmers and Agricultural Research: Complementary Methods, Institute of Development Studies at the University of Sussex, UK, 26-31 July 1987. Presented at International Advisory Committee of ISNAR, meeting on on-farm Client oriented Research at the Hague.

Gupta, Anil K. 1987. *Role of women in risk adjustment in drought prone regions with special reference to credit problems*. October, IIM Working Paper No. 704.

Gupta, Anil K. 1988. *Survival under stress: socio ecological perspective on Farmers' innovation and risk adjustment*. Presented at International Congress on Plant Physiology, New Delhi.

Gupta, Anil K. 1989. *Ecology market forces and design of resource delivery organizations*. Paper prepared for International Conference on Organizational and Behavioral Perspective for Social Development, December 29, 1986 - January 2, Int. Studies in Management and Organization, 18(4):64–82.

Gupta, Anil K. 1990. *Eco-sociology of household risk adjustment and commons: performance in an uncertain world*. Presented at International Conference On Designing Sustainability on the Commons, International Associations for the study of the Common Property. Duke University, USA.

Gupta, Anil K., N.T. Patel, and Rekha Shah. 1988. *Matching farmers' concerns with technologists objectives in dry regions: an exploratory study of scientific goal setting*. CMA, IIM Ahmedabad mimeo.

Gupta, Anil K. and Vora Karma. 1990. *Blending cultural values, indigenous technology and environment: the experience of Bhutan*. Presented at International Conference on Integrated Mountain Development, ICIMOD, Kathmandu, September 10-14, 1990. IIM, Ahmedabad, W.P. No. 883.

Jodha, N.S. 1975. Famine and famine policies: some empirical evidence *Economic and Political Weekly* 10(41):1609–1624.

Jodha, N.S. 1978. *Role of credit in farmer's adjustment against risk in arid and semi-arid tropical areas of India.* ICRISAT, A.P. India, O.P. No. 20.

Jodha, N.S. and A.C. Mascarenhas. 1983. *Adjustment to climatic variability is self provisioning societies: some evidence from India and Tanzania.* ICRISAT, Economic Programme Report No. 48, March. pp 10-12.

Jodha, N.S. 1990. *Sustainable Agriculture in Fragile Resource Zones: Technological imperative.* Presented at International Symposium on Natural Resource Management For Sustainable Agriculture, February 6-10, New Delhi.

Mabugunje. 1979. *Development process: spatial perspective.* London: Hutchinson Press.

Mathai, R. J., U. Pareek, and T.V. Rao. 1977. *Institution building in education and research: from stagnation to self- renewal.* P.S.G. Monograph. IIM Ahmedabad.

Mathur, K. and Anil K. Gupta. 1984. Action research for micro level planning: self appraisal. *International Review of Administrating Sciences* 50(1):60–68.

Munshi, K.M. 1951. *The gospel of the dirty hand and other speeches on the policy and programme of land transformation.* Ministry of Information and Broadcasting, Govt. of India.

Munshi, K.M. 1952. *Land transformation. A philosophy and a faith.* Ministry of Food and Agriculture, New Delhi.

Ostrom, V., D. Feeny, and Picht. 1988. *Rethinking Institutional Analysis and Development Issues, Alternatives and Choices.* International Center for Economic Growth.

Pangare. 1991. Personal Communication.

Peat, F. David. 1987. *Synchronicity-the bridge between matter and mind.* Toronto: Bantam Books.

Perlin, John. 1991. *A forest journey the role of wood in the development of civilization.* New York: W.W. Norton $ Company.

Prigogine, I. and Isabelle Stengers. 1984. *Order out of chaos.* William Heinemann Ltd.

Richards, Paul. 1989. *Agriculture as a performance.* Farmer First Edited by Robert Chambers, Arnold Pacey, and Lori Ann Thrupp. London: Intermediate Technology Pub. pp 39-43.

Runge, Carlisle Ford. 1986. Common property and collective action in economic development. *World Development* 14(5):623–635.

Sen, Amartya. 1967. Isolation assurance and the social rate of discount *Quarterly Journal of Economic* 81:112–124.

Sen, Amartya. 1980. Description as choice. *Oxford Economic Paper* 32(3):353–69.

CHAPTER 5

ENERGY FROM WOOD RESIDUES IN HONDURAS[1]

Jim Wimberly
C. B. Briscoe
Kelton Grubbs
Gregory Morris

ENERGY PRODUCTION AND THE WOOD PRODUCTS INDUSTRY IN HONDURAS

Exports of wood products are a significant source of Honduran foreign exchange. Most sawmills rely on imported petroleum to generate their own electrical power, since grid-connected power from the Empresa Nacional de Energía Eléctrica (ENEE) is unreliable or nonexistent in the remote areas where the mills are located. At the same time, the sawmills generate large quantities of energy-rich wood residues, whose disposal incurs both economic and environmental costs. These residues could meet the energy requirements of all sawmills in Honduras and provide surplus energy in the forms of electricity, process steam, or solid fuels for the national economy. Forest residues, litter from harvesting the forest, and wood residues from sawmills have the potential to become a small, but significant component of the country's total production of energy—either for remote, off-grid applications or for connection to the national grid.

Potential benefits of using forest and wood residues for the production of energy include:

- ***Renewable Energy***—Substitute local renewable sources of energy for imported, nonrenewable petroleum products.

1 We would like to express our sincere appreciation to the following individuals for their support and assistance with this study: Ing. Mateo Molina, Director, COHDEFOR PDF project; Ing. Danilo Escoto, Chief, Technical Assistance Unit to the Industry, COHDEFOR PDF project; Lic. Mauricio Mossi, Chief, Planning Department, ENEE; Dr. Rafael Rosario, Director, Natural Resources Department, USAID/Honduras; Dr. John Warren, Project Officer, USAID/Honduras; Ing. Ramon Alvarez, Forestry Specialist and Assistant Project Officer, USAID/Honduras; Mr. John White, Advisory Team Leader, PDF Project, USDA Forest Service; Mr. Doyle Romans, Industrial Forestry Advisor, PDF Project, USDA Forest Service.

- ***Sustainable Forest Management***—Provide incentives for sustainable forest management by adding energy values to wood residues from harvesting forests and processing timber.
- ***Use of Local Resources***—Use Honduran land, labor, and currency to produce renewable energy, instead of spending scarce foreign exchange to purchase imported petroleum.
- ***Environment***—Reduce environmental degradation by controlling emissions of carbon monoxide and particulates and by limiting the introduction of solid wastes into streams and rivers, drainage ways, or ground water; reduce potential global warming through the use of renewable energy. *As long as forests are managed on a sustained basis, there will be no net contribution to global atmospheric* CO_2.
- ***Independent Production of National Energy***—Sell excess electricity to reduce the costs of planning, construction and construction delays, and environmental degradation that are associated with generation facilities based on petroleum and hydroelectric power.
- ***Operating Margins***—Improve operating margins at sawmills by reducing the costs of energy required for manufacturing.
- ***Employment***—Install and operate wood-based energy systems to create needed employment opportunities, particularly in rural areas.
- ***Rural Economic Development***—Increase the availability of energy to rural populations to enhance rural development.

To realize these benefits, it is necessary to attract investment capital to install generation systems that burn wood residues at the sawmills. In this chapter, we will explore current and future demand, production, and distribution of energy and its effects on forests in Honduras. We will examine the possibilities for the production of energy from wood residues at Honduran sawmills. Are they economically and financially sound investments?

We will compare the costs of installation, operation, and maintenance of three configurations for the production and use of wood-based energy for two representative sizes of sawmills. We will describe the design options, engineering, and equipment for each case, and we will evaluate their operational characteristics and benefits. Then we will present a *pro forma* financial analysis and a sensitivity analysis that illustrate economic considerations and financial performances.

Our study represents a methodology for integrating data on engineering and economics that will provide possible lenders or investors with the information needed to make intelligent decisions while planning for future development of energy production.

An Uncertain Energy Supply for the Future

The five million people of Honduras look to their forests as a principal future source of energy and foreign exchange. Forests once covered almost all of the 112,088 sq. mi. of Honduras, but now they cover about 5,000,000 ha, less than 60% of the country. Still, forests now provide nearly 68% of the country's energy, as fuelwood for domestic use, and 86% of all fuel for household use (DESFIL 1989). In fact, the contribution of energy from the forests of Honduras is proportionately greater than the percentage of land area that the forests occupy.

The World Bank (1980) estimated that by 1995, hydropower would provide 7% of the country's total energy, while petroleum would supply 29%. Biomass would be the largest contributor to the energy supply, with 64% of the total. Fuelwood would continue to be the cheapest source of energy, with 60% for domestic use and 14% for industry. Now, these projections seem overly optimistic. Industry and transportation will continue to rely on imported fossil fuels for energy, since Honduras has no petroleum or coal deposits. Petroleum accounted for 16% of the country's imports in the late 1980s (FAO 1987).

As of 1990, consumption of electrical power in Honduras was about 1,500 million kw/yr. Nearly two-thirds of the total demand for electricity in Honduras is concentrated in its three major cities—Tegucigalpa, San Pedro Sula, and La Ceiba. Even though Honduras has slightly more capacity to generate electricity than it currently needs, its transmission and distribution system remains poorly developed. There is little electrical power available to rural areas. ENEE shoulders increasing debt partly because it is unable to export its surplus electricity to other countries. Therefore, any plans to distribute electrical power to rural areas are uncertain. We may expect that rural areas in Honduras will remain without electricity for the foreseeable future. In the meantime, urban and industrial demands for electricity will absorb Honduras' existing capacity within the next two to four years.

Forest Resources in Honduras

Dwindling Forest Resources

Approximately 80% of existing forested land is suitable for the commercial production of timber, and 8% has the potential to be intensely cultivated for agriculture. The remaining 12% should be natural forests that are managed as watersheds. More than half of the forests in Honduras—2.6 million hectares —are hardwoods, including the well-known Honduran mahogany which once comprised the most significant portion of the country's lumber production. The rest of the forested area is pine, now the primary product in domestic and export markets.

Few efforts are made to reforest harvested timberlands. The forest industry relies heavily on the high rate of Honduran pine's natural regeneration. It takes 30 years for Honduran pine to reach maturity, and many pine forests are now in their third rotation. Table 5.1 shows deforestation rates for both hardwood and pine forests for the past 22 years, with a total estimated rate of 80,000 ha/yr. Without sustained management, Honduras will be barely self-sufficient in forest products in about 15 years.

Table 5.1. Estimated Historical Forest Area

		Inventory			Change, 22 yrs	
Forest Type		FAO 1964	COHDEFOR 1986	ha	%	ha/yr
Pine (commercial)		1,936,500	1,004,500	-932m	-48	-42,364
Pine (immature)		802,300	1,392,300	590m	74	26,814
	Subtotal	2,738,800	2,396,700	-342m	12	-15,550
Hardwoods		4,072,200	2,654,300	-1,418m	-35	-64,450
	Total Forests	6,811,000	5,051,000	1,760m	-26	-80,000

Renewed Efforts to Manage Forests

Although there has been little coordinated and sustained management of forest resources, the government has long recognized forests for their role in soil and water conservation and as sources of valuable products. COHDEFOR's current *Proyecto Desarollo de Forestal* (PDF) is attempting to reverse the effects of negligent practices by introducing and supporting sustainable forest management in Honduras. Its purpose is to improve the management and sustainable productivity of commercial pine forests and the efficiency of industrial conversion and marketing of wood products. PDF plans to control harvesting of the forest and to promote reforestation on an informal basis. It also plans to promote natural regeneration of forests, which should permit perpetual harvesting near the present level while protecting noncommercial forest reserves and major watersheds. The project is financially supported by USAID and receives technical assistance from the USDA Forest Service.

Sustainable forest management can add value to otherwise worthless wood residues. Some of the leftovers from harvesting and the thinnings from pruning—forest residues—may be transported to sawmills for additional fuel. The values of biomass residues would offset the costs of thinning, a desirable silvicultural practice, and, at the same time, reduce the potential of wildfires—a serious problem in Honduras. Additionally, the collection, removal, and transport of such residues could create significant additional rural employment. Rural inhabitants in Honduras are now encouraged to consider these forest residues as additional, potential fuel.

The Need for New Policies of Land Ownership

Recent Honduran laws and regulations mandate forest management, but financial limitations and questionable forest policies have prevented the complete implementation of forest management plans. In particular, the unequal distribution of land ownership in Honduras remains a serious social problem, despite continuing efforts to reform existing policies. In the last census, 64% of the farms occupied only 9% of the land and were, at most, only five hectares. Lack of clear title to land is a great obstacle to sustainable land management. Two or more parties, often the government, claim much of the land. None of the claimants is willing to invest in good land management, especially for forestry, without clear title (FAO 1987).

The division between tree ownership and land ownership handicaps any strategy to manage forests efficiently in Honduras. As of February 1992, all trees in the forest are owned by the government. This policy virtually guarantees antagonism on the part of landowners toward the presence of forests. Two obvious effects of such a policy—overgrazing and frequent forest fires—are apparent throughout the tropics and, especially, in Honduras (FAO 1987, FAO 1968, Tenore and Cosenza 1980).

The Wood Products Industry

Wood products earn the largest amount of foreign exchange for Honduras , next to agricultural products. Table 5.2 lists the major sawmills in Honduras with their 1989 annual production of dimension lumber and estimated production of wood residues (estimated at 50% of log volume) as well as each sawmill's highest annual production data between 1984 and 1989. The uncertain availability of timber in the future, however, implies that any investment in the wood products industry should be evaluated carefully (Chronowski 1985). Many mills are now unable to operate at their full production capacity because there is such tight regulation of the timber supply. The long-term outlook for the wood products industry in Honduras is very uncertain.

Requirements for Electricity at Sawmills

The major form of energy used by sawmills in Honduras is electricity, which runs electric motor drives for saws and other manufacturing equipment. Requirements for electricity at existing sawmills vary significantly from location to location, depending on rated production capacity and remanufacturing. We assumed that remanufacturing would not be a determining factor in the size range of our study. Thus, the size of a sawmill —its production capacity—determines its requirements for energy and the amount of wood residues that it will produce.

Table 5.2. Lumber and Wood Residue Production at Honduran Sawmills

		Annual Production		Highest Production in a Single year	
		1989		1984-1989	
Sawmill	District	Lumber (a)	Residues (b)	Lumber (a)	Residues (b)
Corfino	Colon	17.7	26,547	21.6	32,458
Sansone	Francisco Morazan	11.8	17,732	14.5	21,789
Yodeco	Yoto	11.9	17,822	12.9	19,385
San Jose	Olancho	8.9	13,389	10.4	15,628
Santa Fe	Francisco Morazan	8.5	12,733	10.1	15,177
El Progreso No. 2	El Paraiso	3.8	5,755	6.3	9,467
Sansone No. 2	Comayagua	5.3	7,904	6	9,016
Santa Ana No. 2	Olancho	5.1	7,664	5.1	7,664
San Diego	El Paraiso	3.1	4,613	4.9	7,363
Lumberton	Francisco Morazan	2.3	3,456	4.4	6,612
Arcieri	Francisco Morazan	3.2	4,809	3.8	5,710
Sansone No. 1	Francisco Morazan	1.5	2,254	3.7	5,560
Maderas des Wampu	Olancho	3.2	4,734	3.6	5,410
Santa Elena	Intibuca	2.4	3,637	3.5	5,259
Infozero	Olancho	2	2,945	3.1	4,658
La Laguna	Olancho	2.4	3,622	3.1	4,658
Buena Vista	El Paraiso	2.8	4,223	3	4,508
Teupasenti	El Paraiso	3	4,568	3	4,508
Bijao	Olancho	2.3	3,516	3	4,508
Santa Maria	Francisco Morazan	1.1	1,653	2.9	4,358
Larizabal	Francisco Morazan	1.3	1,954	1.6	2,404
Corralitos	Francisco Morazan	1.4	2,104	1.5	2,254
Quiraguara No.1	Intibuca	1.3	1,908	1.5	2,254
El Platana	Olancho	1.5	2,224	1.5	2,254
Noriega No. 2	Olancho	1.5	2,204	1.5	2,204
Chang No. 1	Francisco Morazan	0.6	870	1.4	2,104
Siria	Francisco Morazan	0.7	1,020	1.4	2,104
Escambray #3	Santa Barbara	1.3	1,923	1.4	2,104
Avelar	Comayagua	0.9	1,286	1.3	1,954
Rio Amarillo	Copan	0.9	1,291	1.3	1,954
Rosales	Yoro	0.1	143	1.3	1,954
La Antorcha	Yoro	1.2	1,833	1.3	1,954

Beatriz	El Paraiso	1.2	1,833	1.2	1,803
Guadalupe	Lempira	0.9	1,358	1.2	1,803
Alvardo	Lempira	0.8	1,162	1.2	1,803
Sacamine	Lempira	1	1,501	1.1	1,653
Tekamblum	Ocotepeque	0.4	615	1.1	1,653
Morazan	Cortes	0.5	730	1	1,503
San Francisco	Olancho	2.5	3,742	2.9	4,358
Catacamas	Olancho	2.7	3,997	2.8	4,208
San Antonio	El Paraiso	2.3	3,411	2.7	4,057
San Pedro	Lempira	1.8	2,690	2.7	4,057
Portillo	Yoro	1.6	2,389	2.6	3,907
Santa Maria	Comayagua	2.1	3,201	2.5	3,757
Noriega #1	Santa Barbara	2.2	3,291	2.5	3,757
Hibueras	Yoro	1.7	2,555	2.4	3,606
Carolina	Yoro	2.2	3,291	2.2	3,306
Naco	Cortes	1.6	2,389	2.1	3,156
Esmeralda No. 1	Francisco Morazan	2	3,005	2.1	3,156
Jesus Bna, Esperanza 2	Olancho	2.1	3,126	2.1	3,156
Guaymas	Yoro	2.1	3,201	2.1	3,156
Jesus Bna, Esperanza	Francisco Morazan	1.4	2,104	2	3,005
IAI Masssu	Comayagua	1.9	2,872	1.9	2,855
Suyapa	Francisco Morazan	0.9	1,398	1.9	2,855
Intibucano	Intibuca	1.6	2,374	1.8	2,705
Patricia	Yoro	1.8	2,705	1.8	2,705
Merendon	Yoro	0	69	1.8	2,705
Cecilia	Francisco Morazan	1	1,473	1.7	2,555
Embalse del Cabon	Cortes	0.4	654	1.6	2,404
El Carrizal	Intibuca	1	1,573	1	1,503
San Juan	Intibuca	0.5	721	1	1,503
Ana	Santa Barbara	0.8	1,250	1	1,503
Promagua	Yoro	0.3	481	1	1,503
La Confiaza	Santa Barbara	0.8	1,129	0.9	1,352
Zuniga	Yoro	0.7	1,067	0.8	1,202
Santa Isabel	El Paraiso	0.7	1,007	0.7	1,052
San Francisco	Intibuca	0.6	887	0.7	1,052
El Esfuerzo	Lempira	0.3	524	0.7	1,052
IAI Santa Marta	Santa Barbara	0.6	845	0.7	1,052
Maya	Copan	0.6	861	0.6	902
IMAHSA	La Paz	0.5	820	0.5	751
El Mochito	Santa Barbara	0.3	451	0.5	751
San Francisco	Copan	0	74	0.3	451
Los Alamos	El Paraiso	0	71	0.3	451
ESNACIFOR	Comayagua	0	68	0.2	301
El Zamorano	Francisco Morazan	0.1	219	0.2	301
Las Mercedes	Lempira	0.2	252	0.2	301
Honduply	Olancho	0.2	286	0.2	286
IAI Corpimwx	Olancho	0.1	161	0.1	161

Sawmills in Honduras typically operate 5 to 5 1/2 days/week for one shift of 8 to 10 hours/day. Some sawmills purchase electricity from ENEE, if a sawmill has access to the grid, but the grid covers only a limited area. The tariff rates for industrial electricity in Honduras, on September 1, 1990, were based on two components: a monthly demand charge of Lempira (Lps.) 20.00/kw, based on 85% of the monthly maximum demand, and an energy charge of Lps. 0.18/kwh.

Most sawmills were built when electrical service from a utility was unavailable in remote areas, so they had to produce their own electricity on-site. Mill owners chose the lowest capital cost alternative—the industrial diesel-engine generator—to supply minimum requirements for power so that they could minimize transportation and storage costs for fuel. Diesel engines are relatively easy to operate, but the sawmills were designed to substitute manpower for electrical power whenever possible. Storage tanks for diesel fuel were designed to hold a supply for one or two weeks.

The fuel requirements for the sawmills covered in our study varied between 500 and 1,200 gallons/week, depending on mill capacity and use of the generators during off-hours. Some mills shut down their diesel generators when the plant is not in operation, while others run their generators at low power levels during nonoperating hours to supply electricity for domestic and other light-duty uses.

The Honduran Government has a long-standing policy of subsidizing the price of diesel fuel to all sectors of the economy. This policy keeps the cost of generating energy at the sawmills below free market levels. During our site visits to sawmill plants in Honduras in September, 1990, the domestic price of diesel fuel was 3.99 Lps/US. gallon—about US$ 0.75/gal. The price for diesel fuel may be compared with Lps. 6.25/gal, US$ 1.18, for regular gasoline. Any proposal to install wood-based energy systems at sawmills must take the Honduran government's subsidization of diesel fuel into account.

Here is a comparison of the estimated costs of electricity produced on-site from diesel fuel with those of power purchased from the utility for sizes of sawmills in our study—large sawmills that produce 9 MMBF/yr and small mills that produce 3 MMBF/yr:

	LARGE SITE	**SMALL SITE**
Diesel Fuel	Lps. 420,000	Lps. 175,000
Utility Electricity	Lps. 120,000	Lps. 70,000

All costs are based on prices of March, 1991, and are representative of typical operations. Costs for energy at a particular location may vary. None of the estimated costs include depreciation, equipment operations, and maintenance. Maintenance and depreciation costs vary greatly, but they are major components of total owning and operating costs. Costs indicated for electricity from the utility are for power purchases only and do not include costs of transformers and other interconnection equipment. The costs for electricity from the utility also assume the availability of a high voltage interconnection. If utility lines in the vicinity of the sawmill are medium voltage lines, the cost of purchased power is approximately 25% higher.

Additional Requirements for Energy at Sawmills

The primary export market for Honduran wood products—the European Common Market—now requires that all lumber imported to Europe be treated against insects. Of the two options for meeting this new regulation—kiln-drying or chemical treatment—kiln-drying is the safer and more environmentally benign. Few sawmills in Honduras in our study kiln-dry their lumber.

However, kiln-drying of lumber could substantially benefit the sawmill industry in several ways. First, it would add value to the wood product. Lumber sold domestically in Honduras is valued approximately 25% higher when it is kiln-dried, and the value might be much higher in Europe. Planing kiln-dried dimension lumber at the sawmill would further up-grade the value of the product for both domestic and export markets. Finally, the cost of shipping would be lower. Kiln-drying alone reduces the weight of the product by as much as one-third. Planing the lumber reduces the weight again by up to 25%.

The recent increases in the demand for kiln-dried products in Honduras' lumber markets suggest that sawmills in Honduras will require additional thermal energy. We designed the energy systems for our study configurations to provide enough electricity and steam for drying 25% of the mill's primary lumber production. We chose this arbitrary value in part because it is large enough to illustrate its impact on the design of the energy systems, but small enough that it does not have a major impact on the economic value of such systems. Perhaps as much as 70% of the lumber could be dried with significant benefit to a sawmill. Such possibilities of kiln-drying and surfacing—with their capital and operating costs as well as enhanced revenues from added value and reduced freight charges—should be investigated in all site-specific feasibility studies.

Production and Disposal of Wood Residues

Exclusive of bark, the amount of wood residues generated during milling operations varies between 40% and 45% of the incoming log volume, depending on the particular sawmill. The actual amounts of wood residues produced at a given site depends on the efficiency of the sawmill operation and the recovery of secondary products. In this study, we assumed that the average production rate of wood residues for sawmills is 43% primary product (lumber). Thus, *the average amount of wastes resulting from milling operations is approximately 50% of the incoming log volume, including 7% bark.*

Most wood residues produced by a typical sawmill are either green sawdust or solid pieces of slabs and edgings. Logs are not usually debarked prior to processing at sawmills in Honduras, and most of the bark, along with slabs and edgings, is waste. Most of the waste volume is green. The assumed heating value of wood residues is 4,975 BTU/lb, and the average moisture content is 45%, wet basis.

Cost of waste disposal methods—incinerators, or hauling to remote dumping sites—have not been considered in our study. In many cases, sawmills in Honduras dispose of their wood residues by dumping them in piles, which are left to continually smolder and burn at the periphery of the property, usually near a river bank. The result is acrid ground-level smoke and haze. In addition, leaching of tannic acids from the decay of sawmill residues contaminates ground water and soil. Water run-off from exposed piles of unburned and partially burned wood residues contaminates nearby streams and impoundments.

The conversion of wood residues into energy have the potential to impart a significant beneficial effect on the environment in Honduras. The magnitude of the benefits to the environment is in direct proportion to the degree in which such systems are adopted by the wood products industry. Controlled burning of sawmill residues will significantly reduce contaminants in the air and water. It could also eliminate the emission of from 200 to 400 tons/yr of carbon monoxide, nitrogen oxide, particulates, and other pollutants that are released into the environment by burning fossil fuels for energy.

Densification—well-known in the United States and northern Europe as fuel for domestic heating and for some industries—is another means of using wood residues. Sawmills that do not have access to the national grid for sale of excess electricity should consider this option. The technology could also be an attractive investment for mills that have access to the grid. Sawdust, or a mixture of sawdust and hogfuel, are extruded at high pressures and temperatures to produce pellets or briquettes, ranging in size from 0.25" to over 3" in diameter. Densified fuels may be made into charcoal to increase combustion efficiency and reduce transportation cost, but this process consumes large amounts of the fuel's heat energy during carbonization.

WOOD-BASED ENERGY SYSTEMS FOR SAWMILLS

The Honduran forest products industry has always used wood residues for fuel, especially when it was a new industry. A wide variety of combustion systems are now available, ranging from the manually fired pile-burning systems that were typical of early sawmills to the highly sophisticated, automated systems found in modern forest products plants. Options now include multiple-stage combustors, suspension burners, and gasifiers.

Several different types of boilers, prime movers, fuel handling systems, and auxiliary systems may be combined to form a wood-based energy system. Careful consideration should be given to such site-specific factors as the quality and availability of both wood residues and water to be used. What types of process energy are required? What capacity is needed? Do employees have the requisite technical skills? Decision-makers must evaluate the trade-offs between the costs of manual labor and the cost of automation, between system efficiency and reliability on the one hand and installation and operating costs on the other.

Selection of Equipment

Our study represents more than three-quarters of the sawmills in Honduras—over 95% of the country's lumber production—which have capacities to produce lumber in excess of one MMBF/yr. These mills can meet all of their demands for energy by burning their wood residues with the lowest capital-cost technology that we describe in this section. The sawmills in our study have production capacities ranging from 1 MMBT/yr to 10 MMBF/yr. The fuel needed will vary from 7,500 MT, wet basis, to at least 25,000 MT of wood residues. The quantities of fuel required indicates that the maximum boiler rating will be less that 30,000 lbs/hr of steam generation. For this size, the most economical boiler would be a fire-tube unit with a maximum operating pressure of less than 200 psig. We have based our economic evaluations on systems with such a boiler.

We have designed the hypothetical energy systems to provide enough steam and electricity for drying 25% of a mill's primary lumber production. Requirements for steam for this process will vary between 5% and 10% of the required boiler capacity. The estimated average electrical load and the fuel requirements of these systems should be adjusted to reflect the impact of the lumber-drying process. Figure 5.1 shows the variation in the estimated electrical power required for sawmill operation, dry-kiln operation, and powerhouse auxiliaries vs. rated mill capacity.

New boiler systems should be selected to operate in conditions similar to those now in operation, without significant pretreatment of the boiler make-up water. We have minimized the consumption of water in our designs. All of the sites that we visited obtained their process water from nearby streams or small surface impoundments. So, the quantity and quality of water available for the production of electrical power varies with the location of the sawmill. Local streams have significant turbidity and color, but most mills have no pretreatment equipment and achieve satisfactory results with some type of internal treatment.

The only combustion system that can use random-sized solid wood pieces for fuel without further processing is a balanced, or negative draft, furnace that has provisions for feeding larger pieces through the furnace door. This furnace can be coupled with a conveyor system that delivers the sawdust and smaller pieces to the furnace through a feed chute in the roof or upper sidewall. An operator may regulate the firing rate by feeding reserve fuel held in an infeed hopper that is connected to the conveyor system. This type of combustion system is not nearly as efficient as a more automated one, but it is much less expensive, and it is compatible with the policy of substituting manpower for mechanical/electrical power whenever practical. We have selected a manual combustion system for this study.

We chose a single-stage turbine for our study. They are less expensive and only slightly less efficient than multistage turbine engines, which are typically used for generating electricity from wood residues.

Two types of condensers are suitable for a single-stage turbine generator, either an air-cooled condenser or a water-cooled surface condenser. For our study, we have chosen both types to make a realistic comparison between a relatively low-cost system that employs an air-cooled condenser and a more expensive but energy-efficient system which includes a water-cooled condenser.

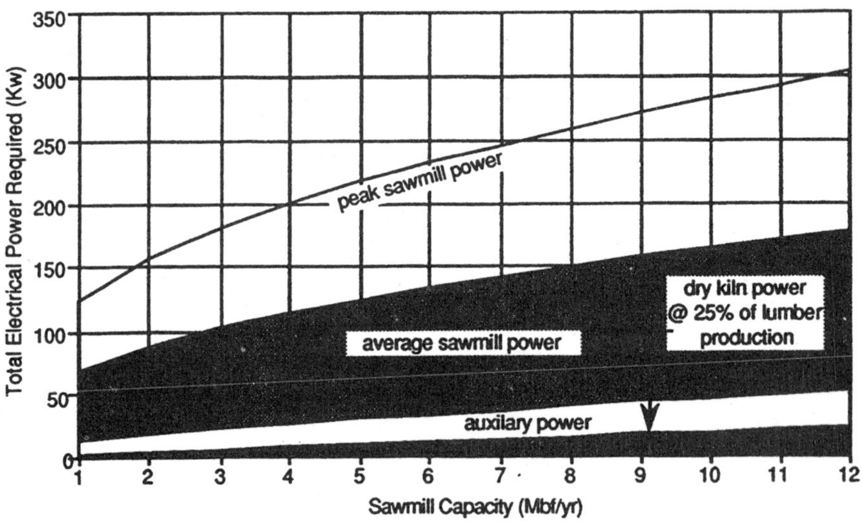

Figure 5.1. Electrical Requirements vs. Mill Capacity.

VARIABLES FOR A CASE-STUDY ANALYSIS

Two Representative Sizes of Sawmills

Our study team, working with COHDEFOR and PDF staff, identified two representative sizes of sawmills:

- a sawmill that produces a **large** volume of lumber: 8 to 10 MMBF/yr
- a sawmill that produces a **small** volume of lumber: 2 to 4 MMBF/yr

We used data collected from actual sites. The San Jose sawmill in Olancho, which produced 8.9 MMBF/yr in 1989, represents a large sawmill; and a small sawmill is represented by the Teupasenti mill in El Paraiso, which produced 3 MMBF/yr in 1989. Accuracy of the site-specific information used in the evaluation is based on the characteristics of these sawmills that were visited by the study team.

From the information obtained from the two mills we estimated the average electrical load at approximately 12.8 kW/MMbf for the larger mill and 29 kW/MMbf for the smaller. However, the data used in this study are representational, not specific. Requirements for electrical power at other locations may be estimated with an equation that relates the magnitude of the specific electrical load (kW/MMbf) to the annual production. A significant error in the estimated load at any specific location is possible with this approach. Many of the calculations in our report are based on general assumptions that need verification on a case-by-case basis before any investment decisions can be made for specific sites.

Two Approaches for Producing Energy

We determined two design approaches to the production of electricity from wood residues at sawmills:

- *Energy self-sufficiency*—is a system that will provide all of the host mill's energy requirements, but no surplus energy for sale to potential off-site customers. This approach will not completely solve the waste disposal problems at most sawmills because the wood residues produced exceed the amount needed to generate energy for the mill's operation.
- *Off-site energy sales*—is a system designed to use all of the available wood residues at a sawmill. This approach provides energy self-sufficiency *and* serves export markets. Two possible export markets exist with this approach. First, sawmills with access to the electrical grid could sell to ENEE, the national electric utility company. Sawmills without access to the grid must identify other markets for their surplus energy. These sawmills have the option to form rural electric cooperatives for nearby towns to encourage the growth of rural or cottage industries. Or, they may convert excess wastes into densified fuels or charcoal for domestic heating and cooking or for energy-intensive industrial markets.

We have assumed that all off-site energy sales of surplus electricity are to the national grid. Sales of *thermal* energy or solid fuels is possible, but not considered.

For energy systems that provide only energy self-sufficiency, the requirement for steam to dry-kiln 25% of lumber production adds a small increase to capital cost relative to having *no* dry kiln. A slightly larger boiler (5% to 10% larger) and turbine-generator are needed. For systems producing power for off-site sales, there is no increase in capital cost. However, such systems would produce slightly less electricity to sell because of the steam and electric requirements of the dry kiln.

Two Wood-Based Energy Systems

The choice of either an air-cooled or water-cooled condenser determines our two proposed wood-based energy systems. In the ***low capital system***, the turbine operates with an air-cooled condenser and has an exhaust pressure, rather than vacuum exhaust conditions. The overall efficiency of the system is reduced, since more fuel is required for a given output of electricity. This configuration is feasible, however, since more than enough wood residues are available on-site to power any given sawmill. Figure 5.2 shows the schematic arrangement of a *low cost system*. It will be designed to operate with an exhaust pressure in the turbine that is sufficient to return condensate to the feedwater storage tank without using condensate pumps. We also decided on a vented feedwater storage tank, a less expensive choice, rather than a deaerating feedwater heater. Such a system gives us a lower installed cost, since we have eliminated a significant amount of condensate handling equipment, but there may be an increased cost for water treatment.

Figure 5.3 illustrates the ***energy efficient system.*** This design incorporates a water-cooled, surface condenser, which requires the installation of condensate and water-cooling pumps, controls for the condensate flow, and a cooling tower. The additional costs of installation and water consumption make this system economically feasible only when excess electricity can be produced for off-site sale.

Technical considerations would not favor one of these systems over the other, but the selection of the most suitable system for a particular site should be based on an assessment of the value of electrical output against the installed and operating costs of the two systems.

Preliminary, expected performance data have been developed for power generating systems incorporating both the ***low capital*** and the ***energy efficient*** designs at sawmill locations in Honduras. We sized each system for the two different approaches to energy production: to produce all of the electrical power required by the sawmill—with its proposed dry kilns and powerhouse auxiliaries—while having wood residues left over; and to use all available wood residues for power generation, with the excess power sold for off-site consumption. Winrock International (1991) has published a more detailed account of this study of sawmills in Honduras.

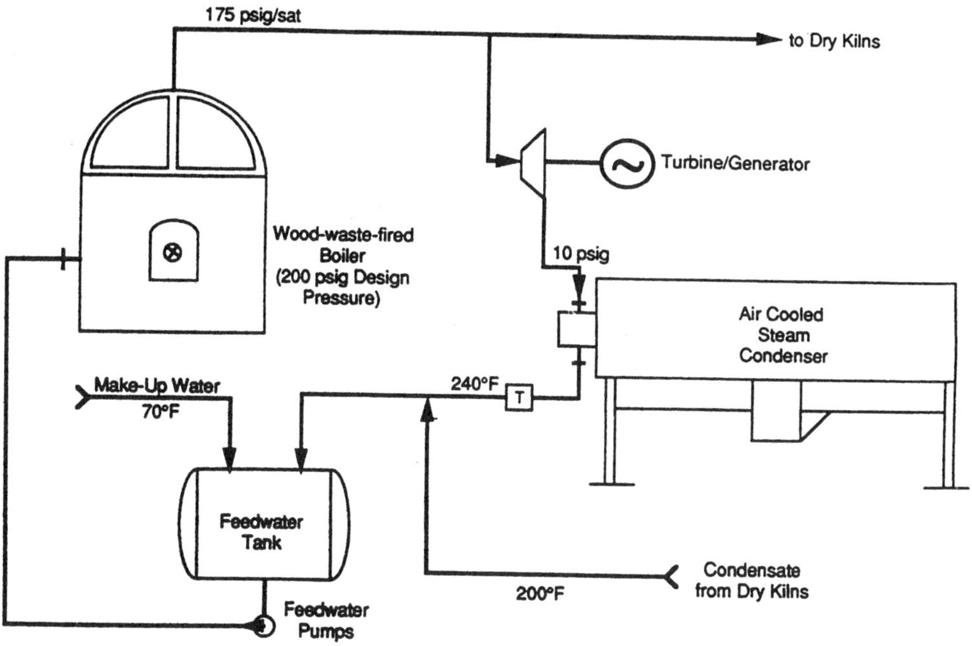

Figure 5.2. Low Capital System - Schematic Diagram.

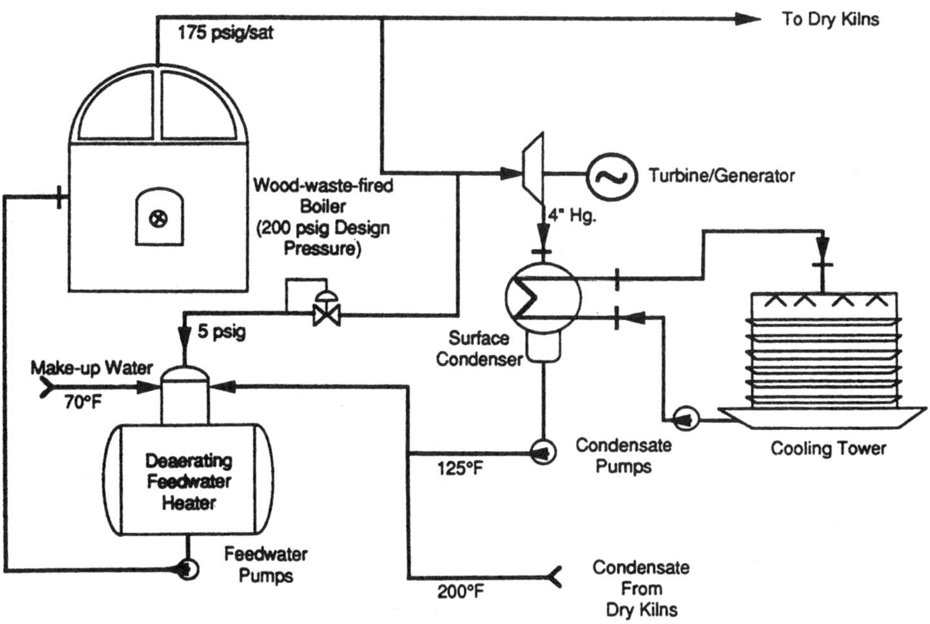

Figure 5.3 Energy Efficient System - Schematic Diagram.

ANALYSIS OF SIX CASE-STUDIES

We have paired our two approaches to the production of energy—***energy self-sufficiency*** and ***off-site energy sales***—with our two wood-based energy systems—***low capital cost*** and ***energy efficient design***. It seems that an energy efficient system would not be appropriate for a sawmill that adopts the energy self-sufficiency approach. All sawmills in the size range of our study produce more than enough wood residues to generate their energy requirements with a low capital system. The additional capital cost of an energy efficient system would not be warranted in such a case.

Three possibilities for production and use of electrical energy emerge:

- ***Low capital cost design, with energy self-sufficiency*** at the sawmill
- ***Low capital cost design, with off-site energy sales*** after meeting internal demands for energy at the sawmill
- ***Energy efficient design, with off-site energy sales*** after meeting internal demands for energy at the sawmill

Our investigations combine these three possibilities with our two representative sizes of sawmills to yield six case-studies.

Mill Size	Approximate % of Residues Required for Self-Sufficiency	Approximate % of Residues Available for Export Power Generation
1.2 MMBF/yr	100	0
"small" (3 MMBF/yr)	60	40
"large" (9 MMBF/yr)	37	63

Energy Self-Sufficiency at the Sawmill

The potential for energy self-sufficient systems becomes apparent when we evaluate low capital systems designed for both small and large sawmill operations. Tables 5.2a, 5.2b, and 5.2c summarize cost information. Figure 5.4 shows the relationship between percentages of wood residues required for energy self-sufficiency and sawmill capacity. Note the summary of figure 5.4 below, which gives approximate percentages of wood residues required for energy self-sufficiency and percentages available to generate surplus energy for the sizes of sawmills in our study:

Energy Produced for Off-Site Sales

We will assume that grid-connected systems may be tied into a distribution network that can consume all surplus electricity produced by the wood-based generating plant. The criterion for the selection of the most suitable design for these systems then depends on the value of the additional electricity versus the installed and operating costs of the system. Compare the information on costs for the low capital, grid-connected system in table 5.2b with that for the energy efficient, grid-connected system in table 5.2c for both large and small sawmills.

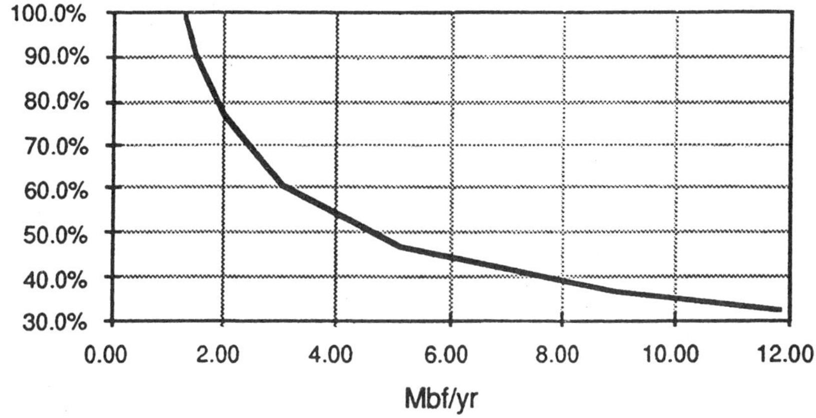

Figure 5.4. Percent of Sawmill Wastes Required for Self-Sufficiency (low capital system design).

Figure 5.5 shows the relationship between generation of electricity and sawmill capacity for ***both*** the ***low capital system*** and the ***energy efficient system***. Note the increase in production of electrical energy using energy efficient systems relative to low-capital systems.

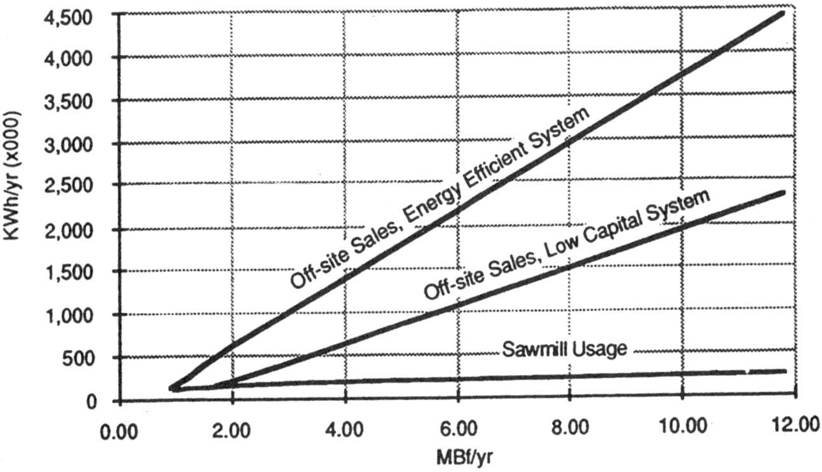

Figure 5.5. Electrical Generation vs. Sawmill Capacity (total electrical generation = sawmill usage + off-site sale).

Table 5.2a. Summary of Installation & Operating Costs for *Energy Self-Sufficient* Systems

Installation Costs	Large Mill	Small Mill
Imported equipment	US$267,075	US$211,500
Locally fabricated equipment	Lps 100,400	Lps 99,550
Installation	Lps 243,950	Lps 202,350
Freight, insurance, nonlocal engineering, supervision, etc.	US$147,289	US$117,467
Import duties and local services	Lps 279,183	Lps 230,174
Total Installed Costs	US$532,012*	US$429,358*
Annual Operating & Maintenance Costs		
Operating Labor	Lps 65,000	Lps 57,500
Operating Supplies	Lps 20,000	Lps 15,000
Maintenance	Lps 35,000	Lps 25,000
Total Annual O & M Cost	Lps 120,000	Lps 97,500

* conversion based on US$ 1.00 = Lps 5.30

Table 5.2b. Summary of Installation & Operating Costs for *Low-Capital, Grid-Connected* Systems

Installation Costs	Large Mill	Small Mill
Imported equipment	US$398,250	US$226,700
Locally fabricated equipment	Lps 103,100	Lps 99,800
Installation	Lps 345,550	Lps 211,050
Freight, insurance, nonlocal engineering, supervision, etc.	US$217,796	US$125,548
Import duties and local services	Lps 395,842	Lps 243,009
Total Installed Costs	US$775,384*	US$456,750*
Annual Operating & Maintenance Costs		
Operating Labor	Lps 87,500	Lps 65,000
Operating Supplies	Lps 35,000	Lps 17,000
Maintenance	Lps 42,000	Lps 27,000
Total Annual O & M Cost	Lps 164,500	Lps 109,000

* conversion based on US$ 1.00 = Lps 5.30

Table 5.2c. Summary of Installation & Operating Costs for *Energy Efficient, Grid-Connected* Systems

Installation Costs	Large Mill	Small Mill
Imported equipment	US$538,150	US$303,600
Locally fabricated equipment	Lps 101,350	Lps 98,750
Installation	Lps 404,300	Lps 232,700
Freight, insurance, nonlocal engineering, supervision, etc.	US$291,458	US$165,735
Import duties and local services	Lps 507,950	Lps 302,313
Total Installed Costs	US$1,020,853*	US$588,913*
Annual Operating & Maintenance Costs		
Operating Labor	Lps 87,500	Lps 65,000
Operating Supplies	Lps 28,000	Lps 14,000
Maintenance	Lps 52,000	Lps 32,000
Total Annual O & M Cost	Lps 167,500	Lps 111,000

* conversion based on US$ 1.00 = Lps 5.30

Case-Study Analysis

With the common assumptions listed in table 5.3, we performed a preliminary financial analysis, using a 10-year *pro forma* cash-flow model for each case study. The analysis determines if a case study will have an IRR (internal rate of return) of at least 15%, assumed by the study team to be the minimum percentage for an investment to be successful. Tables 5.4a and 5.4b give the characteristics of the six cases. Tables 5.5a and 5.5b summarize the analysis. Note that the larger wood-based energy system will have a greater economy of scale for all case studies.

Our analysis is preliminary. It indicates satisfactory investment returns for the six case studies. In cases where our assumptions were general or where our information was incomplete, we erred on the conservative side. To give an example, our estimates of capital costs may be high because we have assumed that most of the major items of equipment would have to be imported. In fact, some of the equipment could be purchased in used condition or may be available domestically at considerable savings. Although we designed the systems to provide steam to kiln-dry 25% of lumber production for each case, we are not crediting any systems with financial benefits associated with the higher value of the dried lumber products. Neither are we including the costs of the kiln-drying system.

Site-specific studies will require a more in-depth assessment of the common assumptions on which these calculations are based and may result in even more attractive investment scenarios.

Low Capital Cost, Energy Self-Sufficiency

Our financial analysis reveals that the calculated IRR for the large sawmill is 21.7% with a payback period of 6.2 years. The small sawmill has an IRR of 16.7% and a payback period of 5.7 years. These energy self-sufficiency systems are therefore acceptable candidates for investment, based on the assumed investment threshold of 15% IRR. Reductions in capital costs from the purchase of used equipment would make them even more attractive. While the large system has a more favorable economy of scale than the smaller, it also uses a smaller percentage of its wood residues to generate energy for on-site needs, and therefore has more utilized wood residues than the small mill. Densification may be an attractive option for utilizing these excess residues at large sawmills that generate their own electrical energy.

Low Capital Cost, Off-Site Energy Sales

Sawmills that are able to generate and sell surplus electricity have distinct economic advantages over those designed for energy self-sufficiency. Revenues received from the sale of surplus electricity more than compensate or the higher capital costs and slightly higher operating costs of the systems. In addition, these systems use *all* of the available wood residues at the sawmill.

Our analysis for these cases shows an IRR of 53.4% and a payback period of 4.0 years for a large sawmill. Small sawmills have a 32.2% IRR and a payback period of 5.0 years. These systems easily meet minimal investment criteria and represent potentially very attractive investments for sawmills that expect to operate through the end of the decade.

Energy Efficient, Off-Site Energy Sales

Sawmills with access to sufficient markets to sell their surplus energy may consider this option. The equipment is more costly than that for low capital cost systems, but the energy efficient equipment produces almost twice as much electricity from the same amount of wood residues. Annual operating costs and skills required are fairly comparable to those of low capital cost systems. The additional revenues earned more than compensate for the increased capital cost. Moreover, it may be easier to negotiate sales contracts for greater amounts of surplus since ENEE has stated it will need additional capacity in the near future.

Our projected financial returns for these systems are extremely attractive, ranging from over 40% IRR for the smaller system to over 75% IRR for the larger system. Simple payback periods are roughly 3.4 years for the large sawmill and 4.4 years for the small. These cases could maximize potential off-site sales of electricity, with substantial contributions to the national electrical grid from the forest products industry.

Sensitivity Analysis

- A sensitivity analysis of our financial analysis will examine major assumptions and variables in our six case studies. We will determine how great a change is necessary in a given variable to cause the IRR to fall to 15%, the minimal return for a successful investment. The five key variables for the sensitivity analysis are: Capital cost of the project
- Operating costs of the project
- Rate of currency devaluation (Lps: US$)
- Rate of inflation in the price of energy in Honduras (%/yr)

Tables 5.6a and 5.6b summarize our results. The first row of data repeats the base-case IRRs from tables 5.5a and 5.5b. The first column of data shows the base-case values of the sensitivity variables. The body of the table shows the values of the variables that cause the IRR to fall to 15%, while all other variables in each case are at base-case levels.

Financial and Operating Costs

Financial performances are more sensitive to increases in capital costs than to increases in operating costs. For example, the large sawmill—in the ***low capital cost, energy self-sufficient*** case—could absorb a 19% increase in *capital cost* before falling to a 15% IRR. It would take a 36% increase in *operating costs* to cause the same fall in IRR. The energy self-sufficiency examples, particularly at the lower end of our size range, are very sensitive to changes in financing and operating costs. All of the case studies that include off-site energy sales could absorb large increases in capital and operating costs before IRRs fall to 15%.

Overall Inflation Rate

Our assumptions in table 5.3 give a 20%/yr rate of general inflation *and* of energy in Honduras, with a devaluation of the Lempira relative to the dollar at 10%/yr. Inflation in Honduras has been on the low side, relative to much of Latin America, but it reached 35% in 1990 when the government implemented major reforms of the currency and the economy. The results of our financial analysis are quite sensitive to any predictions of future inflation rates, which

Table 5.3. Common Assumptions Used for Financial Analysis

Annual Energy System Operating Hours: includes all hours during which boiler is operated	24 hrs/day 300 days/yr
Annual Sawmill Operating Hours: typical total hours of sawmill operations in Honduras	8 hrs/day 250 days/yr
Electricity Sales Price: = based on 0.248 Lps/kWh, ENEE's cost of service in 9/90 plus 20% annual inflation applied over a 15 month period, to arrive at a 1992 price	0.31 Lps/kWh 1992, all hours
Diesel Fuel Price: = 6.94 Lps/gal in 1991; incorporates 20% inflation to 1992	8.33 Lps/gal, 1992
Diesel O&M and Depreciation: conservative estimate based on 2 diesel engines, 250 kW total capacity conservative estimate based on 2 diesel engines 400 kW total capacity	75,000 Lps/yr-- small mill site 85,000 Lps/yr-- large mill site
Currency Exchange Rate: current official exchange rate (3/91)	5.3 Lps/US$, 1991
Honduran Inflation Rate: assumption consistent with opinion of an economic officer at USAID/Honduras	20%/yr, 1991-2002
Honduran Energy Price Inflation Rate: assumed to be same as general inflation rate	20%/yr 1991-2002
Currency Devaluation Rate: assumption consistent with opinion of an economic officer at USAID/Honduras	10%/yr. 1991-2002
Debt Amount: assumption regarding financial structure	80% of total loader capital cost
Debt Terms: assumption based on discussions with lenders and agents	5 year amortization, 1 year grace period
Lempira Debt Rate: assumption consistent with opinion of economic officer at USAID/Honduras	28%
Dollar Debt Rate: assumption based on engineering design	12%
Project Construction Period: assumption based on engineering design	12 months
Discount Rate: assumption used for net present value calculations	15%

Table 5.4a. Characteristics of Wood Energy Systems for Large Sawmills

Sawmill Characteristics	Notes	Low Capital self-sufficiency	Low Capital grid-connected	Energy efficient grid-connected
Annual Lumber Production	a	9	9	9
<u>Generation of Electricity</u>				
Average Sawmill Load	b	114	114	114
Average Dry Kiln Load	c	17	17	17
Generator Capacity	d	273	493	812
Avg. Generator Output-Peak off-site sales (sawmill running)	e	150	448	739
Avg. Generator Output-Peak off-site sales (sawmill down)	f	28	326	614
Avg. Generator Output-Off-peak off-site sales (sawmill down)	g	28	186	342
Avg. Generator Output-Constant off-site sales (sawmill running)	h	0	385	606
Avg. Generator Output-Constant off-site sales (sawmill down)	i	0	271	492
kW to grid, peak sales	j	0	298	580
kW to grid, off-peak sales	k	0	158	308
kW to grid, constant sales	l	0	158	308
Total Off-site Energy Sales	m	0	1,684	3,277
<u>Fuel Use</u>				
Annual Waste Production	n	22,408	22,408	22,408
Annual Boiler Fuel Use	n	8,084	22,408	22,408
Annual Unused Wastes	n	14,324	0	0
<u>Costs</u>				
Total Installed Cost	o	$532,000	$775,000	1,021,000
Annual Operating Cost	p	$110,000	$164,000	$167,500

Table 5.4b. Characteristics of Wood Energy Systems for Small Sawmills

Sawmill Characteristics	Notes	Low Capital self-sufficiency	Low Capital grid-connected	Energy efficient grid-connected
Annual Lumber Production	a	3	3	3
Generation of Electricity				
Average Sawmill Load	b	79	79	79
Average Dry Kiln Load	c	6	6	6
Generator Capacity	d	180	180	180
Avg. Generator Output-Peak off-site sales (sawmill running)	e	99	173	283
Avg. Generator Output-Peak off-site sales (sawmill down)	f	14	88	196
Avg. Generator Output-Off-peak off-site sales (sawmill down)	g	14	53	112
Avg. Generator Output-Constant off-site sales (sawmill running)	h	0	157	238
Avg. Generator Output-Constant off-site sales (sawmill down)	i	0	78	159
kW to grid, peak sales	j	0	74	178
kW to grid, off-peak sales	k	0	39	94
kW to grid, constant sales	l	0	58	140
Total Off-site Energy Sales	m	0	417	1,005
Fuel Use				
Annual Waste Production	n	7,645	7,645	7,645
Annual Boiler Fuel Use	n	4,638	7,645	7,645
Annual Unused Wastes	n	3,007	0	0
Costs				
Total Installed Cost	o	$429,000	$457,000	$589,000
Annual Operating Cost	p	$97,500	$109,000	$111,000

Notes for Table 5.4a

a	in Mbf
b	in kW
c	in kW; electrical load only
d	in kW; includes allowance for sawmill peak loads, i.d. 82% above average load for self-sufficient systems (empirically determined from observations at several sawmills in Honduras), and 10% above average output
e	in kW; = avg. sawmill load + avg. dry kiln load + powerhouse auxiliaries + off-site sales (peak period)
f	in kW; = avg. dry kiln load + powerhouse auxiliaries + off-site sales (peak period)
g	in kW; = avg. dry kiln load + powerhouse auxiliaries + off-site sales (off-peak period)
h	in kW; = avg. sawmill load + avg. dry kiln load + powerhouse auxiliaries + off-site sales (constant amount over 24 hours)
i	in kW; = avg. dry kiln load + powerhouse auxiliaries + off-site sales (constant amount over 24 hours)
j	in kW; for 13 hr/day
k	in kW; for 11 hr/day
l	in kW; for 24 hr/day
m	in million kWh/hr
n	in MT
o	includes consolidated US$ and Lps. costs, @ US$1.00 = 5.30 Lps.
p	in Lps/yr

are extremely difficult to forecast. Most analysts do not expect that the government will be able to bring the annual inflation rate down below 20% in the foreseeable future. They cite a 15% rate as the best that could be achieved.

High general inflation in Honduras actually improves the financial performance of investments in the production of energy because it increases future revenues and savings. The four case studies that include the sale of surplus energy remain attractive investments, even with domestic inflation rates below 10%.

Rate of Currency Devaluation

The devaluation rate of the Lempira relative to the dollar may be even more difficult to predict than the general inflation rate. Until March 1990, the exchange rate had been fixed for a long time at Lps. 2.00 = US$ 1.00. Current policy now seems to allow the Lempira to be exchanged at close to its market price, although some level of control is still in effect. We would have

Table 5.5a. Results of Analysis of Wood Energy Systems for Large Sawmills

	Low Capital self-sufficiency	Low Capital grid-connected	Energy efficient grid-connected
Project net present value, Total* (Lps. x 000)	3,919	9,576	15,548
Project net present value, Total* (US$ x 000)	739	1,807	2,934
Total Capital Cost** (Lps. x 000)	3,010	4,330	5,702
Total Capital Cost** (US$ x 000)	568	817	1,076
Internal Rate of Return (IRR) on Equity Investment (%)	21.70%	53.40%	75.30%
Investment Payback (yr)	5.7	4	3.4

* @ 15% discount rate
** includes Total Installed Cost from Table 4, plus assumed financing and incidental costs

Table 5.5b. Results of Analysis of Wood Energy Systems for Small Sawmills

	Low Capital self-sufficiency	Low Capital grid-connected	Energy efficient grid-connected
Project net present value, Total* (Lps. x 000)	2,752	4,145	6,331
Project net present value, Total* (US$ x 000)	519	2,552	3,291
Total Capital Cost** (Lps. x 000)	2,406	2,552	3,291
Total Capital Cost** (US$ x 000)	454	481	621
Internal Rate of Return (IRR) on Equity Investment (%)	16.70%	32.20%	42.60%
Investment Payback (yr)	6.2	5	4.4

* @ 15% discount rate
** includes Total Installed Cost from Table 4, plus assumed financing and incidental costs

Table 5.6a. Sensitivity Analysis—Large Sawmills: Change in Variables Necessary to Reduce Internal Rate of Return (IRR) to 15%

	Base	Low Capital self-sufficiency	Low Capital grid-connected	Energy efficient grid-connected
Base case IRR		21.70%	53.40%	75.30%
Capital Cost	100.00%	119.00%	203.00%	250.00%
Operating Cost	100.00%	136.00%	299.00%	478.00%
Overall Inflation	20.00%	15.60%	2.80%	n/a
Devaluation	10.00%	17.70%	38.00%	45.70%
Energy Inflation	20.00%	17.20%	7.50%	4.90%

Table 5.6b. Sensitivity Analysis—Small Sawmills: Change in Variables Necessary to Reduce Internal Rate (IRR) to 15%

	Base	Low Capital self-sufficiency	Low Capital grid-connected	Energy efficient grid-connected
Base case IRR		16.70%	32.20%	42.60%
Capital Cost	100.00%	105.00%	149.00%	176.00%
Operating Cost	100.00%	109.00%	188.00%	274.00%
Overall Inflation	20.00%	18.80%	10.10%	5.50%
Devaluation	10.00%	12.10%	26.70%	32.80%
Energy Inflation	20500.00%	19.20%	13.70%	10.70%

an exactly correcting devaluation rate of 15%, if we assume a 20% inflation rate in Honduras and a 5% inflation rate for the dollar. In fact, most analysts expect the rate of devaluation to be lower than 15%. We have chosen 10% for our base case. For those sawmills selling surplus energy to the grid, only devaluation rates exceeding the assumed 20% general inflation rate would seriously impair their attractiveness for investors.

Rate of Inflation for the Price of Energy

We have assumed this rate to be the same as that of the overall, general inflation for our base case. However, the rate of inflation for energy prices in Honduras varies from the general rate. In response to the Persian Gulf situation, the prices of diesel fuel increased by about 75% in the last quarter of 1990. With recent, successful roll-back and stabilization of world oil prices, the Honduran Government announced a policy that would place a special tax on fuel to stabilize prices at their level of March, 1991. If these two conditions continue, then prices for diesel fuel in Honduras could increase at rates below those of general inflation, at least over the short-term.

CONCLUSIONS AND RECOMMENDATIONS

Given the tenuous situation of forests and the increasing demand for energy in Honduras, the long-term outlook for the forest products industry is very uncertain. However, our study shows that the economic feasibility to invest in wood-based energy systems for sawmills in Honduras appears attractive. Rates of return are within acceptable limits for all of the six case studies. Payback periods are well within the 15-year period of reasonably assured supplies of timber.

Our assessment of demands for energy and the potential to produce energy from wood residues from forestry and, more especially, from sawmill residues, yields three points:

- There are no significant technical issues that would limit the use of wood-based energy in Honduras.
- Investments in wood-based energy systems at sawmills in Honduras would meet or exceed reasonable criteria for investment performance. Such investments at larger sawmills and at sawmills that have markets for the sale of surplus energy would be especially attractive.
- Energy generated from wood-based systems could significantly increase the national energy supply in Honduras, and the use of these systems could substantially reduce the problem of disposing of wood residues.

Potential for Production of Energy from the Forest Industry in Honduras

If sawmills in Honduras in the size range that we considered in this study were to achieve only energy self-sufficiency with wood-based energy systems, the potential savings to the country in fuel costs alone would be greater than 1,000,000 gal/yr of diesel fuel. Such systems would provide nearly 10 million kWh/yr of electricity to the sawmills and enough steam to kiln-dry approximately 25% of the nation's lumber output, increasing the value of wood products.

Energy self-sufficient systems, however, would provide only a partial solution to the problem of disposing of wood residues at Honduran sawmills. If all of the mills were to install self-sufficient energy systems, there would still be over 150,000 tons/yr of wood residues at the plants. Less than half of the wood residues available to generate energy still remains as wastes for large sawmills.

We can calculate the theoretical amount of *excess* power that could be generated from these remaining wood residues in Honduras, but it is difficult to estimate the amount of actual, salable power that could be used. We need to know if a sawmill has access to the grid or to other markets that can use its surplus energy. If the sawmills in our study chose the low capital cost systems, they would have collectively more than 20,000,000 kWh of power to sell, in addition to the approximately 10,000 kWh of power produced for their own consumption. Energy-efficient systems would yield nearly 53,700,000 kWh of power to sell for off-site consumption, in addition to the mills' own requirements.

Our calculations of surplus power potentially available from Honduran sawmills do not include the enormous potential generation of power from forest residues. Collection and combustion of forest residues on a sustainable basis in Honduras could bring social, silvicultural, and environmental benefits to the country that appear to warrant further investigation of this option.

Technical Conclusions and Recommendations

There are no significant technical limitations to installing wood-based energy systems at sawmills in Honduras. The most practical systems seem to be those that combine fire-tube boilers and single-stage steam turbine-generators. These may be designed to require no additional fuel preparation and to operate without sophisticated auxiliaries and control systems. Similar technology was used in sawmills in the United States in the first half of this century. Ample information is available on the design parameters that should be considered. In fact, such technology is already in use at a few sites in Honduras.

The low capital cost scheme in figure 5.2 represents the least-cost alternative for energy self-sufficiency. We recommend the energy efficient system depicted in figure 5.4 for generating maximum electrical power to sell. Note that more efficient equipment—in terms of kilowatts of electricity produced per unit of fuel, or wood residues—is available. Such equipment is twice as efficient as the "energy efficient" design that we described in our study. These systems are much more expensive, but they should be considered in site-specific analyses of large sawmills that have the potential to sell as much surplus electricity as they can generate.

Site-specific feasibility studies should explore kiln-drying of lumber, since it represents a significant additional benefit to sawmills in Honduras. We have designed the wood-based energy systems in our case studies to provide enough steam and electricity for drying 25% of the mill's primary production. Some sawmills in Honduras could profit by kiln-drying as much as 70% of their lumber. The costs of kiln-drying and, possibly, surfacing—including capital and operating costs for each—and the potentially increased revenues from added value to the product and reduced freight costs must be evaluated.

Densification of wood residues for sale to off-site markets—such as energy-intensive industrial operations or domestic cooking and heating—is another way to dispose of wood residues. This option may be economically attractive to Honduran sawmills that do not have access to national, local, or micro- grids. Densified fuels are far more economical to transport than loose sawmill residues, and they can be used relatively easily as a fuel source.

Financial Conclusions and Recommendations

Investments in wood-based energy systems appear to meet or exceed acceptable financial performance criteria. Those systems able to sell surplus electricity to national, local, or micro- grids show higher returns than systems designed only for energy self-sufficiency. Larger systems show higher economic returns than smaller systems. The purchase of used equipment, or perhaps the purchase of locally fabricated equipment, could increase the value of the investment even more. Such investments are also attractive because they improve overall operations at the sawmills, especially if they include kiln-drying and/or increased manufacture of secondary products. In addition, there are the potential returns from densification and off-site sales of *excess* sawmill wastes that we have not analyzed in our study.

The analysis for the *energy efficient, off-site energy sales* cases shows internal rates of return of 70% and a simple payback period of 3.4 years for the large mill. The small mill has a 40% IRR and a simple payback period of 4.4 years. We are assuming one-year grace periods on the loans. The large and small *low capital, off-site energy sales* have IRRs of 53.4% and 32.2% and simple payback period of 4.0 and 5.0 years. The large and small *low capital, energy self-sufficient* cases have the least attractive, but most certainly acceptable, results. IRRs are 21.7% and 16.7%, for the large and small sawmills, with simple payback periods of 5.7 and 6.2 years.

Investments in the energy-efficient systems would be even more attractive if all surplus power can be sold at prices near Lps. 0.31/kWh in 1992 (US$ 0.052/kWh). This base price is low, relative to international rates, and reflects ENEE's current costs of power generation—almost entirely from hydropower. The rate may be adjusted slightly upward for peak periods. ENEE's peak power demand is expected to exceed capacity by about 1992 or 1993. After that time, costs of generation for peak power will no longer be based on low-cost hydro. ENEE will need to purchase surplus power, if it is available, by the early 1990s. The possible increase of revenue from the sale to the utility of surplus power for peak periods have not been considered in our study.

Governmental Support

All operating sawmills in Honduras are now privately owned. However, the support of the government, while not required, is an important factor for investments in wood-based energy systems. In fact, it is probably essential in order for sawmills to sell surplus electricity to the government-owned ENEE. To date, the utility has encouraged the use of wood-based energy. ENEE has indicated that it would be willing to purchase all available surplus electricity generated by wood residues from forest management and sawmills. ENEE has also supported the establishment of rural electric cooperatives. Such cooperatives would be the markets for surplus electricity produced at sawmills in remote locations. Our study has not investigated purchase agreements for electrical power between private generators and ENEE. Details of such agreements should be established prior to investment in equipment intended to generate surplus electricity for sale to national, local, or micro- grids.

Sources for Financing

The availability of capital is a crucial factor for investment in wood-based energy systems. The number of private loans available have been extremely limited, and they have been difficult to get. Interest rates from Honduran banks to purchase equipment have typically been in the range of 15% to 20%. Recently, government policies on consumer interest rates have become more liberal, making more funds available, but rates have climbed to the mid-20% level and are expected to approach 30%.

Possible funding from international sources include loans from bilateral agencies—either directly or through local banks. Multilateral lending institutions, such as the Inter-American Development Bank—with funding available through local banks—and other, private sources such as venture capital firms, may be other alternatives for funds. For example, USAID, a bilateral source, has provided part of the funding for the PDF project. This money is intended to support capital investments, specifically for improvements in the operations of Honduran sawmills and could be used, presumably, for wood-based energy systems. Such funding would be obtained through local banks. At the time of our study, however, the mechanisms to establish such investment support had not been completed. The single largest consideration for funding from private, international sources will be repatriation of capital. Those sawmills that export their products, and therefore have access to hard currencies, will find it easier to attract such capital. Guarantees from the Honduran Government would be very helpful for such cases.

Overall, wood-based energy systems at sawmills in Honduras appear to be both technically and economically viable. The results of our analysis should prompt investigations of more site-specific opportunities and stimulate investments in wood-based energy systems throughout Honduras. Our suggestions for further studies include:

- **Site-specific analysis of a Honduran sawmill *with* access to the grid**. This is a detailed analysis of the technical and economic feasibility of a wood energy system designed for *maximum energy production*, with sales of all surplus electricity to the grid. Such a study could also compare the benefits of densification of excess wood residues with those from selling surplus electrical power generated by residues.
- **Site-specific analysis of a Honduran sawmill *without* grid access**. This is a detailed analysis of the technical and economic feasibility of a wood-energy system designed to generate only

enough energy required for *self-sufficiency*. This study could also include a feasibility assessment of densifying excess wood residues as part of the energy system.

- **Site-specific analysis of a sawmill without grid access, but with the favorable circumstances to establish a local or micro-grid market.** This study focuses on a sawmill in a remote location, with no access to the grid, but near villages that do not have electricity. If the nearby village(s) is large enough and has some purchasing power, it could establish its own grid and the sawmill(s) could sell surplus electricity to the village cooperative. Such a project would yield numerous benefits—rural electrification, increased revenues for the sawmill(s), and consumption of excess wood residues.
- **Site-specific analysis of an *independent* power station using residues from sawmills—and, possibly from the forest—for fuel.** This is a situation in which the generating plant would purchase wood residues from surrounding sawmills and forest management projects and sell electricity to the grid. Some of the facility's electricity and/or steam could also be sold directly to nearby sawmills or other industries. An independent power project should be located close to a group of sawmills since the costs of transporting wood residues is high.

Literature Cited

Anon. 1984. Aserradero el, utilizacion de los residuos de madera del aserradero el rincon con fines energeticos.

Anon. 1981. Determinacion del volumen de residuos de madera en cortas de *Pinus oocarpa* con destino a serrio.

Anon. n.d. Species descriptions, *Pinus caribaea*, Caribbean pine, and *P. oocarpa*, Ocote pine. Includes specific gravity.

Baker, A.J. 1983. New findings on wood fuel values. *American Forests* 2(83):45–47, 51-52.

Beckwith, C.L. and M. Godfrey. 1990. CARE INTERNACIONAL en Honduras, agricultura y recursos naturales (ARN) Estrategia 1990-1995.

Chronowski, R.A. 1985. Honduras-energy assessment mission, UNDP/World Bank Fuelwood and Forestry Sector report.

COHDEFOR (Corp. Hondureno de Desarrollo Forestal). 1990. Anuario Estadistico, Seccion Industrias Forestales, Departamento Forestal 1989.

COHDEFOR. 1988. *Plan de manejo forestal, area no. 1 La Union.* (excerpts; original loaned from Danilo Escoto, COHDEFOR).

Curtis, A.B., Jr. 1976. How to calculate wood fuel values. *Forest Industries* 12(86): 44–45.

DESFIL (Development Strategies for Fragile Lands). 1989. *Perfil Ambiental de Honduras 1989.* (excerpts; original loaned from John White, PDF).

ENEE. 1988. Datos estadisticos de la Empresa Nacional de Energia Electrica, 1987. Honduras, C.A. (Enclosed are Boletin Estadistico Junio 1990, and Proyeccion de Ventas Tegucigalpa, Escenario Base (GWH)).

FAO. 1987. FAO Report 6476-HO, Honduras: issues and options in the energy sector.

FAO. 1968. Survey of pine forests, Honduras. FAO Report FAO/SF:26-HON 50. (excerpts; original loaned from John White, PDF).

FIAFSA (Forestal Industrial Agua Fria, SA). 1984. Estudio para una central de energia a vapor, potencia 2500 kw utilizando residuos de madera.

Groothausen, Cornelio. 1989. Tablas de rendimiento para *Pinus oocarpa* en la region forestal de Comayagua, Honduras, C.A. Centro Nacional de Investigacion Forestal Aplicada (CENIFA), Escuela Nacional de Ciencias Forestales (ESNACIFOR), Apdo 2, Siguatepeque, HON.

Harwood, J.H. 1981. Lenha e restos de madeiras como fonte de energia na Amazonia. *Acta Amazonica* 11(3):553–559.

Howes, K.M.W, and R.A. Rummery, eds. 1979. *The impact of changes in energy costs on the rural sector of the Australian economy*. Proc Workshop by CSIRO, Western Australia Dpt of Agri, & Murdoch Univ; Bunbury, W.A., Oct 1979.

Hunt, John. 1989. Assessment of the Empresa Nacional de Energia Electrica bulk power supply system. USAID/H.

Instituto Geografico Nacional. 1966. Mapa general, Republica de Honduras. 1:1 000 000.

Landsberg, H H, Chairm. 1979. *Energy: the next twenty years. Ford Foundation & Resources for the Future study group*. Cambridge: Harper & Row.

Tenore, F.M., and L. Cosenza. 1980. *Feasibility study of power generation from wood waste, ENEE*. 2 vols. (Charles T. Main report).

USAID/HON. 1988. USAID y Honduras, Asociados para el desarrollo 1942-1988.

USDA. 1980. *Cutting energy costs. The 1980 Yearbook of Agriculture*. USDA, WDC.

Weinberg, C.J., and R.H. Williams. Energy from the sun. *Scientific American* 9(90):146–155.

Winrock International. 1991. Energy from sawmill wastes in Honduras. Report No. 91-05. Report for Office of Energy. Washington: U.S. Agency for International Development.

World Bank. 1980. *Energy in the developing countries*. World Bank, WDC.

CHAPTER 6

A PREFEASIBILITY ASSESSMENT OF THE POTENTIAL OF WOOD-WASTE POWER SYSTEMS FOR THE INDONESIAN WOOD PRODUCTS INDUSTRY[1]

Jerome M. Weingart
Peter A. Jezek
Gregory Morris

WOOD WASTE IS A RESOURCE FOR ENERGY[2]

Wood waste discarded by the forest products industry in Indonesia can constitute as much as 50% of all raw timber that enters a plywood factory or sawmill. This waste is burned, used for landfill, or dumped in rivers or the ocean. Plywood plants and sawmills depend on diesel fuel that is substantially subsidized by the government to supply their on-site captive electrical power because there is no local power grid in most forest industry locations outside of Java. The plywood industry requires reliable power, so that even when grid connections are available, a plant must have full backup capability. Yet the wood waste, already available at the site, and essentially free, represents a potential power source, not only for the wood products industry but also for local rural communities. Wood waste from Indonesian plywood plants could produce approximately 150 MWe (MWe = megawatt-energy) of on-site electrical power, which could benefit the country economically, socially, and environmentally. There is a potential market sufficiently large to allow mass production of basic, standardized generator units that burn wood waste exists.

1 Abstracted from a 1988 report of research sponsored by the Bioenergy Systems and Technology Program of the U.S. Agency for International development in cooperation with private industry and USAID/Jakarta to investigate the possibilities of substituting this wood waste for diesel fuel to produce on-site electrical power at plywood factories. The original report is available from the Director, BEST Project, Winrock International, 1161 North Kent Street, Arlington, Virginia 22209. The authors thank Dr. Sam Schweitzer, USAID Office of Energy and Infrastructure, for his support and encouragement of this work.

2 In the six years since this study was conducted, diesel price subsidies have decreased substantially and the Government of Indonesia, now a net importer of diesel fuel, is actively supporting environmentally sustainable means to decreasing diesel fuel use. The project described here appears even more attractive and feasible for 1994 than it did in 1987. [JMW].

We foresee a market for equipment suppliers in the United States of several million throughout the 1990s.

U. S./Indonesian joint venture companies could build, operate, and sell electrical power from wood waste and establish a vital new industry in rural Indonesia to stimulate the country's economy on local and national levels. No longer entirely dependent on shifting or increasing diesel-fuel prices, the wood-processing industry could become more self-sufficient, prosperous, and stable with efficient use of raw timber. Once the technology to generate electricity from wood waste becomes established in the Indonesian forest products industry, its success could stimulate PLN (Perusahaan Listrik Negara—the state electrical power company) or private initiatives to construct and operate similar power plants. If used to provide 24-hour electricity for rural communities, this electricity could invigorate local economies in Indonesia, enabling rural people to increase their incomes and improve the quality of their lives.

Plywood plants in Indonesia now use wood waste to generate process steam in relatively primitive boilers that emit noxious gases because they have no flue-gas controls. Unacceptable environmental conditions—dust, smoke, floating debris in rivers, and smoldering trash heaps—are the result of inadequate means to dispose of wood waste. Air-borne sawdust and smoke from burning wood in the open air cause respiratory diseases and increase the danger of explosions at the plants. Wood ash from the uncontrolled burning of waste combines chemically with bacteria in the soil to create toxic matter that poisons both the soil and the water beneath ground level. Burning the wood waste for electricity would eliminate such air, water, and soil pollution, and it would improve the working conditions at the plants. Generators that use wood waste for fuel would be properly equipped with combustors and boilers that would recycle particulate matter back to the boiler to achieve more complete combustion, eliminating a source of air pollution. Disposal of wood ash would be monitored, and waste water from the cooling towers would be treated before emptied into rivers and oceans to prevent environmental damage.

The purpose of this study is to determine the financial and commercial attractiveness of constructing and operating standardized modular power plants that burn wood waste in Indonesia. Specifically, we focus on 2 MWe to 4 MWe generators that are well suited for private U.S./Indonesian joint ventures to build, own, and operate. We visited 28 plywood factories in Java, Kalimantan, and Malaku to determine technical, operational, and financial criteria for using wood waste as a fuel. Focusing on site-specific wood-energy balances, we reviewed the present status of the plants' generation of electrical power and process steam. We estimated a range of costs for proposed power-generation systems. We made certain technical and financial assumptions to create a hypothetical base case which we subjected to a *pro-forma* financial analysis, using six combinations of diesel-fuel and wood-waste prices, to see if these power-generating systems could be commercially and financially attractive for potential investors.

A MARKET ASSESSMENT: THE POTENTIAL FOR WOOD WASTE AT PLYWOOD PLANTS

The Indonesian Plywood Industry

The forest products industry is a major source of foreign exchange for Indonesia. Until the early 1970s, the country exported wood primarily in the form of raw logs. In 1981, the government imposed a quota to limit log exports to approximately 25% of total production. In addition, to increase the value of exported timber products, the government required that logging companies build integrated facilities to manufacture products that would earn more foreign exchange than logs.

Beginning in 1973 with two mills and a total production of 19,000 m^3, the plywood industry grew to 29 mills with a combined production of slightly over one million m^3 by 1980. Most plants were built after 1981. At least 115 plywood factories and integrated wood products plants operated in Indonesia by the late 1980s. Typically, about 10 plywood factories form clusters along riverfronts or coastlines in sparsely populated areas. Raw logs arrive as rafts, on barges, or on coastal freighters, and the finished product is loaded onto ships for the international market. A typical factory is a sprawling complex with a sawmill, several plywood production lines, and housing and cafeteria facilities for several thousand workers.

Plywood plants operate continuously, day and night, six days a week, with workers on 12-hour shifts. In 1985, the industry produced over 3.8 million m^3 of plywood, earning $811 million in a world market where both price and demand were low. In 1987, plywood production surged to a level of 7 million m^3. Log production for 1987-88 was estimated to be 37 million m^3, of which 65%, or 13 million m^3, would be processed by sawmills and the remaining 35%, or 13 million m^3, would be designated for plywood.

To control such rapid expansion and, hopefully, to prevent deforestation, the Indonesian government and the Indonesian Wood Panel Association have decided to issue no more permits to build new plywood plants for the near future. Current production will be held constant. The plywood plants are based on logging concessions with a life span of 20 years or more, and the industry claims that these holdings are sufficient to sustain current production levels during this period. In addition, the industry is beginning to diversify. Under the guidance of the Wood Panel Association, it is starting to produce such high value products as paper overlay, paneling, moldings, and furniture.

Most timber comes from Kalimantan, with plywood plants in Java, Sumatra, and Malaku. It remains to be seen how continued logging activity will affect Kalimantan's tropical forests, despite efforts to manufacture labor-intensive wood products, and how these new directions for the industry will affect its demands for electrical power. Responsibility for reforestation should be a consideration of any detailed assessment of power generation systems that burn wood waste in the plywood industry. As of the late 1980s, loggers were paying Rp 8,000 (the local equivalent of $4.00) for each cubic meter of wood used to the Government of Indonesia for reforestation, and the government was responsible for reforestation programs.

Electricity at Indonesian Plywood Factories: Present and Potential Use of On-site Wood Wastes

Total demand for electricity at plywood plants in Indonesia is about 126 MWe. PLN or on-site steam turbine-generators supply only about 26 MWe, while the rest is produced by diesel generators installed at the factories. Diesel fuel is readily available, and 50% of its cost has been subsidized by the government. The annual electricity demand by the industry of 728,000 MWh/yr (MWh/yr = megawatt-hours per year)—represents $73 million/yr in fuel costs at today's world-market prices ($0.20/liter).

In 1987, 80% of the plywood factories used diesel fuel to produce electricity, and the remainder burned wood waste. However, all plywood plants use wood waste to generate steam for dryers and hot presses. In these cases, the wood waste provides the extremely reliable electrical power demanded by the manufacturing process. As little as three minutes' power interruption will ruin any plywood in the presses at that time.

The average recovery rate—or fraction of logs in the finished product—at the plywood mills is typically about 50%; the rest is waste. Three million cubic meters of annual wood waste could be used to generate 200 MWe at a 70% annual average capacity factor, equivalent to twice the present installed diesel-generator capacity of the industry.

SITE VISITS

Before visiting the plywood plants, we sent the plant owners a questionnaire to obtain information on plant operations and wood processing, electricity production and consumption, and process-steam production and consumption. We obtained additional site-specific information on the balances of wood supply and energy production during the visits. The status of the plant's operations and management's views on installing steam turbines and on using wood waste to fire boilers and steam turbines was of particular interest. In some cases, we also held discussions at company headquarters in Jakarta to review relevant commercial and financial issues, in particular, the viability of biomass energy based on wood waste. Our results are expressed as averages for typical conditions during the period of our study in 1987.

Plant Operation and Wood Processing Data

At the plants visited, wood consumption ranged from 12,000 m^3/mo to 100,000 m^3/mo, and plywood production was between 4,000 m^3/mo and 40,000 m^3/mo. Actual recovery rates for all of the plants averaged 54%.

Available wood waste ranged from 47,000 m^3/mo/plant to 5,500 m^3/mo/plant. One plant used all of its wood waste—18,000 m^3—and only two had any wood waste remaining—a total of 20,000 m^3. Most of the waste—we estimate from 2,000 to 29,000 m^3—was used to fire boilers for steam production.

Electricity Production and Consumption

Diesel-generation capacity installed at the plants varied from 0.45 MWe to 12.7 MWe. Four of the plants supplemented this power with PLN connections, from 1.4 MWe to 7.9 MWe. Three of the plants had no installed diesel-generation capacity, but relied totally on steam. Only seven of the plants had steam-generation capacity, ranging from 2.2 MWe to 6.8 MWe. Peak demand was as high as 11.0 MWe, in the largest plant, and as low as 1.0 MWe; most averaged 2 MWe to 5 MWe. The average peak electricity demand for plywood production was 70 kWhe/m^3 of plywood produced (kWhe = kilowatt-hour-electric). Electrical consumption ranged from 580 MWhe/mo to 580 MWhe/mo (MWhe = megawatt-hour electric). Diesel fuel consumption ranged from 28,000 l^3/mo, costing $4,000, to 760,000 l^3/mo, costing $138,000 ($l^3$/mo = cubic liters per month). The largest plant had as many as nine diesel generators, but the majority of the plants (18) had anywhere from three to five units. Five of the plants had only one unit.

Process Steam Production and Consumption

Most of the boilers were rated between 20 t/hr and 30 t/hr (t/hr = ton per hour). Since steam is used primarily for drying and hot presses, the steam-pressure requirements are low, around 12 kg/cm^2 to 15 kg/cm^2. Average temperatures needed are around 170 C to 180 C. We estimate demand for steam at 12 t/hr to 25 t/hr.

Owner Interest in Wood-Waste Technology

Almost all of the plant visits experienced moderate to severe difficulties in operating electrical power-generation and steam production systems. Plywood-plant owners and operators typically have neither the technical background nor the interest in such operations. However, many expressed strong interest in the possibilities of contracting with a third party who would assume these responsibilities and wood waste to generate power. To lower their risk of investment, owners need the answers to many questions before they decide to invest in the technology that burns wood waste. Table 6.1 indicates the

Table 6.1. The Concerns and Questions of Plywood Plant Owners

Costs and Operations	Financing	Government Policy
What are the capital and installations costs, both fixed and variable	What are the various alternative financing mechanisms and financial variable?	What is the Indonesian government's policy regarding power generated from wood waste?
What are the maintenance requirements relative to present operations?	How would the transition to the use of wood waste be financed? Would it be financed by owner equity, debt financing , or a combination of the two?	Would wood-waste electrical power be independent or connected to a grid?
What is the reliability of wood-waste power generation relative to diesel power?	What are the options for ownership arrangements?	What position will the PLN take relative to the commercial diffusion of the technology for electrical power generated by wood waste—will it be for or against?
What is the salvage or recovery value of existing diesel equipment?	What is the payback time and the return on equity for different methods of financing?	Would the investment be secure in the event of sudden changes in government that would affect the operation of the Indonesian wood products industry or the generation of privately owned electrical power?
What is the avoided cost of on-site diesel-generated electricity or of PLN-supplied electricity?		
	Are commercial and multi-lateral development banks and other institutions able to assume debt and equity positions in such a project?	
What is the anticipated increase or decrease in the above costs?		
What is the value of alternative uses of wood waste that could be consumed for power generation?		

primary areas of owners' concerns and their questions about the proposed power-generation units. We have concluded that owners of plywood plants need to see a clear demonstration of the technical, economic, and financial feasibility of wood-waste technology before they will invest in its use. An alternative approach would be for an energy services company to assume the risk and contract for the supply of electricity and steam, as well as contract for exclusive no-fee access to the wood waste produced on site.

TECHNOLOGICAL REQUIREMENTS AND RECOMMENDATIONS

The Fuel

The average moisture content of wood waste used as fuel in steam boilers is 23%, according to the Combustion Research Laboratory near Jakarta, and energy content is about 6,600 BTU/lb. We have found that sufficient wood waste is available to provide electricity and steam at levels above those currently required in the plants that we visited. The capacity of wood waste to generate power in these plants is about 48 MWe. Thus, the diesel fuel saved in one year would be 68 million liters, with potential foreign-exchange export earnings of about $12 million (1987$).

The Equipment

From our study of these plants' peak demands for electricity (2 MWe to 11 MWe), we have identified the performance requirements for steam turbines and boilers that use wood waste. Prepackaged power-generation systems—"standard plants"—of 3 MWe net would be well suited for most of the plants' operations. There are several types of steam turbines manufactured in the United States that could be suitable for the plants. Options include straight-condensing, steam-extraction, and back-pressure turbines. Since the plywood plants require reliable electricity and steam at the same time, the latter two types of turbines would be good choices for efficiency, but their capital costs and complexity of operation might favor the use of simpler equipment such as straight-condensing turbines.

We have made no engineering analyses of boiler requirements and their fabrication. However, our on-site inspection of plants that use steam turbines indicates that 12.1 lb to 12.2 lb of steam are required to generate one kWhe. We estimate that one pound of wood chips would produce about 4.5 lb of steam at the desired pressure of 34 kg/cm^2. Subsequent studies will determine the required steam flows for each type of turbine under consideration to design, manufacture, and price appropriate boilers. However, we have found that a wood-waste-fired boiler with a capacity of up to 30 tons of steam per hour, at 725°F and 600 psig, could generate 5 MWe net with straight-condensing turbines.

Recommendations for Manufacturers

The demands of the market, rather than available technology, have guided this study. Thus, the business environment would determine the requirements that any new wood-waste technology for on-site generation of electrical power should meet in order to be an attractive investment for those in the plywood industry. Any equipment manufacturer or supplier must take cost considerations into account. From a manufacturer's viewpoint, the major obstacle to wood-waste technology is the initial cost of design. However, as we have noted before, the overall potential of 150 MWe—and, in our study, of 48 MWe—indicates a substantial market for the manufacture of standardized, modular generator units that burn wood waste. Rather than employing conventional, individually designed, engineering approaches, suppliers must consider the manufacture and sale of multiple units that would capture the cost-reduction effects of the industrial learning curve and, therefore, recover initial design costs in a shorter period of time. Production of modular units would ensure quality control during manufacturing, and standardization of training, operations, and maintenance.

Subsequent studies must identify the most effective way to package available off-the-shelf hardware that could best meet the specific requirements of the plywood industry. Given buyers' concerns, designs that can minimize capital costs, assure simplicity of operation, and provide opportunities for local employment will prove the most commercially desirable. The generator units would be predesigned and standardized to minimize costs of operation, maintenance, and replacement. In addition, manufacturers could employ the latest technological advances to control atmospheric pollution. Off-the-shelf equipment could be easily shipped via barge to a plywood production site. The system could then be erected in the field and installed as prebuilt modules, or the parts could be factory-assembled into skid-mounted modules, to minimize installation costs and the need for skilled labor at the site.

A SAMPLE WOOD WASTE-FIRED POWER GENERATION SYSTEM

Once we have determined the preliminary specifications for a proposed power-generation system, we must consider the significant technical, economic, and financial parameters that influence its feasibility. We have made conservative assumptions to describe a base case (12-year financing) to determine how varying fuel prices affect payback time and rate of return on equity. Our results demonstrate the project's financial appeal and provide a framework for future discussions of more defined projects with potential investors, lenders, and equipment vendors.

Technical Assumptions

Fuel

About 15 tons of steam/hr are required to operate a straight-condensing turbine of 2.75 MWe. For this amount of steam, wood waste must be consumed at the rate of 4 green ton/hr. The fuel will be air-dried (23% moisture content), and its energy content is 6,622 BTU/green ton. Heat rates would be 19.264 BTU/kWh gross and 21,190 BTU/kWh net.

Equipment

We have selected a straight-condensing turbine for the base case analysis. Separate boilers would burn wood waste to produce steam for the hot presses. Therefore, our generating system would provide only electrical power, and the rate of steam extraction would be zero. Once this technology has been accepted in Indonesia, we would recommend more complex systems employing cogeneration and steam extraction that would be more efficient both in use of fuel and in total costs.

Maintenance

Maintenance will be scheduled for Sundays, when diesel backup-fuel will be used. If several steam generators that burn wood waste are available at a plant, managers could stagger maintenance schedules to provide continuous electricity, with no reliance on diesel fuel. Our experience predicts that regular maintenance should require 5% of a year's total hours of operation, or 438 hours (36.5 hr/mo). For our analysis, we have allowed 104 hr/mo of scheduled maintenance, a period more than adequate for thorough care. In addition, our experience suggests that a well-maintained plant would have a forced outage, or disruption of operations, of 6% of total operating hours, or 42 hr/mo. Total operating hours, the number of hours in a month minus scheduled and forced outage, are 584 hr/mo, or 7,008 hr/yr.

ECONOMIC AND FINANCIAL ASSUMPTIONS

Contractual Arrangements

Several contractual arrangements could make the use of modular power-generation systems that burn wood waste attractive to the Indonesian plywood industry. In one approach, a U.S./Indonesian joint-venture firm could offer the predesigned units on a packaged "turnkey" basis. Plywood mills would provide wood waste as fuel to the power company under a long-term contract, while the supplier would own, operate, and maintain the generators. Under a long-term leasing arrangement, a company would install, operate, and maintain the facilities on a contract basis, and it would also remove, refurbish, and sell some of the present on-site diesel generators to reduce the overall cost of the installation. In another case, one or more firms would build, operate, and transfer the power plants to those plywood mills that are interested in ownership. The supplier would be retained as a consultant, training local people to operate the plants and assisting in maintenance and repairs. For our analysis we will assume third-party ownership of the power plant with leaseback and with on-site operation and sale of electricity. The project will be owned independently of the plywood mill and will pay the mill for the lease of the project.

Capital Costs

A 2.5-MWe-net ($960/kWe) power system would cost approximately $2.84 million on an installed "turnkey" basis, including fuel-handling and storage facilities. Fully burdened capital costs of the project would be about $4 million. A 10% contingency fee ($240,000), another $200,000 for miscellaneous construction needs not covered by the contractor, and startup of $150,000, plus owners' costs during construction and startup ($250,000) as well as closing costs ($100,000) and a developer's fee ($125,000) are based on typical costs in the United States. Working capital of $36,000 would be needed for the project to cover cash costs with a two-month lag in the payment of accrued revenues. Debt-service reserve, calculated for six months' annual debt, is $258,000.

The project would be built with 100% construction financing at 10% interest with a 2% commitment fee. These rates are typical of those charged in the United States. With financing costs of $277,000 ($69,000 commitment fee and $208,000 interest during construction), total loaded-capital costs are $4,036,000.

Financing

Once operations begin, we would obtain permanent project financing with a debt structure of 85% of the total loaded cost and 15% equity. The amount of debt is $3,431,000, and the amount of equity is $605,400. The term of the debt is 12 years, with an interest rate of 11%. Annual debt payment is $516,000, and the interest on the debt-service reserve fund (half of the annual debt payment) is 7%. An investor who provides the equity receives two-thirds of the distributable cash flow, while the remaining one-third is paid to the developer.

Operations and Maintenance

We are conservative in our assumptions of operations and maintenance costs. In our case, $10,000/yr will be paid by the power plant to the plywood mill for the lease of the project site. In addition, the project will be charged

$40,000/yr as a management fee. We estimate $62,000/yr for property taxes and insurance. Direct expenses of the power plant would total $110,000/yr. We need a staff of 15 to 20 people, earning a monthly income of $50.00, to operate the plant. Thus, we may assume salaries to be $10,000/yr. We have assumed world-market prices for operating supplies at $50,000 and annual routine and contract maintenance at $40,000. Ash disposal and administration and office expenses are estimated at $5,000/yr each. Since on-site backup-diesel power would be available, we would not expect any purchased or backup power from PLN. Total costs of operation and maintenance for the project's first year would be $222,000. We assume this total cost will escalate at the rate of 4%/yr.

AN ECONOMIC AND FINANCIAL ANALYSIS: THE SUBSTITUTION OF WOOD WASTE FOR DIESEL FUEL

Our financial analysis was developed as a *pro-forma* model and used by Future Resources Associates, Inc., of Berkeley, California for the study of private power projects. The complete analysis includes six parts: statement of income and expenses, overall project summary, first year system performance summary (by month), project revenues, project operation and maintenance costs, and capital costs and financing. For simplicity and clarity, we will explain our process briefly and then present our results.

Fuel Costs are the Variables

Using our previously assumed parameters as constant, we have investigated six scenarios of diesel and wood-waste price combinations: The actual cost of wood-waste fuel for our project depends on the alternatives available for wood waste produced at plywood plants. There is a limited market for this waste in Indonesia. In practice, only 5% to 30% of the wood waste from plywood mills is sold for firewood or for building huts. A "high" price of $3.00/m^3 is paid for the small fraction of waste that is sold. This price may be converted into a cost of $5.43/t for a wood-burning power plant at a plywood mill. At the other extreme, it is likely that most or all of the wood waste required to generate power would be free, and its use would save the plywood mill the cost of its disposal. Recall that we need 4 green ton/hr of wood waste to produce the 15 tons of steam/hr necessary to operate our 2.75 MWe straight-condensing turbine. Knowing hours of operation and energy needs at the plant, we may calculate the consumption and costs of wood waste for fuel. These figures are a significant variable in the cost of operations and income and expense analyses of our study.

Our cost of diesel fuel is based on 1989 quotes, which were 50% subsidized by the government, of $0.47/gal or $0.13/l. Thus, the world-market price was $0.26/l. We need one liter of diesel fuel to produce 3 kWhe. Consumption rates of diesel fuel needed for plant operations are multiplied by diesel-fuel costs. Since we expect that the Indonesian government might begin to phase out its subsidy of the plywood industry over a five-year period, beginning in 1990, we have included two cases which illustrate this development for our proposed project. These results form a significant factor for project revenues, which are based exclusively on avoided diesel-fuel costs in our study. Thus, the money that we save by burning wood waste, and not buying diesel fuel, is money that our project has earned.

For our six cases, we have compared varying costs of operations, based on cheap or free wood waste for fuel, with varying project revenues, based on avoided diesel-fuel costs. We expect that all fuel costs will rise at a rate of 5%/yr. Remembering also that we have various project capital and financing costs, as well as routine maintenance and expenses, described in the previous section, we balance these costs and expenses against our revenues to determine annual cash flows before taxes for the six cases of our project. See figures 6.1 and 6.2 for the cash flows during the 12-year term of our debt payment.

WHEN WOOD WASTE-FIRED GENERATORS ARE AN ATTRACTIVE INVESTMENT

Two of the most important financial indicators for a proposed project are payback time and the return on equity. The results of these calculations are shown in figures 6.3 and 6.4.

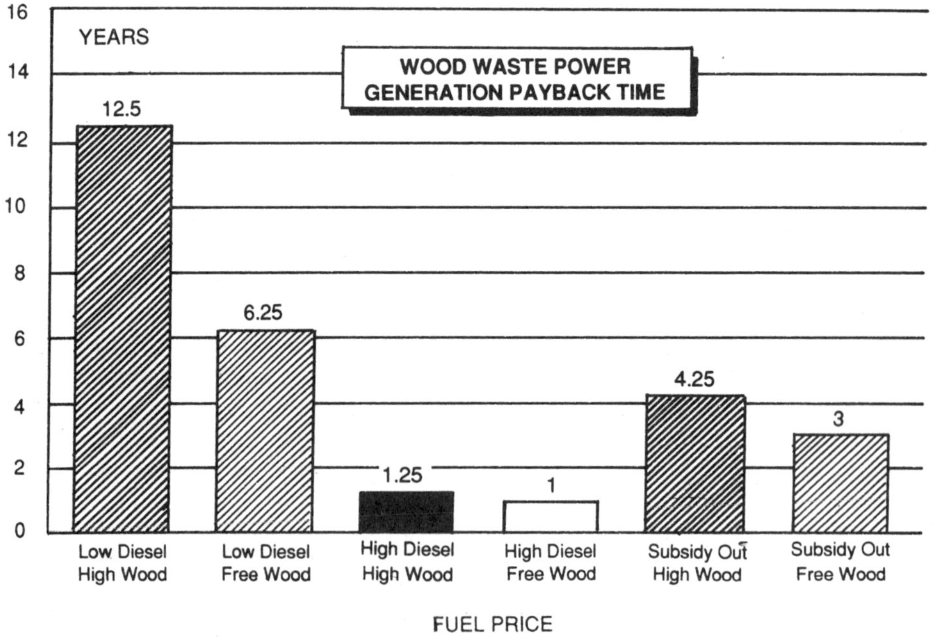

Figure 6.1. Payback Time for an Investment in Wood Waste-Power Generation Systems in Indonesia.

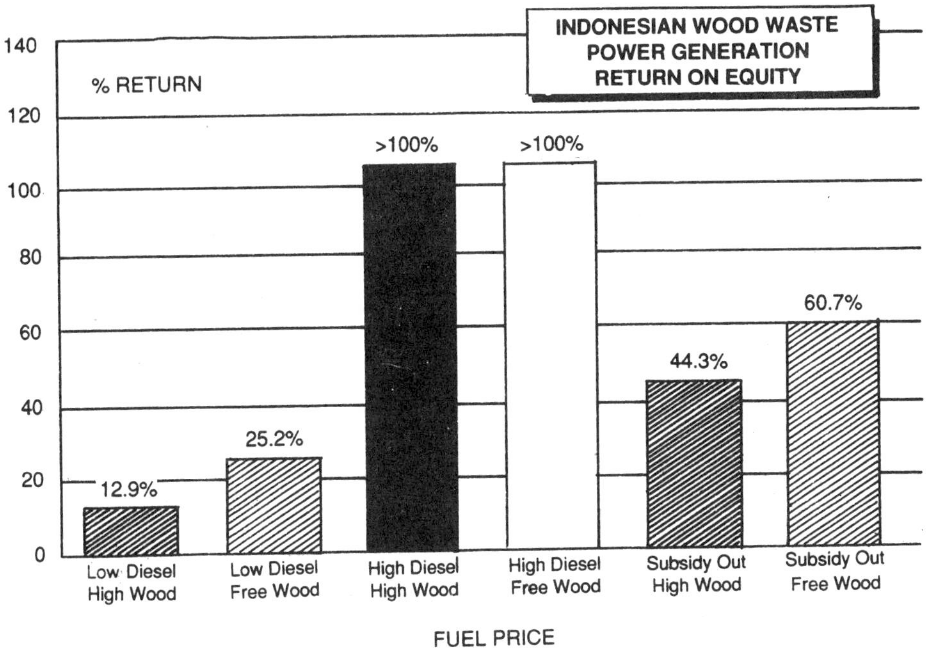

Figure 6.2. Return on Equity for an Investment in Wood Waste-Power Generation Systems in Indonesia.

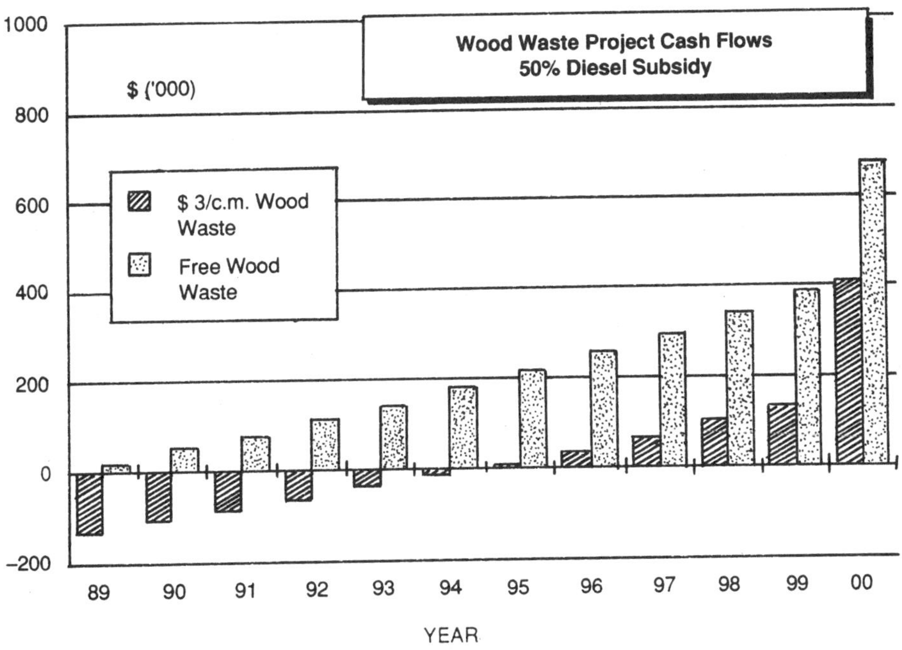

Figure 6.3. Wood Waste Power Generation Project Cash Flows (50% Domestic Diesel Fuel Subsidy).

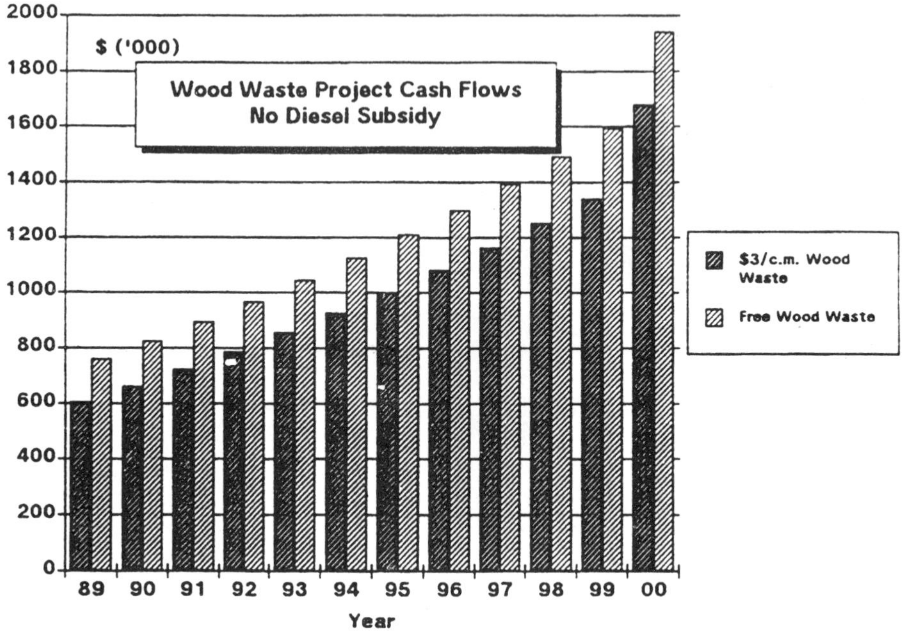

Figure 6.4. Wood Waste Power Generation Project Cash Flows (No Domestic Diesel Fuel Subsidy).

Our conservative assumptions, a price subsidy for diesel fuel and a charge of $5.43/t for wood waste, yields a negative cash flow for the first six years of our project. For Case 1, the investment would be unattractive, with a payback time of 12.5 years and a 12.9% return on equity. The investment picture looks better for Case 2. Even with the 50% subsidy of diesel fuel, it would take only 6.25 years or less to fully pay for a 2.5-MWe wood-waste-fired power plant. In addition, there would be a 25% return on equity, which would be earned by savings in diesel fuel costs alone if wood waste were freely available. If we include operation and maintenance costs as well as the short life of diesel generators in our analysis, the effective payback time would be shorter.

On the other hand, Cases 3 and 4 indicate that if the diesel-fuel prices for the Indonesian plywood industry are at world-market levels of $0.26/l, then the savings on fuel prices are so great that the price of wood waste, whether free or not, becomes relatively insignificant. Payback time is 1 yr to 1.25 yr, and the before-tax return on the investment is well over 100% in both cases. The owner's equity would be repaid within the first year.

It is reasonable to expect that the diesel-fuel subsidy might be phased out gradually in Indonesia. For the purposes of our analysis, we have assumed that prices would gradually rise to world-market prices during a five-year period that would begin in 1990. Cases 5 and 6 illustrate such conditions for wood waste at $5.43/t or free. In the first instance, payback time would be 4.25 years with a return on equity of 44.3%. However, if the wood waste is free, the payback time would decrease to three years and the return on equity would consequently increase to 60.7%.

SUMMARY AND RECOMMENDATIONS

The results of Case 2 would attract investors if this project were in the United States. However, financial conditions in Indonesia dictate short payback-periods, often two years or less, and high rates of return, well over 25%. Therefore, attracting financing for the initial power plants will require incentives and guarantees beyond those typically required in the United States.

Given owners' numerous concerns about costs and performance, which we have discussed above, we have found several persuasive arguments in the favor of wood-waste technology from our preliminary study. First, the costs of power generation using on-site wood-waste power systems will be stable and predictable, relative to diesel power-generation, over the 30-year life of the equipment since the equipment is comparatively simple and, most importantly, since the on-site wood-waste fuel supply is assured and virtually free. By contrast, the future price of diesel fuel is uncertain. Indonesia has rather substantial foreign-exchange and domestic problems and may remove the 50% subsidy on diesel fuel in the coming years.

We have already mentioned several contractual and financing possibilities that could introduce wood-waste technology to the Indonesian wood products industry. Given the uncertain investment climate in the country, we recommend a more detailed feasibility analysis of the problem to develop guidelines for both tight project development and management and to carefully structure financing plans that are specific to Indonesian financial variables. Such a study would establish detailed engineering specifications and breakdowns of costs for the power systems and identify the most suitable locations for the first plants.

Ultimately, a demonstration project will refine our preliminary specifications to determine the technical and economic performance of power-generation systems that burn wood waste to potential owners, investors, and lenders. The current perceived risks of investment will be substantially reduced, and the project's value to the Indonesian wood products industry and to the country as a whole will be recognized.

CHAPTER 7

EXPERIENCES WITH ENERGY FROM WOOD RESIDUES IN THE WESTERN UNITED STATES

O. Bender

AVAILABLE RESIDUES FROM WOOD PROCESSING VARY BY PRODUCT

Sawdust, chips, edgings, and other wood residues accumulate at wood processing plants. These residues create fire and environmental hazards, but they can be converted into valuable energy for internal use or for sale to utilities. The percentage of wood residues available to produce energy ranges from 10% to 60% of the timber that enters a plant. Actual volumes of residues depend on the type of wood processing, the efficiency of the manufacturing process, the scale of the operation, and existing markets for wood by-products.

Veneer and sawmill plants generate the greatest volumes of wood residues, and the pulp and paper industry generate the lowest on a percentage basis (figure 7.1). Sawmills that cut large logs for quality lumber can have from 50% to 60% of timber raw material as waste by-products. Some of these residues are saleable to pulp mills, if they need chips, and some are sold to fiberboard or particleboard mills or to plants that make compressed fuel-logs. In sawmills in the United States, perhaps 30% of the original, unprocessed log is saleable to other forest products industries, so about 20% is potentially available for the production of energy. Veneer and plywood mills produce, on average, less wood residues than sawmills, but their percentages of waste can vary from nearly 0% to 65% of their log-input volume.

The volume of wood residues available at a plant also depends on the plant's size. Sawmills may have a higher percentage of wood residues per processed log, but the small size of mills in many developing countries often precludes the installation of economically viable wood-based energy systems. In contrast, pulp and paper plants, despite their low percentage of wood by-products, have high fiber through-put. Particleboard and fiberboard plants, along with pulp and paper mills, have an average of 20% of their raw material input available as fuel to generate energy. In developed countries, large pulp and paper plants switched rapidly to energy self-sufficiency based on wood residues when prices for fossil fuel rose (Klass 1988).

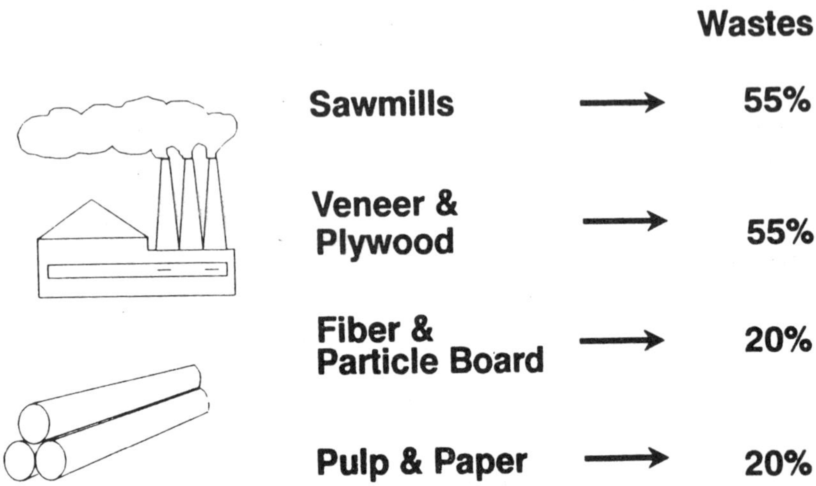

Figure 7.1. Average Wood Wastes from Forest Products Industries (percent unprocessed log).

THE UNITED STATES ILLUSTRATES A VARIETY OF PATTERNS FOR PRODUCING WOOD-BASED ENERGY

The forest products industry consumed 66% of the wood used for energy in the United States in 1987. The percentage used by the industry is declining because of increasing wood use for energy in the domestic sector. Regionally, use of wood-based energy was greatest in the South, followed by the West, and then the Northeast and North Central regions.

Over 3.7% of the energy produced in the United States was derived from biomass, principally from wood, in the mid-1980s (NFPA 1987). The amounts of wood residues used for energy rose from 82 million tons in 1972 to about 138 million tons in 1987. In 1972, the forest products industry used 60 million tons of wood residues, over 70% of the energy derived from wood, to generate steam and electrical power. By 1987, the industry used 91 million tons, a 30% increase since 1972 (NFPA 1987). Industrial and residential use of wood-based energy appears to have plateaued in recent years because markets have probably been saturated and prices for fossil fuel have generally declined in real terms. The government of the United States no longer collects data on the production of wood-based energy, however, so this observation is speculative.

Wood residues have been sold as by-products for a long time, and there are price differentials among residues of varying quality. Integrated facilities that manufacture solid wood products and pulp have used residues from sawmills as feedstocks for their pulp mills since the early 1900s. Soon after

World War II, many independent sawmills also began to sell their wood residues to the pulp industry. By the late 1970s, the energy market became a new option for the use of wood by-products from the forest products industry because prices of traditional fossil fuels were rising and enforcement of environmental standards was becoming stricter. Larger, more modern sawmills used wood residues to produce energy for their internal needs, thus reducing the amount of power that they purchased from the local utility company. Pulp and paper factories also converted to wood-based energy. Over 85% of the forest products industry produced energy from on-site or purchased wood residues by the early 1980s (Klass 1988).

New federal legislation in the 1970s, which promoted the sale of private power to utilities, made the economics more favorable for the industrial production of wood-based energy. Congress passed the Public Utilities Regulatory Policy Act which ensures that if these industries could meet the utilities' avoided costs, they could sell any excess energy to the power companies.

In 1980, the largest forest products companies in the United States produced 70% of their energy requirements from wood residues, recycled gas, and other by-products (NFPA 1987).

The 14 largest companies using their wood residues for energy included:

Boise Cascade
Champion International
Crown Zellerbach
Temple Eastex
Koppers
MacMillian
Bloedel
Masonite
Potlatch
Georgia Pacific
Kimberly-Clark
Weyerhaeuser
Willamette Industries
Southwest Forest Industries

An average of only 18% of the fuel used by the forest products industry was natural gas or fuel oil, and the industry purchased only 10% of the electricity that it consumed in 1980 (NFPA 1987). The industry sold 2,500 MWh of surplus electricity in 1978, and this amount jumped to 29,400 MWh, a 10-fold increase, a decade later. Conversely, use of natural gas by the paper industry in the United States fell by 34.9% from 1972 to 1986, and its use of oil declined from 21.3% in 1972 to only 8.5% in 1986.

THE WESTERN UNITED STATES

Three western states illustrate typical patterns for the industrial production of wood-based energy. Similar patterns are observed in the South and Northeast. In these different states, the amount of energy produced and the size of the energy systems at the plants depend on several conditions. First, there should be a commitment at the state level to encourage the development of alternative sources of energy. Pricing formulas that govern the sale of private power to the grid and regional tariffs for electricity should be established. Finally, the cost of alternative fuels for boilers at the processing plants and the environmental costs that are internalized at the energy facilities must be examined.

California

In an effort to keep consumer rates for utilities low and to encourage local use of nonfossil fuel, California developed, in the early 1980s, an aggressive program to promote the private production of electrical power. Four standard-offer contracts for private producers were developed to encourage such production of electrical power. *Standard Offer No. 4* was the catalyst for stimulating the development of a large number of projects for wood-based energy because it established high payments for energy (fuel) and for a fixed capacity for power production. This latter provision replaced the need to invest in utilities by giving the private producer a payment for his installed baseload capacity.

By 1987, the cumulative capacity of facilities that generated power from biomass in California equaled 673 MW, with a projected capacity by 1990 of 1,152 MW, according to the database released by the California Energy Commission on November 18, 1987. Residues from the forest products industry accounted for 80% of the potential for generating power from biomass, while the remaining 20% came from agricultural residues (pits, shells, prunings), urban wood residues, and animal manure.

Of the 151 biomass plants listed by the California Energy Commission, 100 were facilities installed by forest products plants. Seventy-four plants producing wood-based energy were built and operating. Only 31 were capable of generating power, with the remainder producing steam. Eight were shutdown or inactive, and 16 were in the permit, planning, or construction stage.

Sizes of plants, measured by their capacity to produce energy, ranged from a minuscule 50 kW to 23.5 MW, with 10.6 MW as an average in the state. Total installed capacity has increased partly because the California Public Utility Commission has changed its policies regarding pricing for the purchase of electricity. The commission has provided capacity payments to companies that produce electrical power. For example, the 16 plants in the "planning, permit, and construction" phase are larger, on average, than those built earlier. Their sizes range from 8.5 MW to 42 MW, with an average of 23 MW. Three of the plants that were shut down were obsolete and probably remained closed. Three plants that were inactive were closed temporarily because they had problems with pollution. These plants averaged 8 MW. Of the 31 plants that generate power, 27 are cogenerating. One of the four operating plants that are not cogenerating is undecided, and three generate only power (no steam) for direct sale. The plants are using no new or innovative systems, with the exception of one that has a fluidized bed.

The management and operating conditions at each plant determine the lower economic limits for its facilities. Of the seven plants with extremely small generating capacity, all but one were cogenerators:

Wood Power Plant/Location	Capacity (MW)
Jeld-Wen, Susanville	0.99
Susanville Forest Products, Inc. Susanville	2.50
Sierra Pacific Industries, Quincy	4.50
Desert Plant Industries, San Bernadino	0.05
California Cedar Products, Stockton	0.85
Roseburg Lumber Company, Anderson	3.00
Roseburg Lumber Company, Anderson	1.70

Oregon

Total production of wood-based energy by the forest products industry in Oregon was slightly higher than in Washington, but power from burning wood had not received the same support as it had in California by the late 1980s. There was enthusiasm for wood-based energy in the early 1980s, but some companies switched back to the use of oil when the price of diesel fuel dropped in the late 1980s and when the utility tariff became competitive with the cost of energy from wood.

In 1985, there were 137 industrial facilities for wood combustion that were listed in the Oregon State Department of Energy's *Directory of Oregon Biomass Energy Facilities* (Sifford 1987). Of this group, only 18 were cogeneration facilities, and eight were direct generation facilities. The other 111 plants burned wood wastes to generate steam for internal use.

Design capacities of wood-based power plants in Oregon range from 1 MW to 50 MW. The average capacity is 7.2 MW, which is smaller than that of the typical facility in California. The actual generation of wood-based power at the time of the 1985 survey was much less than the capacity. Eight mills reported no generation. The smallest generation was 256 kW, and the largest was 32.4 MW, averaging 3.5 MW—well below the average design capacity for Oregon plants. Sale of power by the forest products industry is depressed because there is a surplus of relatively cheap power in Oregon with its proximity to the hydroelectric plant at Bonneville. Some forest products companies now buy electricity off the grid because it is cheaper to buy it than to produce it from their wood residues.

Washington

The forest products industry in Washington was the least responsive of the three western states to the idea of installing wood-based energy systems for cogeneration or generation of electric power to sell. This response appears to be caused by the low rates available from the Bonneville Power Administration and by the lack of state support for alternative fuels.

Of the 52 industrial facilities for wood combustion that are listed by the Washington State Energy Office, only nine have cogeneration plants. The smallest plant has a design capacity of 600 kW, and the largest has a 35-MW capacity. The average design capacity of the plants is 14.4 MW. In contrast, the actual production during 1986, the year of the last survey, averaged only 8 MW. Most of the generation was for steam, only 56% of design capacity.

VOLUME OF WOOD RESIDUES BY PRODUCT

Sawmills Yield the Highest Volumes of Wood Residues

Sawmills, veneer mills, and plywood mills produce the largest quantities of wood residues on a per-unit basis in the forest products industry. Figure 7.2 diagrams the process at sawmills and illustrates that, on average, the recovery of the original log is only 50%, excluding the bark (logs are generally measured or "scaled" without the bark). The rest of the log becomes the slabs, edgings, and chips, as well as sawdust, that could be used as fuel or as raw material for particleboard or fiberboard plants or for pulp mills.

FAO statistics of developing countries suggest that sawmills consume about four times as much industrial roundwood—wood in log form—as do plants that manufacture wood-based panels. Much of the waste materials in sawmills—30% of the debarked log—is bought by other wood processing plants that are located near the sawmill. Roughly 20% of the original log is still available for fuel. Particleboard, fiberboard, or pulp mills typically buy solid wood residues, especially those from sawmills. It is much cheaper to process these residues than it is to convert whole logs into chips.

If a processing facility for secondary wood products, such as a particleboard or fiberboard plant or a pulp mill, is not within an economic hauling distance, a sawmill has no alternatives. It must give away, burn, dump, or sell its wood residues for a minimal price. If that is the case, then 50% of the debarked logs could be available for fuel.

Since it is easy to enter the wood processing business, the scale economies of sawmills vary widely in A.I.D.-assisted countries. Many entrepreneurs or farmers are in the lumber business because the capital costs of starting a rudimentary sawmill are low. Small sawmills are run by farming families, who balance their sawmill operations with their planting and harvesting schedules on a seasonal basis. Logs may only be available during the dry season, or the mill may operate only when there are orders for sawnwood. Small mills sell lumber right off the saw, or they may stack and dry-air the lumber if there are no ready customers.

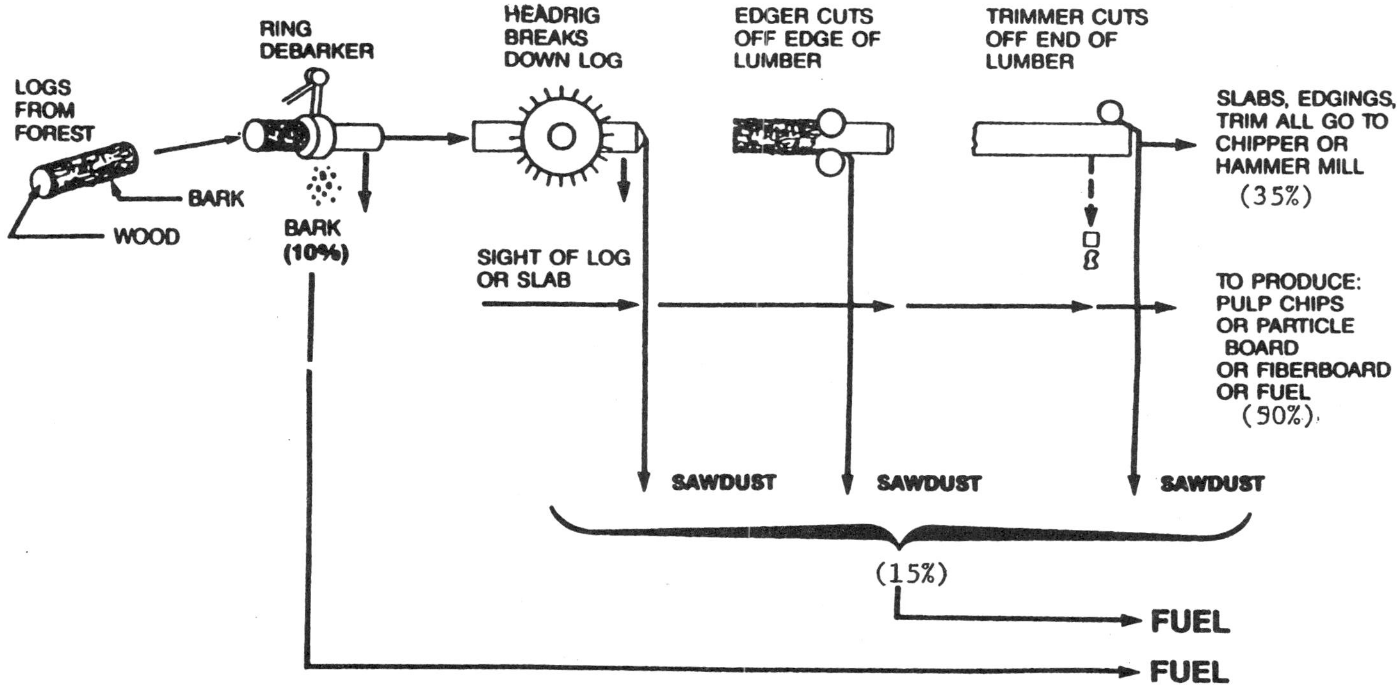

Figure 7.2. Sawmill Residues Available for Energy Use (* % of debarked logs).

The ability of sawmills to produce thermal-electrical energy depends upon a variety of factors, but the most important consideration is the size of the mill. Size determines the quantity of available wood residues, the demand for internal thermal-electrical power, and the likelihood of a plant's having excess power or steam for sale. A sawmill generally needs to produce over 10,000 m^3/yr of lumber in order to have sufficient residues to meet the requirements of a 500-kW generator. Smaller sawmills have neither the capital nor the thermal-electrical demand to be interested in generating energy from their wood residues. They want to sell or give them away because burning or dumping them represents an additional cost.

Veneer Mills Have the Highest Percentage of Wood Residues

Veneer mills are usually larger and more complex than sawmills, so their scale economies favor production of energy. There are some exceptions, however, such as some mills in Brazil, where the veneer is hung on a washline to dry, and the practice in Nepal, where veneer is spread on the grass. The quantity of wood residues from veneer mills exceeds that from sawmills. Figure 7.3 graphically describes the manufacturing process of veneer and plywood and shows the percentages of the original log that become wood residues at each stage. The process involves a greater through-put and results in a higher percentage of unusable wood per log. Large veneer mills with greater needs for steam and electricity than the typical sawmill are logical sites for installing wood-based energy systems, if the framework of policies and the economic incentives are favorable.

Once a mill has decided to kiln-dry its lumber or veneer, it will have to generate steam or hot air. Consequently the mill must acquire a boiler and other equipment to dry the wood, which increases the cost of producing these primary wood products. Burning wood residues is the most efficient means for providing steam.

Until oil prices rose during the 1970s, the cost for electricity or for a stand-alone diesel generating system was often lower than that for any system burning wood residues. Small—under 10 MW—commercially competitive wood-based systems were not available. Large turbines were too expensive. If installed, they presented management with the problem of marketing excess electricity, since sale of private power was prohibited in most countries. Government subsidies made tariffs for electricity in rural areas generally low, often far below the actual costs of production.

Other Wood Products Industries also Yield Wood Residues

Energy is a crucial element in the processing of plywood, particleboard, and fiberboard. Figure 7.3 illustrates the manufacture of plywood, with the percentages of the debarked log that are available for fuel at each stage of the manufacturing process. Since steam is required for drying and pressing, a

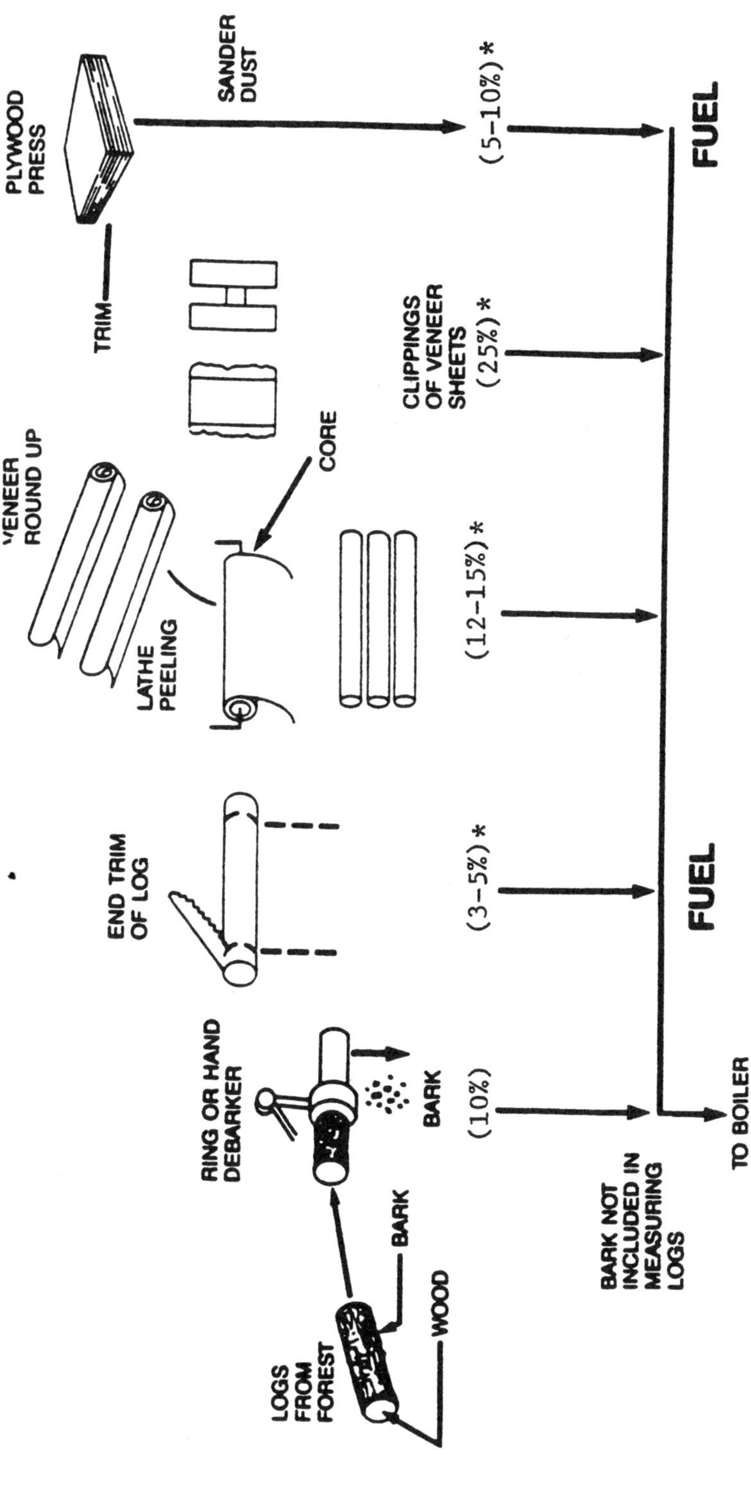

Figure 7.3. Veneer/Plywood Residues Available for Energy Use (* % of debarked logs).

boiler is a necessity. For some mills, it is cheaper to burn their wood residues for power than to buy electricity from the rural grid. If electrical power is not available from the grid, the mills provide their own with diesel power plants at their facilities.

In most developing countries, electrical power generated from diesel fuel is the source of power in the forest products industry. As long as diesel fuel was cheap and environmental standards nonexistent, mills that were not located on the grid could afford to generate the power they needed with diesel fuel, while they used their wood residues for steam or simply burned them or dumped them in landfills. That situation changed in the 1970s when real prices for diesel fuel rose four or more times.

Since higher quality wood residues are needed to make particleboard and fiberboard, these wastes have a high opportunity cost to such wood processing industries. Many particleboard and fiberboard plants live off the wood residues produced by sawmills and plywood plants. Pulp, particleboard, and fiberboard manufacturers discovered that it was much cheaper to buy the better quality wood residues from nearby sawmills, veneer mills, and plywood mills than to buy logs directly, debark them, and process them into the sizes of chips needed to make their products.

Figure 7.4 illustrates that particleboard and fiberboard plants have a low percentage of waste recovery, on average only 15%-20% of the whole unprocessed log. After selling their higher quality wood residues, about 10%-15% of the original log remains to produce energy at pulp and paper mills. These residues are composed of sawdust and bark (10%), pieces of wood that would go through chippers, and the fines (5%)—screen particles that are too small for the machines.

Pulp mills and particleboard and fiberboard mills that are not able to purchase all of their raw material from sawmills will need to purchase and process their logs directly. Under such conditions, the mills use around 90% of the log, leaving 10% or less available for the production of energy. Their wood residues include the bark—few particleboard or fiberboard mills use bark—unchippable material, and screenings.

Figure 7.5 shows that pulp mills usually make better use of their wood residues than mechanical wood processing plants, such as sawmills, because they have higher capital and operating costs that ensure efficient and cost-effective management of their resources. The demand for thermal and electrical power at pulp mills often far exceeds their on-site supply, which they normally obtain by recovering their waste liquors and their rejected wood residues from pulping—the bark, knots, fines. Hence, pulp mills are net purchasers of power unless they are integrated with mechanical wood processing plants. Even wood-based panel plants require more capital and technical expertise to operate than a sawmill, so there is more incentive in these plants to reduce their costs by using wood residues to produce power.

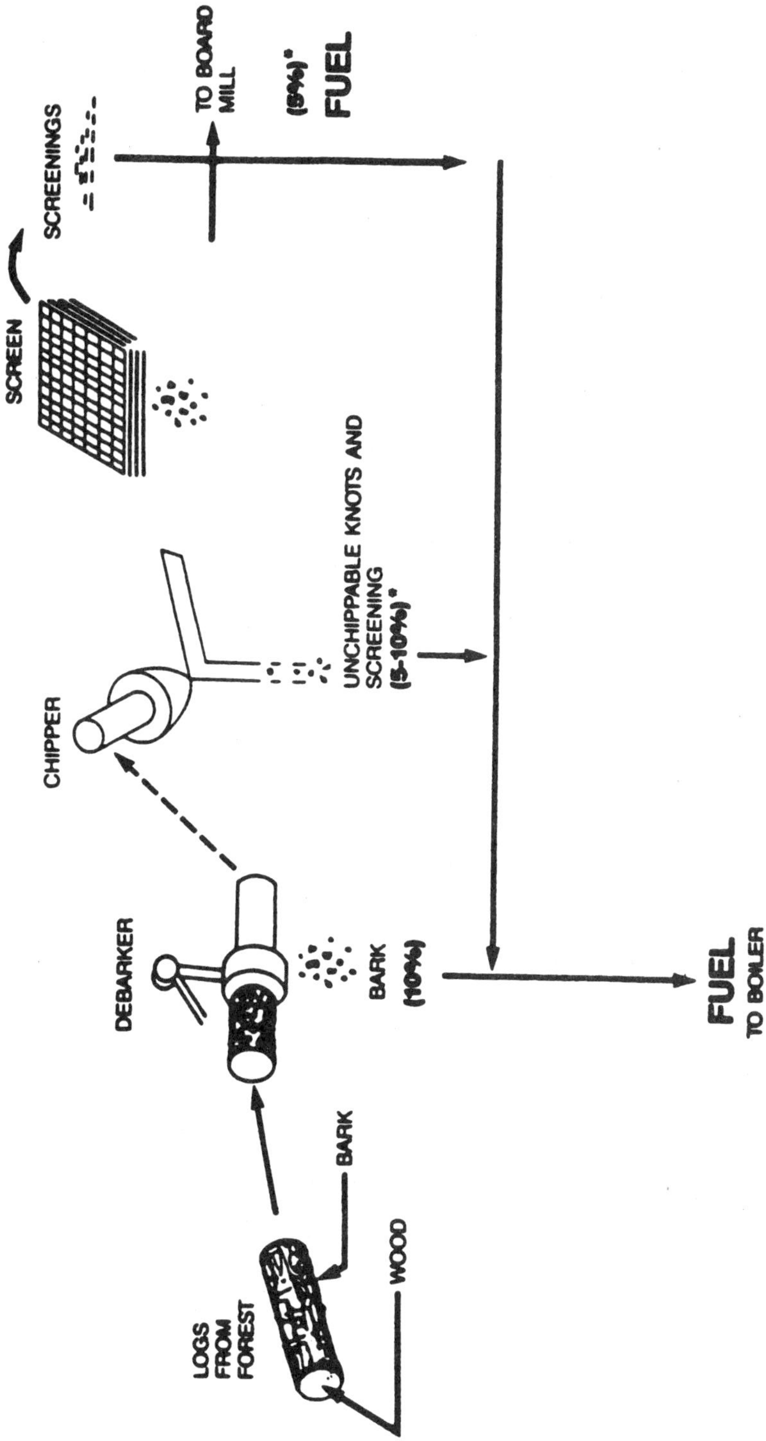

Figure 7.4. Particle/Fiberboard Residues Available for Energy Use (* % of debarked logs).

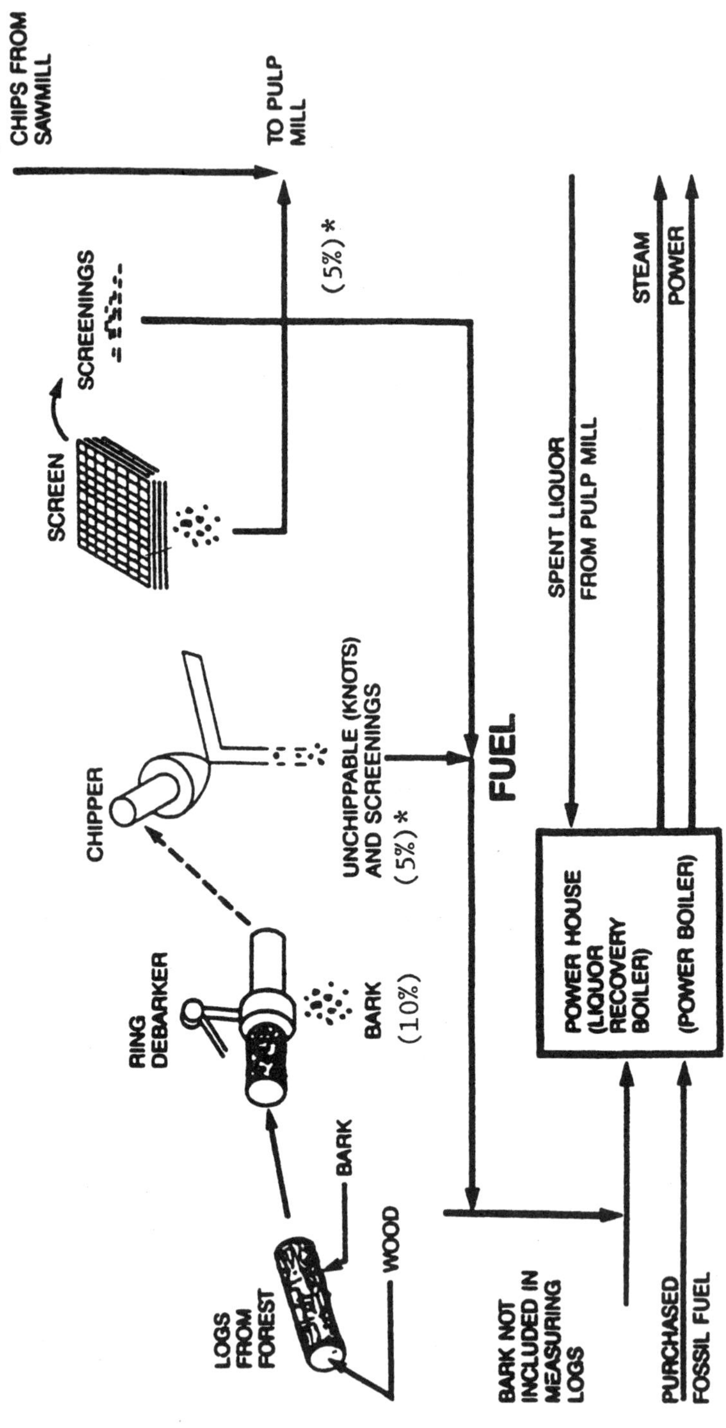

Figure 7.5. Pulp Mill Residues Available for Energy Use (* % of debarked logs).

TECHNOLOGIES FOR USING WOOD RESIDUES

Availability of equipment for handling and burning wood to generate heat or power was limited prior to the oil crises of the 1970s. Higher prices for fossil fuels stimulated manufacturers of equipment to improve their technologies for conversion of wood to energy. Burning of wood residues in wigwam or teepee burners is not efficient. Economically viable energy systems that are designed for wood combustion range, with some exceptions, from three MW to 20 MW in design capacity.

The minimum economically viable size for any system that burns wood for steam or electricity is around that of a 3-MW design capacity. These systems require about 23,000 oven dry metric tonnes (ODMT) of fuel per year for direct generation of power. Such a demand for feedstock would require a sawmill that processes over 50,000 m^3 of lumber, assuming a specific gravity of 0.45 and a waste recovery of 50% in the sawmill.

According to industry experts, the largest size for a wood-based energy system to be economically competitive is one with a design capacity of about 20 MW. A 20-MW plant, with no internal power consumption, would need 152,000 ODMT/yr. This demand would require a sawmill with a capacity to produce 338,000 m^3/yr of lumber. Such a capacity is equivalent to a large mill on the west coast of the United States which cuts 140 million board ft/yr. Georgia Pacific's energy plant at Crossett, Arkansas, is the world's largest biomass energy plant based on wood residues, with a design capacity of 42 MW. It was designed to take advantage of residues from several processing plants as well as fuelwood from external sources (Dominguez 1993).

Prices for alternative energy or electricity that would be available to wood processing plants could make 3-MW wood-based energy systems an economic option. At the higher end of the scale, the size of the wood processing mill and its ability to import additional feedstock are the factors that may preclude the installation of a 20-MW system.

Larger wood-based energy systems in the United States, those with a capacity of over 40 MW, often require large hauling distances to ensure a sufficient supply of wood. Burlington, Vermont, recently switched back from its large wood-based energy plant to some use of diesel fuel. Lower oil prices and the need to haul wood from a three-state radius were given as critical reasons for this change. In addition, environmentalists have interfered with the development of wood-based energy in Connecticut, where they have blocked the construction of two generation plants that burn wood residues.

Literature Cited

Dominguez, Jules. 1993. A case history comparison of wood versus fossil fuel as an energy source. In *Forests and wood-based biomass energy as rural development assets.* (W.R. Bentley and M.M. Gowen, eds.). New Delhi: Oxford & IBH Publishing Co. Pvt. Ltd.

Klass, Donald L. 1988. The U.S. biofuels industry. Paper presented the International Renewable Energy Conference (September 18-24, 1988), East-West Center, Honolulu.

NFPA (National Forest Products Association). 1987. Energy consumption of fifteen reporting companies by energy source—lumber and wood products industry. Washington, D.C: NFPA.

Sifford, Alex. 1987. *Directory of Oregon biomass energy facilities*. Prepared for the Bonneville Power Administration under U.S. Department of Energy Grant. Salem: Oregon Department of Energy.

CHAPTER 8

A CASE HISTORY COMPARISON OF WOOD VERSUS FOSSIL FUEL AS AN ENERGY SOURCE

Jules Dominguez

INTRODUCTION

In this chapter, we develop a case study of the Georgia-Pacific Corporation's decision to build a major wood-based biomass energy facility at Crossett, Arkansas. This study will illustrate the factors surrounding a choice between biomass and fossil fuel sources of energy. As we will see, the relationships among these factors can change rapidly and unpredictably.

Georgia-Pacific Corporation is a large forest products company which manufactures pulp and paper and building materials, such as lumber and plywood. The manufacture of pulp and paper is an energy-intensive process, demanding large amounts of thermal energy in the form of steam as well as electricity. Wood products manufacturing is less energy-intensive, but also uses thermal energy to dry the wood and electricity to run the machines.

Both types of manufacturing processes generate wood waste, which Georgia-Pacific uses as fuel to supply a major part of its energy needs. Since we have the facilities to handle and burn wood waste, we have opportunities to purchase additional waste materials from others for use as fuel. Georgia-Pacific uses waste material for fuel at 97 locations across the United States. Approximately 63% of Georgia-Pacific's *total* energy needs are supplied by waste materials, some of which are wood-based. Some of the waste is generated internally and some of it is purchased.

AN OPPORTUNITY TO REDUCE FUEL COSTS

Let me share our experience with a boiler that we converted to the use of wood waste for fuel at a unique time in the mid-1980s, when the prices of petroleum-based fuel changed radically. Our story begins in 1983, when the cost of natural gas and fuel oil was in the range of $4 to $5 million BTUs.

Our manufacturing complex in Crossett, Arkansas, included two plywood mills, a paper mill and a chemical plant. In 1983, we were using two wood-waste boilers in the plywood mills. Our paper mill had one wood-waste boiler, a black liquor recovery boiler and two gas/oil-fired boilers, all of which produced steam. In addition, the paper mill employed three steam turbine generators to produce approximately 75 megawatts of electricity. Two of the turbine-generators were double extraction/condensing units, and the third was extraction/back-pressure. We purchased the additional electricity from the local electric utility company to make up the balance of our demand.

At this time, a new black liquor recovery boiler had just been completed for the paper mill. Since we were planning to rebuild and relocate the old one to another paper mill, it was still in place, "laid up" to protect it from deterioration. Nothing happened with these plans. We had to choose between finding a buyer for the boiler or demolishing it. Recovery boilers are characterized by certain design features that accommodate specific properties —moisture and ash content—of the fuel, and that facilitate the recovery of chemicals. These features, a large furnace and wide spacing of convection tubes, made us realize that our old boiler would be ideally suited to the burning of wood. Such a conversion had been made successfully in other mills.

A FEASIBILITY STUDY

Since the cost of steam produced by natural gas and oil was so high, management studied the economic feasibility of converting this boiler to burn wood. In addition to the production of steam, we hoped to have the potential to generate more electricity at a low cost, if we could make more steam by burning wood. To gain this potential, we needed to rewind one of the generators to increase its capacity. In addition, we needed to make "condensing" power, since all of our heat sink was spent already for "backpressure" power.

We included in our feasibility study the possibility of our making additional electricity, which would enhance the economic viability of the boiler conversion, even though we anticipated an insufficient supply of wood waste within a reasonable hauling distance to fully fuel the boiler. We planned to harvest whole-tree chips from scrub wood to make up our fuel deficit, realizing that such an operation would cost more than wood-waste fuel. However, the cost would still be less than for gas or oil. We reasoned that we would improve the growth of commercial timber from such harvesting, and we factored this intangible benefit into our evaluation.

We calculated capital and operating costs from all of the factors and circumstances discussed above. Since the projected return on our investment exceeded the corporate "hurdle rate" in effect at that time, we decided in favor of the project.

HIGH PERFORMANCE DESIGN

A consulting engineering firm was hired, under the direction of a team of Georgia-Pacific operating and engineering personnel. Specifications were prepared and proposals solicited from boiler manufacturers to design the boiler modifications and supply the materials. Additional specifications were prepared and bids solicited for the boiler auxiliaries, stack gas scrubber, and the fuel and ash handling systems and equipment, and the project was implemented. The technical details of this boiler have been described previously in a paper by Barsin, Proterra and Stewart (1987).

Our result is a high-performance boiler, and we believe that it has the largest wood-burning capacity and the lowest maintenance requirements of any such boiler in North America. All of the companies that bid on the conversion predicted a capacity to generate steam in the range of 525,000 lb/hr, but they were using normal parameters based on a conservative design. After we put the boiler into operation, we found that it could easily exceed the estimated rating. The contractor was asked to make another evaluation of the converted boiler's ultimate capacity, based on its actual operation. Its rerated capacity is 650,000 lb/hr.

High performance is achieved by a dual-arch configuration of the front and rear furnace walls above the grate and by the ability to deliver a large percentage of the total combustion air as overfire air at the arches. These features provide good penetration and mixing of the overfire air, which increases the combustion temperature and completeness of combustion in the zone below the arches, resulting in good load-following ability and low unburned carbon losses.

LOW OPERATIONS AND MAINTENANCE COSTS

Features which contribute to the low maintenance requirements of this boiler are: water-cooled vibrating grate in lieu of air-cooler traveling grate; no mechanical dust collector; minimal cinder-rejection system with no sand classifiers; wet scrubber for final particulate removal from stack gas in lieu of baghouse or precipitator; and a clean induced-draft fan. The water-cooled grate also contributes to high performance by permitting high combustion air temperature to be used.

We selected natural gas as the only back-up fuel for the boiler so that we would avoid cold-end corrosion and convection pass plugging, problems inherent with the burning of No. 6 fuel oil. The fuel-handling system includes a rotary-dumping railcar unloader, a truck dumper, and an automatic stacker/reclaimer which simultaneously builds and feeds from a doughnut-shaped pile which is also the short-term fuel storage. Bottom ash is handled by a wet system which sluices to an ash pond. Flyash is bled to the ash pond from the venturi scrubber recirculating pumps.

Based on emission testing, the state Environmental Protection Agency agreed that it would assume carbon monoxide levels to be within allowable limits, provided that we control oxygen levels within the narrow range of 3% to 4.5% of the flue gas content. We monitor and report oxygen levels instead of those of carbon monoxide. Emission limits for nitrogen oxide for this boiler are 0.5 lb/million BTUs for wood only, 0.3 lb/million BTUs for burning a combination of wood and gas, and 0.2 lb/million BTUs for burning gas only.

The boiler meets these limits with no problems. We have gas burners with secondary air ports to stage combustion in order to control nitrogen oxide levels, but, so far, we have not found it necessary to use them.

SOME ECONOMIC CONSIDERATIONS

While the boiler's operating performance has been an enormous success, meeting or exceeding Georgia-Pacific's expectations for operations and maintenance, its economic performance has been disappointing. Recall that we based our economic predictions on 1983-84 prices for natural gas and fuel oil. The boiler began to operate in the late summer of 1985. Fuel oil prices dropped precipitously in 1986 and have never recovered. Prices for natural gas fell as well, and they are now less than half the 1984 prices. Indeed, whole-tree chips used for fuel are more expensive than natural gas. Currently, we fuel the boiler with wood waste and gas, rather than with wood waste and whole-tree chips.

Our experience on recent boiler projects indicates that the capital costs are about four times as much for a facility that burns wood for fuel than for one that burns natural gas. Therefore, we would need a large differential in fuel prices to justify the burning of wood in a purely economic evaluation. In general, wood-fired facilities require more maintenance and more attention to operations, and the combustion efficiency of wood is inherently lower than that of gas. These latter two factors further reduce any savings in fuel and increase the price differential required to realize any economic rationale for burning wood.

Wood is a good fuel, and the technology to use it is well-proved. However, economics and the ready availability of wood at a particular factory will determine the extent of its use. Each situation of fuel type and condition of its availability, whether it is wood waste generated internally at the plant, wood waste purchased for fuel, or whole-tree chips, has its own unique economic factors. As of 1992, the conditions and our own economic projections do not favor the installation of additional wood-burning boiler capacity at this particular site. Georgia-Pacific will continue to evaluate all situations at our numerous plants as we consider future prices and availability of all types of fuels.

Literature Cited

Barsin, J.A., J. Proterra, and G. Stewart. 1987. *Conversion of Recovery Boiler to Bark Burning.* Presented to TAPPI Engineering 1987, New Orleans LA.

CHAPTER 9

EFFECT OF A WOOD-FUELED POWER PLANT ON ATMOSPHERIC CARBON DIOXIDE[1]

M. J. Ducey
P. M. S. Ashton
B. C. Larson
L. M. Vernegaard
K. Dalla Rosa
G. Knoettner

INTRODUCTION

The influence of human activity on global climate has emerged as a major environmental concern. In particular, human-incurred increases in greenhouse gases such as carbon dioxide (CO_2) have raised questions about long-term changes in climate. Atmospheric CO_2 has increased by 25% since the industrial revolution to a concentration of 350 ppm today (Gammon et al. 1985) and is expected to reach 500-700 ppm within a century (Trabalka et al. 1985). Possible effects include changes in temperature and precipitation as well as direct physiological effects on plant growth. These effects are poorly characterized, and significant uncertainties remain. Avoiding the predicted negative effects and the risk brought by a changing atmosphere has become an important goal at national and international levels. However, strategies for ameliorating atmospheric change must be implemented on the ground at local scales.

Recently, several researchers have explored forest management as a way to offset the release of CO_2 to the atmosphere (Sedjo and Solomon 1989, Vitousek 1991, Wright et al., in press). Woodwell (1987) estimates that approximately 21 billion tons of carbon are stored in the biota and soils of terrestrial ecosystems. Since forest ecosystems store 90% of all terrestrial carbon (Graham 1990), manipulation of this carbon sink through forest

1 The authors gratefully acknowledge the assistance of Dr. William H. Smith. This research was sponsored by the U.S. Department of Energy, Office of Energy Research, Environmental Sciences Division, Office of Health and Environmental Research, under appointment to the Graduate Fellowships for Global Change administered by Oak Ridge Institute for Science and Education.

management could provide a strong regulatory influence over levels of atmospheric CO_2. Wright et al. (in press) suggested four strategies of forest management to control or offset increasing atmospheric CO_2. The first three strategies are mechanisms for expanding carbon sinks on the landscape, while the fourth reduces rates of release of atmospheric CO_2 from fossil fuels:

- Planting new forests
- Planting trees in urban areas and shelterbelts
- Increasing net annual timber growth on existing timberlands
- Using wood, a renewable energy source, to replace fossil fuels

All four strategies are subject to uncertain information and to limitations. For example, Vitousek (1991) noted that to offset the annual release of carbon into the atmosphere from combustion of fossil fuels, all previously forested lands, including cities and farmland, would have to be reforested. Sedjo and Solomon (1989) discussed the possibility of offsetting current CO_2 emissions through increased forest productivity. They concluded that the rate of growth in the world's existing forests would have to double, an unlikely increase in productivity given reasonable social, economic, and biological constraints.

Unlike the first three strategies, the fourth does not result in a net decrease of atmospheric carbon. Instead, wood is substituted for fossil fuel to produce energy so that the rate of release from fossil carbon sinks to the atmosphere decreases (Wright et al., in press). Although combustion of wood releases CO_2, reforestation of the harvested area can recapture much of the carbon that has been released. The carbon flux of a wood-to-energy system theoretically approaches a steady state, while a fossil-fuel system represents a continuous release of carbon from nonatmospheric carbon sinks.

Biofuel systems are not new. Currently, about half the world's population relies on biomass to supply their most basic energy needs, such as cooking and heating (Bormann et al. 1991). At a higher technological level, biomass fuels many small power plants. In New England alone, approximately 400 such plants power sawmills or produce modest amounts of commercial electricity (Killingly Energy Limited Partnership 1988). While such facilities represent a potential asset for rural development, they may also form part of a comprehensive strategy to manage levels of atmospheric CO_2. Our case study tests the viability of substituting biomass for fossil fuel at a proposed wood-burning generator in northeastern Connecticut by quantitatively examining its effects on the local carbon sink.

THE PROPOSAL FOR THE KILLINGLY WOOD-FUELED ENERGY PLANT

The Killingly Energy Limited Partnership (KELP 1988) submitted a proposal to construct a 32-MW electrical power generating facility in northeastern Connecticut (figure 9.1). The plant, with a planned operating life of 50 years, would be fueled by 450,000 dry tons/yr of wood chips. Much of the fuel would come from management within the surrounding forests, which are predominantly of the species-diverse oak-hickory type (Eyre 1980). The remaining chips would be derived from wastewood, such as demolition wood and debris from land clearing.

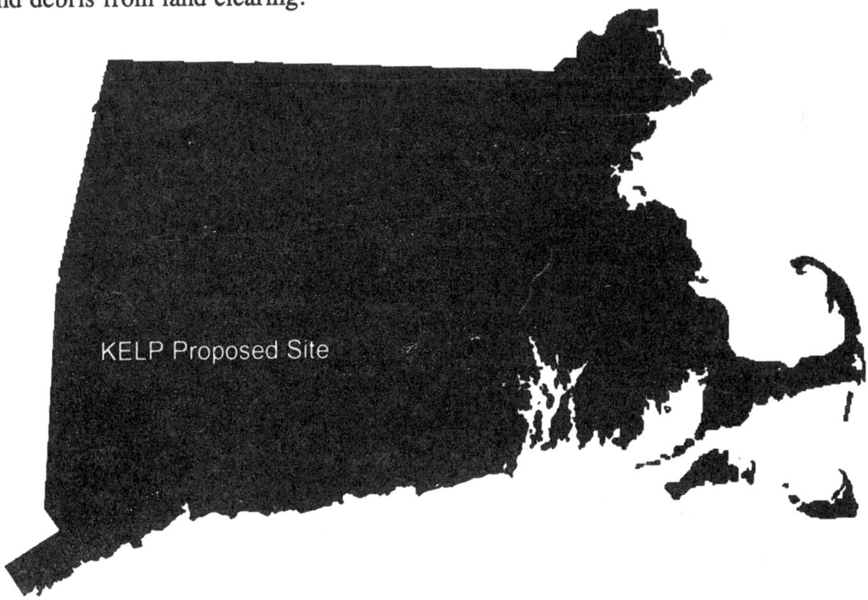

Figure 9.1. Location of the Proposed Generating Facility in Northeastern Connecticut.

In their proposal, KELP asserted that a wood-fueled energy system would have several benefits:

- Reduced CO_2 emissions from burning wood rather than fossil fuel
- Improved silviculture and forest management through the creation of a market for previously unsalable materials
- Reduced demand for landfill space
- Increased local autonomy in the production and supply of energy

The effect of the Killingly plant on atmospheric CO_2 depends upon the ratio of biomass to fossil fuel used and on the ultimate fate of the carbon from these two sources. A prerequisite to determining the effects of the plant's operations on the CO_2 balance is to ascertain whether or not current or probable systems of forest management in southern New England could support KELP's projected demand for wood chips. KELP's (1988) stated goal was to procure between 10% and 30% of its requirements for wood chips, or from 45,000 to 135,000 dry tons/yr, from silvicultural sources at a cost of \$17.32/green ton delivered to the mill. Their estimated prices reflect the anticipated cost of demolition wood from the nearby Boston and Hartford metropolitan areas. KELP's management could be expected to minimize costs by exploiting inexpensive sources of wood chips before considering alternative sources. Thus, the comparative economics of all potential sources of wood chips would determine buying decisions and the resulting dynamics of the carbon flux.

METHODS AND ASSUMPTIONS

Woodshed Analysis

The availability of different fuel sources, especially of silvicultural wood, is critical to estimating the effects of the plant's operations on atmospheric CO_2. We define silvicultural wood as that produced by the harvest of standing trees, excluding site conversion from forest to other land uses. The distinction is important because a harvested forest will regenerate and continue to grow, while land converted to other uses does not support continued accumulation of carbon.

Since the Killingly plant is expected to have a long operating life, sustainable yield, rather than the current harvesting rate, is the criteria for estimating the production of wood chips from the surrounding woodshed. Determining sustainable yield involves three main components: the biological yield of the woodlands; the maximum economic hauling distance; and the harvesting behavior of landowners within the woodshed.

Biological Yield

For this case study, we defined biomass that is available for fuel as the tops and limbs of all sawtimber trees as well as all rough culls and salvable dead trees. We made this assumption for three reasons. First, since there is a large difference in value between sawtimber and chips, few boles large enough to be classified as sawtimber would be chipped for fuel. Second, short-rotation biomass crops do not provide optimal returns in oak and hickory forests because the present value of a potential sawtimber crop always exceeds its value as fuel (Larson 1986). Also, early thinnings for biomass in this complex forest type are uneconomical because it costs too much to select and mark trees for harvest. Third, logging operators in northern New England report that any harvest operation involving only chipping is unlikely to be profitable (Huyler 1989). If conditions in southern New England prove similar to those farther north, it is unlikely that low thinnings of mature stands will produce substantial quantities of silvicultural fuel. Thus, the majority of wood chips must be a byproduct of harvesting sawtimber, where the material which would otherwise be left in the stand is brought to the landing and chipped. This scenario would yield the maximum amount of wood chips. In reality, however, green trees and snags may be retained for wildlife or aesthetics in many silvicultural operations, and the yield of wood chips would be reduced.

To determine the biologically sustainable yield of wood chips within the potential woodshed area, we must also know the rate of production. Unfortunately, reliable yield tables for biomass are unavailable in southern New England. Estimating biomass yields using a modeling approach is difficult because typical stands may contain over 20 overstory species. However, a regional imbalance in stand classes allows a semiempirical approach. The forest stands across the potential woodshed are generally stratified, even-aged stands that date from harvests of old-field white pine stands and abandonment of fields and pastures in the early part of the century and from a major hurricane in 1938. We used current standing volume/acre, based on U.S. Forest Service inventory data, as a proxy for site quality in each county (Dickson and Bowers 1976, Peters and Bowers 1977a, 1977b). The expected yield for typical stands in Windham County, Connecticut, is 50 green tons/acre of tops, limbs, culls, and salvable dead trees after an 80-year rotation. Annual yield/acre of timberland in each county was estimated as:

$$Y_i = (Y_w V_i) / (R V_w)$$

where Y is yield/acre of timberland, V is current standing volume/acre, and R is rotation length (80 years). The subscripts i and w denote respectively the county of interest and Windham County. Use of this formula involves two major assumptions: first, that the distributions of age within the stands are similar among the counties, as discussed above, and second, that yield is a linear function with respect to site quality (e.g. MacKinney and Chaiken 1939).

We converted green weight to dry weight using an estimated moisture content of 0.57, based on species values in the U.S. Forest Products Laboratory (1987). Carbon content of wood was calculated as 49% of dry weight (Haygreen and Bowyer 1989).

Hauling Distance and Woodshed Area

The network of roads in southern New England does not provide equal ease of transportation in all directions. Local landforms—and, therefore, roads—tend to be oriented north to south. Several rivers, bridged only at a few points, also obstruct transportation. A simple, circular woodshed, as a consequence, would be inappropriate for the case study. We estimated the area of woodshed within hauling distance in each county by using a grid overlay with 5-mile (8-kilometer) increments. We used a linear relationship between maximum hauling distance and harvest and transportation costs:

$$P = H + Ts$$

where P is the price at the Killingly plant ($17.32/green ton), H is the cost of harvesting from stump to roadside, T is the transport cost/ton mile, and s is the road distance to Killingly in miles. While a more complex expression for transportation cost might be desirable, the data are insufficient to support a more intricate model in southern New England. We selected an upper bound of 65 miles to reflect anticipated competition from other generating facilities and pulp markets to the north and west of the proposed site. We converted harvesting targets to a percentage of maximum biologically sustainable yield in the woodshed as a function of harvest and transport costs to provide a basis for comparison with present and anticipated landowner harvesting behavior.

Effects of the Killingly Plant on Atmospheric Carbon Dioxide

We calculated the effect of the Killingly electrical power generating plant on atmospheric CO_2 over its 50-year life by using a static accounting approach that is subject to a number of constraints and assumptions. We assumed that the limited increase in stumpage price for hardwoods because of the wood chip market would have no effect on rates of sawtimber harvest. While increased rates of decomposition following a harvest might result in substantial loss of carbon stored in the forest soils (Birdsey, in press), we assumed that the difference in carbon loss in the soil following harvesting of only sawtimber and harvesting of both sawtimber and biomass would be insignificant. Based on the data of Wharton and Raile (1984), approximately 20% of the harvested biomass would come from living trees in the pole and fuelwood size classes, whose carbon was assumed to have remained in either the above- or below-ground carbon sinks of the forest. However, these suppressed trees are typically incapable of response, or respond only slowly, after their release from competition (Marquis 1981a, 1981b). Therefore, we assumed no change in forest growth following their harvest.

We calculated emissions from alternate fossil fuel sources for a power plant of the same size as the Killingly facility by using the results of Marland (1982), as recalculated by Nordhaus and Yohe (1983). Since landfill space is increasingly scarce in southern New England, we assumed that wastewood consumed under the biomass scenario would be incinerated or left to decompose under the fossil fuel scenario.

RESULTS

Portions of 18 counties in three states lie within the 65-mile maximum radius of woodshed area (table 9.1). Maximum sustainable harvest as a function of size of the woodshed is shown in figure 9.2. A decreased rate of growth in biomass supply begins at 45 miles from the Killingly plant, as the boundary of the woodshed approaches Long Island Sound on the south and the Boston metropolitan area on the east. The nature of the road network has a severe effect on the pattern of supply. Assuming a circular woodshed would overestimate the area by 44% to 80%.

Table 9.1. Characteristics of Counties within the Killingly woodshed

County	Area (acres x 10^3)	Median Distance (miles)	Maximum Annual Sustained Yield (Green Tons)
Hartford, CT	472.8	57	128,000
Middlesex, CT	237.8	54	96,000
New Haven, CT	386.8	>65	112,000
New London, CT	427	33	157,000
Tolland, CT	266.1	34	104,000
Windham, CT	330.1	14	137,000
Bristol, RI	15.9	41	1,700
Kent, RI	110.4	23	30,000
Newport, RI	73.6	46	10,000
Providence, RI	265.9	20	73,000
Washington, RI	205.6	33	56,000
Bristol, MA	354.6	45	110,000
Hampden, MA	397.8	56	153,000
Hampshire, MA	338.8	>65	144,000
Middlesex, MA	528.1	>65	117,000
Norfolk, MA	252.2	51	55,000
Plymouth, MA	418.7	>65	135,000
Worcester, MA	968.5	49	362,000

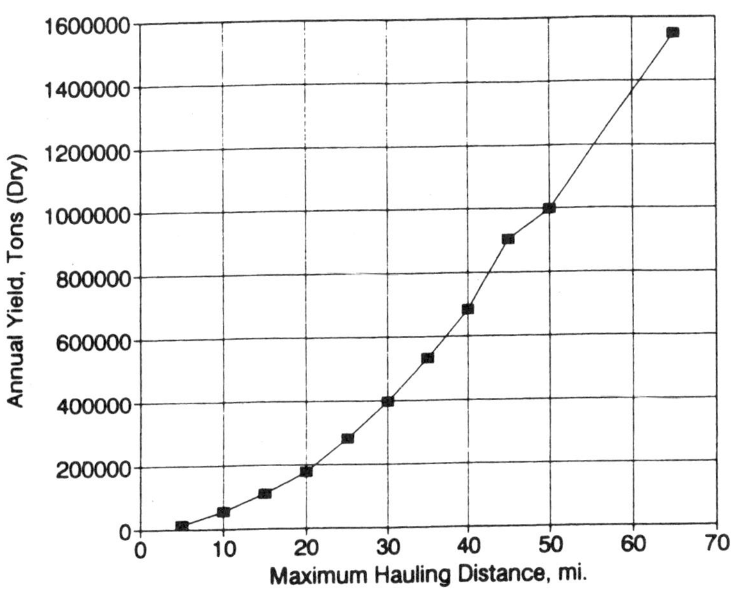

Figure 9.2. Maximum Annual Sustained Yield of Biomass vs. Maximum Hauling Distance for the Woodshed.

The percentage of maximum sustainable harvest required to meet Killingly's upper target of 30% of its fuel from silvicultural chips is shown as a function of harvest cost and transport cost in figure 9.3. In figure 9.3, demand exceeds biological supply only at high levels of both harvest and transport cost, above and to the right of 100%. However, demand exceeds 25% of biological supply over much of the range. The percentage required to meet the lower target of 10% of fuel from biomass is shown in figure 9.4. Here, demand is always less than biological supply, generally less than 15%.

Contributions to atmospheric CO_2 for the scenario of 10% of Killingly's fuel from silvicultural wood are shown in figure 9.5. The contributions include fuel combustion, decomposition of unused woodly material, and disposal of demolition wood and other wastes. While stack emissions are always higher for the biomass case, reduced losses from decomposition and waste disposal lead to a net benefit relative to fossil-fuel combustion. Depending on which fossil fuel is replaced, the strategy of substituting biomass for fossil fuel has a marginal benefit of 10,900 to 42,000 tons of carbon/yr. The benefit is similar in magnitude but slightly less than the fossil carbon consumed in each scenario.

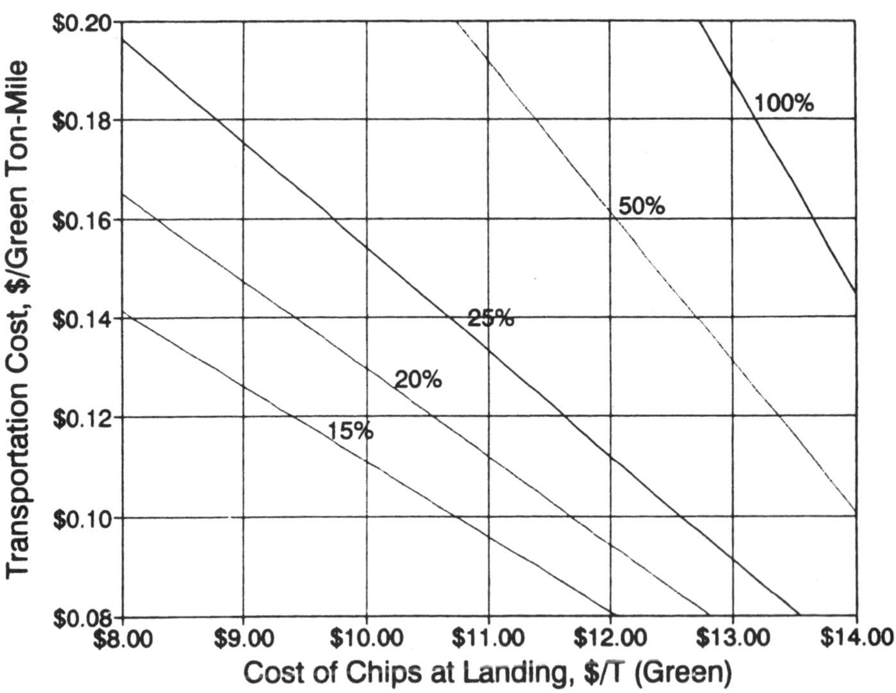

Figure 9.3. Percent of Maximum Sustained Yield Required to Meet Killingly's Lower Objective of 30% silvicultural Wood, as a Function of Hauling and Transportation Costs.

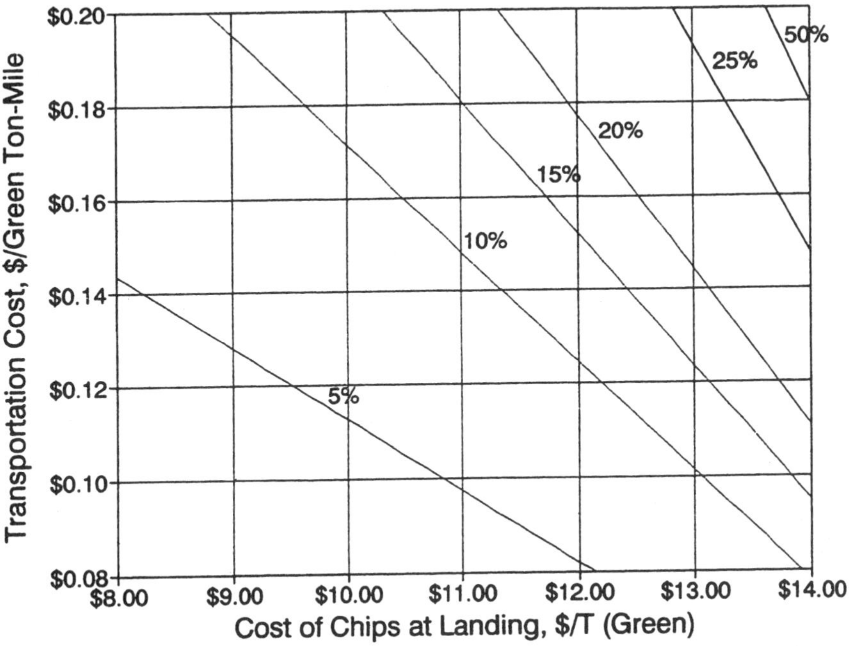

Figure 9.4. Percent of Maximum Sustained Yield Required to Meet Killingly's Lower Objective of 10% silvicultural Wood, as a Function of Hauling and Transportation Costs.

DISCUSSION

The results of the woodshed analysis indicate that southern New England is biologically capable of supporting the Killingly plant's demand for silvicultural wood chips. However, a comparison of the required levels of harvesting with the behavior of landowners suggests that the potential for electricity from biomass in this region may be limited by social factors. Ninety percent of the forest land in southern New England is distributed among 184,000 private owners with diverse values and attitudes toward harvesting. Approximately 45% of the land is owned by those who are opposed to harvesting of any sort; another 25%, by owners who employ silvicultural systems, such as uneven-aged management, which do not lend themselves to chipping operations (Kingsley 1976). Therefore, 30% of the biologically sustainable yield should be considered the absolute maximum supply of chips.

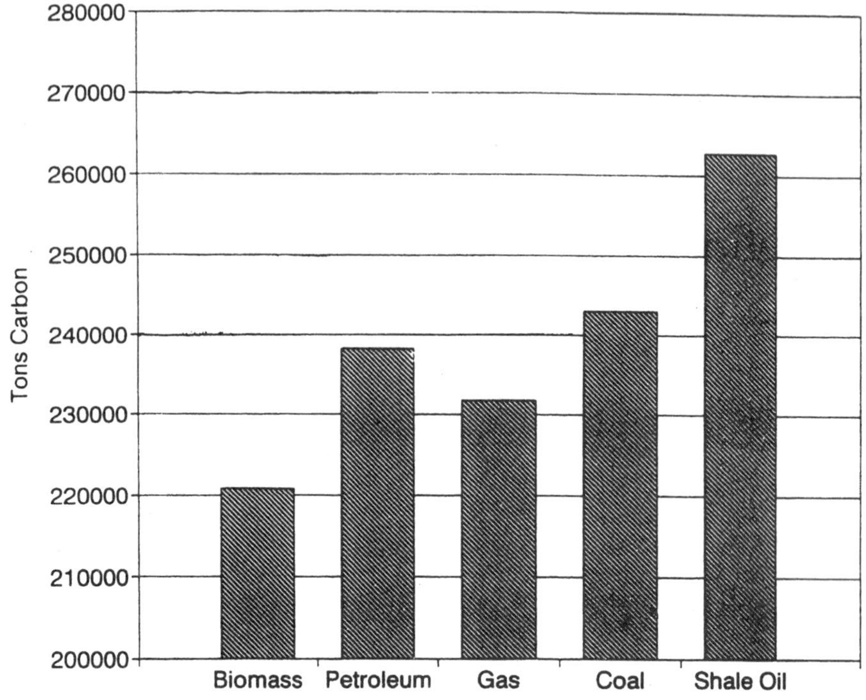

Figure 9.5. Carbon Contributions to the Atmosphere by Pathway for Biomass and Fossil Fuel Scenarios.

Actual rates of harvesting may be somewhat lower. Between 1971 and 1984, only 25% of the annual increment of sawtimber was harvested in Connecticut (Dickson and McAfee 1985). Similarly, Lindsay and Gilbert (1991) report that less than 25% of forest owners in Vermont are willing to sell in current fuelwood markets. Given reasonable harvesting costs of between \$9 and \$13/green ton (Huyler 1989) and transport costs of \$0.10 to \$0.14/ton mile (Young et al. 1988, 1991), Killingly's optimal goal of meeting 30% of its demand for fuel by using silvicultural chips would be difficult to attain. The lower goal of 10%, however, appears reasonable.

It is important to ask how any change from present conditions might influence the supply of wood chips. Over short time scales, this supply may be somewhat unstable. Killingly's silvicultural supply depends directly on the harvest of sawtimber, which varies markedly with changes in the regional and national economy (Neuel and Wharton 1988). While waste and demolition wood sources may compensate for lost silvicultural supply, construction and development may be positively rather than negatively correlated with changes in sawtimber harvest. If so, then the cost of fuel could fluctuate substantially.

Over longer time scales, changes in costs of harvesting and transportation costs may have a significant effect. With increasing affluence and urbanization, changes in the behavior of landowners, as discussed by Binkley

(1981), could threaten the silvicultural portion of Killingly's fuel supply. In addition, as Nyland (1992) points out, much of the harvest of sawtimber in New England has consisted of unsustainable cutting, with limits on the diameters of trees to be cut, in even-aged stands. This practice could degrade both the biological productivity of the resource and the economic incentives for harvesting sawtimber in the region. Since the supply of wood chips from silviculture is closely linked with the harvest of sawtimber, the effect on the supply of wood chips could be significant.

Results of the CO_2 analysis suggest that the marginal benefit of using wood fuel in the Killingly plant is caused by the utilization of carbon which would otherwise decompose or be incinerated. These results represent a "least benefit" case, because there is little or no opportunity to increase afforestation or rates of forest growth. In addition, since early thinnings for biomass are uneconomical in the oak-hickory forest type (Larson 1986), there is little opportunity to shift biomass production toward sawlogs and other high-quality products with long lifetimes. Thus, biomass harvests have little effect on the "wood-in-use" sinks for carbon.

In cases where shorter rotation lengths are possible, however, it might be economically advantageous to grow biomass crops to generate electrical power either by afforesting marginal croplands or as part of an integrated agroforestry strategy. An aboveground carbon sink would be created. While this sink would be transitory at the stand level, appearing and disappearing with each rotation, an equilibrium of carbon stored in woody biomass would develop at the landscape level. Biomass crops would reduce erosion, protecting inorganic carbon in the soil. The effects on organic soil carbon would depend on decomposition rates. In addition to its direct economic benefits, production of biomass in a context of agroforestry would ameliorate CO_2 by providing a substitute for fossil fuels, creating a landscape-scale carbon sink, and reducing soil carbon losses. These factors suggest that substituting biomass for fossil fuels can provide a number of benefits in a context of rural development while reducing atmospheric CO_2.

CONCLUSIONS

The Killingly case study yields several insights about using biomass energy as a strategy to reduce atmospheric CO_2. Our analysis of the woodshed for the Killingly plant suggests that social factors, including ownership patterns of forest land and their effect on harvesting behavior, form a critical element in determining both the sustainability of the resource and the strategy to reduce CO_2. Both short-term economic fluctuations and long-term social and biological trends may have a strong influence on biomass supply and should be considered during the planning phase of any similar development. The Killingly plant offers a limited, marginal benefit by capturing carbon that would otherwise decompose or be incinerated, and substituting it for fossil fuels. In an integrated strategy for rural development, however, biomass energy could offer greater rewards. Afforestation using short-rotation biomass crops would provide additional benefits by creating an additional terrestrial carbon sink and by protecting soil carbon. Biomass energy represents both an asset for rural development and a potential weapon in the struggle against global warming.

Literature Cited

Binkley, C. S. 1981. *Timber supply from private nonindustrial forests*. Yale University School of Forestry and Environmental Studies Bulletin No. 92.

Birdsey, R. In press. Changes in forest carbon storage from increasing forest area under timber growth. In *Forests and global warming*. (R.N. Sampson and D. Hair, eds.). Washington, D.C.: American Forestry Association.

Bormann, F.H., K.R. Smith, and B.T. Bormann. 1991. Earth to hearth: a microcomputer model for comparing biofuel systems. In *Biomass and bioenergy*. 1(1):17–34.

Dickson, D. R. and T. M. Bowers. 1976. Forest statistics for Connecticut. U.S.D.A. Forest Serv. Northeast For. Exp. Sta., Resource Bull. NE–44.

Dickson, D. R. and C. L. McAfee. 1986. Forest statistics for Connecticut—1972 to 1985. U.S.D.A. Forest Serv. Northeast For. Exp. Sta., Resource Bull. NE–103.

Eyre, F. H. (ed.). 1980. *Forest cover types of the United States and Canada*. Washington, D.C.: Society of American Foresters.

Gammon, R. H., E. T. Sundquist, and P. J. Fraser. 1985. History of carbon dioxide in the atmosphere. In *Atmospheric carbon dioxide and the carbon cycle*. (J. R. Trabalka, ed.). Washington, D.C.: U. S. Dept. of Energy, Office of Energy Research. DOE/ER-0239. pp 26-62.

Graham, R.L., M.G. Turner, and V.H. Dale. 1990. How increasing CO_2 and climate change affect forests. In *Bioscience* 40:575–587.

Haygreen, J. G. and J. L. Bowyer. 1989. *Forest products and wood science: an introduction*. Ames: Iowa State University Press.

Huyler, K. K. 1989. Fuel supply structure of wood-fired power plants in the Northeast: a loggers' perspective. U.S.D.A. Forest Serv. Northeastern For. Exp. Sta., Research Paper NE–624.

Killingly Energy Limited Partnership (KELP). 1988. Application of the KELP to the Connecticut siting council for a certificate of environmental compatibility and public need for the construction of a 32.2 MW (Net) wood-burning electric generating facility. ERL Project No. 6322-C11–87.

Killingly Energy Limited Partnership. 1990. *Energy and environment*, 2d ed.

Kingsley, N. P. 1976. The forest-land owners of southern New England. U.S.D.A. Forest Serv. Northeast For. Exp. Sta., Resource Bull. NE–41.

Larson, B. C. 1986. Extensive coppice production in natural upland hardwood stands of the southeast. Final Report. U.S. Department. Contract #18–873.

Lindsay, J. J. and A. H. Gilbert. 1991. The availability of fuelwood from Vermont's nonindustrial private forest lands (NIPF). *Northern J. Appl. For.* 8:57–59.

MacKinney, A. L. and L. E. Chaiken. 1939. Volume, yield, and growth of loblolly pine in the mid-Atlantic coastal region. U.S.D.A. Forest Serv. Appalachian For. Exp. Sta., Tech. Note 33.

Marland, G. 1982. The impact of synthetic fuels on global carbon dioxide emissions. In *The Carbon Dioxide Review: 1982.* (W. Clark, ed.). New York: Oxford University Press. pp 406-410.

Marquis, D. A. 1981a. Removal or retention of unmerchantable saplings in Allegheny hardwoods: effect of regeneration after clearcutting. In *J. Forestry* 79:280–283.

Marquis, D. A. 1981b. Survival, growth, and quality of residual trees following clearcutting in Allegheny hardwood stands. U.S.D.A. Forest Serv. Northeast For. Exp. Sta., Res. Paper NE-477.

Neuel, R. L. Jr. and E. H. Wharton. 1988. The timber industries of southern New England—a periodic assessment of timber output. U.S.D.A. Forest Serv. Northeast For. Exp. Sta., Resource Bull. NE-101.

Nordhaus, W. D. and G. W. Yohe. 1983. Future paths of energy and carbon dioxide emissions. In *Changing Climate*. (National Resource Council, ed.). Washington, D.C.: National Academy Press. pp 87-152.

Nyland, R. D. 1992. Exploitation and greed in eastern hardwood forests. In *J. Forestry* 90(1):33–37.

Peters, J. R. and T. M. Bowers. 1977a. Forest statistics for Massachusetts. U.S.D.A. Forest Serv. Northeastern For. Exp. Sta., Resource Bull. NE-48.

Peters, J. R. and T. M. Bowers. 977b. Forest statistics for Rhode Island. U.S.D.A. Forest Serv. Northeastern For. Exp. Sta., Resource Bull. NE-49.

Sedjo, R.A. and A.M. Solomon. 1989. Climate and forests. In *Greenhouse warming: abatement and adaptation.* (N.J. Roseberg, W.E. Easterling, P.R. Crosson and J. Darmstadter, eds.). Washington, D.C.: Resources for the Future. pp 105-119.

Trabalka, J. R., J. A. Edmonds, J. Reilly, R. H. Gardner and L. D. Voorhees. 1985. Human alterations of the global carbon cycle and the projected future. In *Atmospheric carbon dioxide and the carbon cycle.* (J.R. Trabalka, ed.). DOE/ER-0239. Washington, D.C.: U.S. Department of Energy, Office of Energy Research. pp 247-287

U.S. Forest Products Laboratory (USFPL). 1987. Wood handbook: wood as an engineering material. U.S.D.A. Forest Serv. Agric. Handb. 72.

Vitousek, P.M. 1991. Can planted forests counteract increasing atmospheric carbon dioxide? In *J. Environ. Qual.* 20:348–354.

Wharton, E. H. and G. K. Raile. 1984. Biomass statistics for the northern United States. U.S.D.A. Forest Serv. Northeast For. Exp. Sta., Research Note NE-318.

Woodwell, G.M. 1987. Forests and climate: surprises in store. In *Oceanus*. 29:71–75.

Wright, L.L, R.L. Graham, A.F. Turhollow, and B.C. English. In press. Opportunities for short-rotation woody crops and the potential impacts on carbon conservation. In *Forests and*

Global Warming. (R.N. Sampson and D. Hair, eds.). Washington, D.C.: American Forestry Association.

Young, T. M., D. M. Ostermeier, and D. W. Domenech. 1988. Average wood chip trucking costs for the southeastern United States. *Southern J. Appl. For.* 12:267–269.

Young, T. M., D. M. Ostermeier, J. D. Thomas, and R. T. Brooks Jr. 1991. Computer model simulates supply, cost of chips. In *Forest Industries* 118(8):20–21.

Part Three

Managing the Tropical Forest: Can It Be Done?

INTRODUCTION

Future use of wood energy revolves around our capacity to construct and enforce sustainable forest management and use policies. Over the past decade, scientists and policy makers have come to better appreciate the intricacies involved in creating truly sustainable tropical forest management schemes. The biological relationships that will prevent or foster long term multiple use of tropical forests are discussed at length in this section. Nutrient balances, ecosystem relationships, and mixed uses of various types of forests are presented. A primary finding of past and present research into sustainable management of tropical forests is that we must proceed very slowly in continuing use of these resources. Our limited knowledge on how these ecosystems—the forest, soils, fauna, and hydrological systems—actually relate and replenish themselves should make us cautious. Multiple use forestry, while perhaps the long run goal in tropical forest management, cannot be construed to mean wood energy is always an appropriate component of a local economy's base. The papers in this section show that we need to move towards forestry use systems that biologically mimic or are based on the historic indigenous species and local use patterns, and wood-based energy may not fit.

More intensive management of some forest lands for timber production would relieve the biophysical and economic impacts on tropical forests. This is especially true of opportunities for forests growing high-quality timber and fiber plantations. Many of these opportunities are in the temperate conifer regions. Consequently, this spatial relationship cannot be ignored in formulating tropical resource policy solutions. The global and regional relationships among timber supplies and wood or fiber consumers, as a consequence, interact with ecological relationships at the microsite and landscape levels.

CHAPTER 10

NUTRIENT BUDGETS OF YOUNG PLANTATIONS WITH NATIVE TREES: STRATEGIES FOR SUSTAINED MANAGEMENT

Florencia Montagnini
Freddy Sancho

INTRODUCTION

Our research focuses on a specific area in the humid lowland tropics of Central America in Costa Rica where banana plantations are predominant. Deforestation and the consequent degradation of land from mismanagement—shifting agriculture and cattle grazing—are common problems. Unpalatable native grasses and ferns reclaim abandoned land. A 4000 mm/yr rainfall hastens erosion of soil that has been depleted of nutrients from repeated harvesting.

Restoration of the soil to its former productivity and ecological balance has become a predominant concern of the government. There are local low- to medium-income farmers who are willing to risk some part of their land to raise trees, for they consider trees as a form of savings or investment under conditions of high inflation. They may reserve 10% to 15% of their land for trees as a future cash value. Regional forest services encourage farmers to plant exotic species—*Eucalyptus deglupta*, *Gmelina arborea*, and *Cordia alliodora*. Fungus usually infects the eucalyptus, and *Cordia alliodora* is a 30-year rotation species that grows on fertile soil, but it is not suitable for degraded lands. Such governmental policies need to be reconsidered. Local people should have a role in policy-making to insure appropriate choices of land use, species, and management, for they have specific knowledge about conditions in which they live.

We advocate a strategy of gathering criteria about each local region: the economic value of the regional species; the availability of seeds or seedlings; the growth rate data of regional species; and the indications of species' effects on soil nutrients. We have found that certain native tree species have a potential to grow well on poor, degraded soils. These trees are fast-growing and contribute organic matter to the soil within three years. Their litter and small branches may be managed to obtain nutrients: the litter may be used as mulch for agriculture (for maize crops) or left on the forest floor to return nutrients to the surface layers of degraded soils as well as prevent soil erosion. Our discussion describes these species and their combined nutrient recycling capacities.

BENEFITS OF TREE PLANTATIONS

Currently, a number of governments and private efforts in rural development throughout the tropics are promoting tree planting, including tree-crop combinations as practiced in agroforestry. Trees are regarded as a source of cash, savings, and assets for the rural poor (Chambers and Leach 1990). Since tropical plantations managed for high yields can be at least four to ten times more productive than unmanaged natural forests (Wadsworth 1983), they also can help to meet the growing global demand for timber. This type of management is expected to help decrease the pressure on natural forests (Evans 1987).

The sustained management of tree plantations becomes a biologically and socially feasible alternative on soils that are unsuitable for the continuous practicing of agriculture that uses prevailing local technologies (Gladstone and Ledig 1990). Particularly, tree plantations and tree-crop combinations represent productive alternatives for the uses of deforested lands that have poor regeneration of natural forests since they are either intensely degraded or distant from sources of propagules. Low soil fertility, soil compaction from cattle grazing, and invasion by grasses and other aggressive vegetation—all evidences of site degradation—can be serious obstacles to both forest regeneration and conventional agriculture. As the area of degradation expands, there is increasing emphasis on the planting of tree species that can grow in such poor conditions to yield potentially profitable products (timber, fuelwood and other) as well as environmental benefits (soil conservation, watershed protection) (Evans 1987).

The choice of appropriate tree species for plantation forestry or agroforestry is influenced by knowledge of a species' performance as well as by economic and environmental benefits. In local situations, the choice of a tree species is determined by the availability of its seeds or seedlings and by information on its silvicultural characteristics and management—for instance, fast growth and the possibility of intercropping during early establishment. Most reforestation or tree-planting programs and subsidies promote the use of well-known, often exotic, species. About 85% of plantation forestry in the tropics is dominated by three genera: *Pinus*, *Eucalyptus*, and *Tectona* (Evans 1987), while there may be thousands of indigenous species suitable for similar purposes. Native trees can be more appropriate than exotics because they are better adapted to local environmental conditions, seeds and propagules are locally available, and farmers are familiar with them and their uses. In addition, the use of indigenous trees in productive systems helps to preserve genetic diversity and fosters a better balance with the local flora and fauna than exotics.

Ecological Consequences of Fast-Growing Tree Plantations

The factors that influence the productivity of tropical forest plantations are poorly understood. Some studies report yield or biomass productivity in relation to climatic factors, but few refer to site characteristics such as elevation or soil type (Lugo et al. 1988). Trees can influence site characteristics through their recycling of nutrients and their interactions with the microenvironment.

The most important beneficial effects of trees on soils can include improvement of soil structure and increased availability of nutrients (Fassbender 1984, Nair 1989, Sanchez et al. 1985, Sanchez 1987). Symbiotic nitrogen fixation by trees often results in increased soil nitrogen availability (Alpizar et al. 1986, Montagnini et al. 1986, Dommergues 1987). On the other hand, young tropical tree plantations which incorporate considerable amounts of nutrients in their biomass over a relatively short period of time are rapidly aggrading ecosystems (Bruijnzeel 1991). In early stages of growth, the amount of nutrient uptake from the soil usually exceeds the amount of nutrients supplied to the soil by litterfall and wash from the canopy (Bruijnzeel 1991).

Decline of site fertility can be a serious limitation to sustained plantation forestry in tropical regions: soil fertility can be decreased through excessive removal of living biomass, particularly if nutrients in tree crowns are lost through harvesting or site preparation (Perry and Maghembe 1989). On the other hand, Wadsworth (1983) suggests that, with the possible exception of phosphorus, repeated harvests usually would not result in serious deficiencies of soil nutrients.

Lundgren (1980) proposed that the ameliorating effects of plantation forests on soils occur only during the five-to-ten-year period immediately following canopy closure (the "fallow enrichment phase"). During the maximum-production phase, actual deterioration of site quality can occur: mineral nutrients are absorbed by the trees while litter accumulates on the forest floor, but conditions are unfavorable for the decomposition of organic matter (Lundgren 1980). Sanchez et al. (1985) concluded that deleterious effects on soils occur only during plantation establishment, but they also stressed that the extraction of nutrients by harvesting and the leaching losses incurred prior to canopy closure lead to a depletion of key nutrients, particularly potassium, that should be replaced if yields are to be sustained on subsequent rotations.

Tree species vary in their rates of nutrient uptake and in their capacity for nutrient recycling. The possibility of using particular species to accumulate nutrients was suggested by Sanchez et al. (1985) from their observations that certain species have the ability (e.g., *Gmelina arborea*) to accumulate calcium and magnesium, while others favor potassium and phosphorus. Little data is available on second and third rotations, so we do not have enough clues about what the critical nutrients for maintaining site productivity are. Data on different plantation species' rates of nutrient acquisition and recycling capabilities will help to design the best management strategies that either can take advantage of the ameliorating effects of trees on site fertility or avoid site deterioration at harvest.

Assessing the Impacts of Trees on Site Nutrients: An Example from Costa Rica

Local and regional trials of tree species for reforestation often reveal outstanding performances from indigenous trees. For example, out of 13 native tree species in an experimental plantation at La Selva Biological Station of the Organization for Tropical Studies (OTS) in the Atlantic lowlands of Costa Rica, Central America, at least four—*Stryphnodendron excelsum*, *Vochysia hondurensis*, *Vochysia ferruginea*, and *Hyeronima alchorneoides*—had growth rates equal to or greater than the exotic species that were currently recommended for the region (Espinoza and Butterfield 1989). This research demonstrates the potential of many native trees for commercial purposes. Most interestingly, it shows how certain indigenous species are able to grow well on degraded sites with poor, acid soils that could not sustain conventional agriculture, and how some of the species may have positive effects on soil fertility at an early stage (Montagnini and Sancho 1990a, 1990b, Montagnini et al. 1991). Results of our studies at the same site showed that after 2.5 years, these species contributed to the restoration of soil fertility by increasing organic matter, nitrogen, and soil cation levels to values close to those considered adequate for agricultural crops (Montagnini and Sancho 1990a, 1990b).

In the following sections, we present results from our studies of nutrient budgets in the same experimental plantation as an example of how biomass measurements, coupled with information on tissue nutrients, can be used to understand the potential environmental impacts of trees on site nutrients. We compare biomass and nutrient content among the species as well as among the trees' above-ground parts, forest-floor litter, understory vegetation, and soil nutrient pools. This information can be used to design management strategies to take advantage of the trees' ameliorating effects on soils and to avoid depletion of site nutrients at harvest. These strategies should be valuable for promoting the use of systems—mixed or pure plantations, agroforestry—that include these fast-growing timber species in the area and in other tropical lowland regions with similar ecological characteristics.

THE STUDY SITE

The experimental plantation was established in December, 1985, on abandoned pasture at the OTS La Selva Biological Station (10°26'N, 86°59'W, 50 meters mean elevation, 24°C mean annual temperature, 4000 mm mean annual rainfall, with maximum in July and minimum in March) (La Selva Biological Station weather reports). Soils are Fluventic Dystropepts, derived from alluvially deposited volcanic materials; they are deep, well drained, and free of stones, with low or medium organic-matter content, moderately heavy textured, and generally acid and infertile (Sancho and Mata 1987). The area had been cleared in the 1950s, and cattle had grazed there until 1984. The dominant species in the pastures were grasses (*Olyra latifolia*, *Melinis minutiflora*), ferns (*Pteridium* sp.), and bushes (*Psidium guajava* and *Piper culebranum*). In the pastures, there were patches of approximately 20-year-old adjacent forest with: *Pentachlethra macroloba*, a mimosoid, nitrogen-fixing legume dominant in the primary forest at La Selva; *Piper culebranum* and species of the elastomataceae family; ferns (*Pteridium* spp.); and tree seedlings in the understory.

The site was cleaned manually before planting. The tree species were planted randomly in a block design with five replicates, each plot containing seven rows of seven trees (14 m x 14 m each), with two meters between trees. Five similar 14 m x 14 m plots were also established in an adjacent open area with grass and in a nearby patch of secondary forest. During the first year, weeds were cut manually four times. Thereafter, weeding was done mechanically until canopy closure made it no longer necessary. The grass was also similarly weeded to keep it free of trees at the same time.

The Tree Species

The criteria for species selection for this study were good growth, as determined during the first three to four years of the plantation (Espinoza Camacho and Butterfield 1989, Gonzalez et al. 1990), presence of root nodules in the leguminous species (field observations), and economic value (Gonzalez et al. 1990, Chudnoff 1984, Holdridge and Poveda 1975). *Stryphnodendron excelsum* Harms (Leguminosae, subfamily Mimosoideae) ("vainillo") is found only in Costa Rica, although representatives of this genus are native to all of tropical South America, i.e. Brazil, Costa Rica, Guiana (Allen and Allen 1981). This species grows in low elevations with very humid climates and apparently adapts to alluvial soils as well as to slopes and abandoned pastures with degraded soils (Gonzalez et al. 1990). Its timber is primarily used for general construction, and, because of its ability to take high polish, small furniture and turnery (Allen and Allen 1981). Its fruits are eaten by many species of wildlife, mostly small mammals.

Vochysia ferruginea Mart (Vochysiaceae) ("botarrama") grows in lowland forests from Nicaragua to Brazil (Whitmore and Hartshorn 1969). It is found usually on acidic, well drained, and low-fertility soils, although it can adapt to a variety of soils (Gonzalez et al. 1990). It is a self-pruning pioneer species that forms uniform, even-aged stands in abandoned fields, and its wood is used for plywood and construction.

Vochysia hondurensis Sprague (Vochysiaceae) ("mayo") is found from Mexico to Panama, at elevations up to 900 m (Whitmore and Hartshorn 1969). It usually grows in humid, low altitude areas on either alluvial or residual (less fertile) soils. Considered a substitute for mahogany, its timber is highly desirable for carpentry, plywood, and furniture.

H. alchorneoides (O) (Euphorbiaceae) ("pilon") ranges from southern Mexico to southern Brazil (Chudnoff 1984). This species grows well on hills and on abandoned grasses, but not much is known about its edaphic requirements. Its timber is used for heavy construction, furniture, cabinet work, decorative veneers, and turnery (Chudnoff 1984). Botanical characteristics of the species were described by Holdridge and Poveda (1975), Hartshorn (1983), Standley (1937-38) and Hartshorn and Hammel (unpublished). Detailed studies on seed and germination characteristics are presented by Gonzalez (1991).

METHODS

Sampling procedures and chemical methods are described in Montagnini and Sancho (1990a, 1990b), and Montagnini et al. (1991). The soils under the species, in a grassy area free of trees and in a 20-year-old secondary forest, were sampled for soil fertility, and their nitrogen availability was also measured. Tree biomass and nutrient content from stems, branches, and leaves were measured at the time of plot thinning when the plantation was four years old. The biomass and nutrient concentrations of the understory vegetation and the forest floor under the trees were also measured. Nutrient budgets were calculated by multiplying the biomass of each compartment times its nutrient concentration (nitrogen, calcium, magnesium, potassium, phosphorus).

RESULTS AND DISCUSSION

Tree Biomass

The values of whole-tree biomass reported here (table 10.1) are greater than of those reported for four-year-old *Albizia lebbek* (Parrota 1989) and of five-and-a-half-year-old *Leucaena leucocephala* (Wang et al. 1991), both growing in dense plantations for biomass production in Puerto Rico. Values shown here of the total tree above-ground net primary productivity (calculated by dividing whole-tree biomass by tree age) lie within the ranges reported elsewhere for monospecific plantations in the humid tropics. The value for *V. hondurensis* is close to that reported for *Gmelina arborea* (12.8 tons/ha/yr) in the Brazilian Amazon (Russell 1987) as well as to *Gmelina arborea* (12.7 tons/ha) and to *Albizia falcataria*, both found in the Philippines (11.3) (Kawajara et al. 1981). However, the increments shown here are lower than those reported for some of the fastest-growing trees in the humid tropics, such as *Acacia mangium* (15.5 to 18.0 tons/ha in Malaysia) and *Leucaena leucocephala* (20.0 to 30.0, and even up to 80.0 tons/ha in Hawaii and in other tropical sites, Young 1989).

Table 10.1. Means of Tree Diameter at Breast Height (dbh), Height, Aboveground Biomass and Annual Increment

	Dbh	Height	Above-ground live biomass				Mean annual increment	
			Stem	Branches	Leaves	Total	Total	Stems
	(cm)	(m)	(kg/ha)				(t ha -1 yr -1)	
S. exc.	12.0a	8.9b	35,250a	15,250a	4,325a	54,825	13.7	8.8
V. fer.	10.3a	8.1b	24,750b	14,250a	5,925a	44,925	11.2	6.2
V. hon.	10.8a	12.0a	41,750a	6,500b	7,250a	55,500	13.9	10.4
H. alc.	10.8a	9.0a	26,250b	12,250a	5,350a	43,850	12.0	6.5

Note: In this and in following tables, differences between sites for a given parameter are statistically significant when means are followed by different letters.

Annual increments of stemwood biomass for broadleafs range from 1 to 28 tons/ha/yr. Fast-growing species such as *Gmelina arborea* and *E. saligna* range from 10 to 20 and from 8 to 28 tons/ha, respectively, and relatively slower-growing species such as *Swietenia* sp. and *Tectona grandis* range from 1 to 4 and 3 to 12 tons/ha, respectively (Wadsworth 1983). Other values for fast-growing trees in tropical humid regions include various eucalyptus species grown in the Americas and in Asia (from 7.2 to 11.9 tons/ha); *Gmelina arborea* in Costa Rica (11.8 tons/ha) (Lugo et al. 1988); 1.3- to 5.3 *Leucaena leucocephala* growing in premontane and lowland humid sites (2.8 to 15.9 tons/ha); *Prosopis juliflora*, on a moist site in India (9.4 tons/ha), and *Populus deltoides*, on a subtropical site in India (6.4 tons/ha) (Lugo et al. 1990). Thus, the mean annual increments of stemwood biomass for the species in this study also fall within the ranges reported for other fast-growing tree species in the humid tropics.

Nutrient Accumulation in Above-Ground Tree Biomass

Nitrogen

Consistent with trends found in nitrogen (N) tissue concentrations (Montagnini et al. 1991), *S. excelsum* had the highest stem, branch, and total above-ground tree biomass N. Although *S. excelsum* had higher N-leaf concentrations than the other species, it had similar N-leaf biomass because its leaf biomass was relatively lower (figure 10.1). Taking leaves and branches together, approximately 200 kg/ha, or 60% of *S. excelsum*'s above-ground tree biomass

N (figure 10.1) would be left on the site at the time of harvest if only stems were removed. *V. hondurensis* had a similar proportion of N to that of *S. excelsum* in its leaves and branches portion. Again, over 50% of above-ground tree biomass N could be recycled if left at the site at the time of harvest. *V. ferruginea*, with relatively lower stem biomass, had proportionally more N in leaves (52.9%) and branches (42.1%), while *H. alchorneoides* showed a more even distribution of above-ground tree biomass N in stems (39%), branches (27%), and leaves (33.8%).

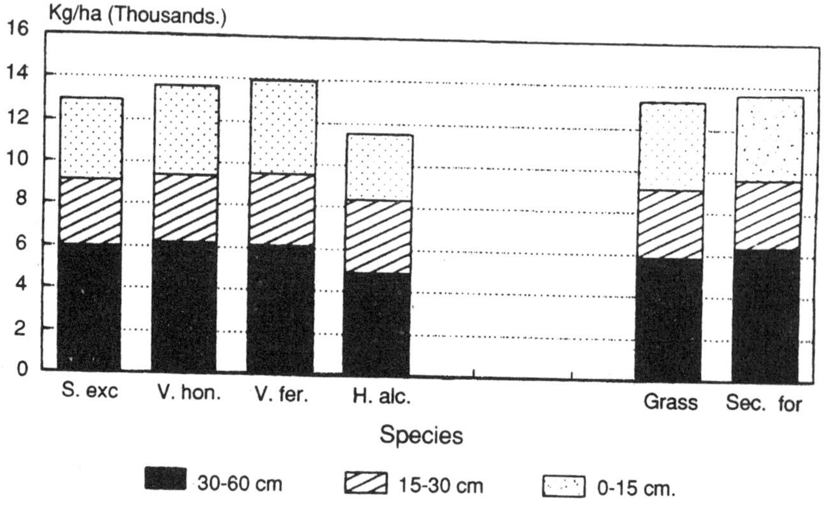

Figure 10.1. Soil Nutrient Balances - Nitrogen.

Calcium

V. hondurensis, with the highest stem biomass and calcium (Ca) concentration, also had the highest stemwood Ca (over 600 kg/ha, or 84% of above-ground tree biomass Ca), approximately twice as much as either *S. excelsum* or *V. ferruginea*, and several times more than *H. alchorneoides* (figure 10.2). Therefore, the harvest of *V. hondurensis* trees could substantially reduce the amount of Ca in the site. However, while the *V. hondurensis* trees are living, relatively large amounts of Ca could be recycled because, although it represented only 16% of above-ground tree biomass, the amount of Ca in leaves and branches together was over 100 kg/ha.

The proportion of Ca in stems relative to above-ground tree biomass was similar for *S. excelsum* and *V. ferruginea* (76.6% and 70.8% respectively) (figure 10.2), but the absolute amounts were less than half those of *V. hondurensis*. *H. alchorneoides* again showed a more evenly distributed quantity of Ca in stems, leaves, and branches, with above-ground tree biomass Ca slightly over half that of *S. excelsum* and *V. ferruginea*.

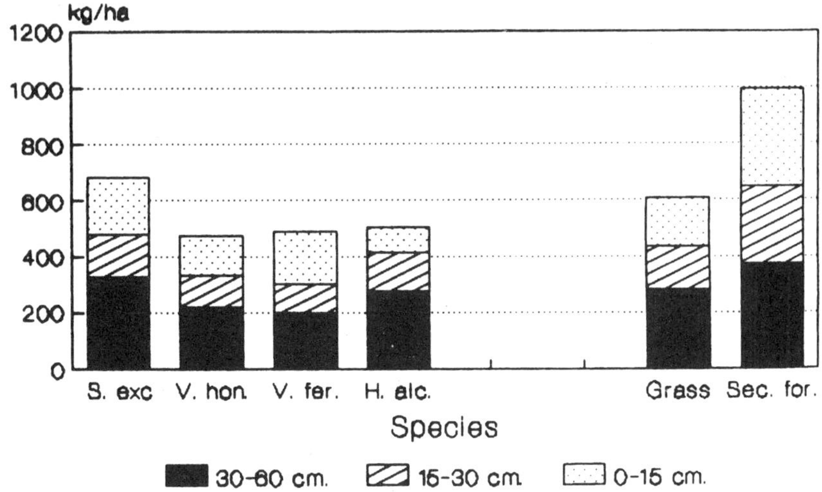

Figure 10.2. Soil Nutrient Balances - Calcium.

Magnesium

V. hondurensis, with its high stem biomass and high magnesium (Mg) concentration, also had the highest Mg in stemwood (55% of above-ground tree biomass, or approximately 30 kg/ha) (figure 10.3). Again, removal of *V. hondurensis* stems would affect the site's Mg budget more dramatically than for any of the other species, especially if the whole tree is harvested. The other species showed lower proportions of above-ground tree biomass Mg in their stems (30% to 46%) (figure 10.3).

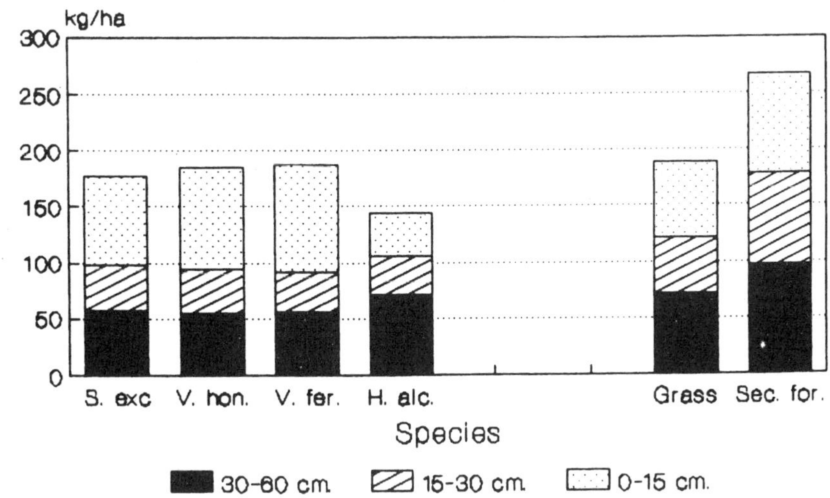

Figure 10.3. Soil Nutrient Balances - Magnesium.

Potassium

For potassium (K) the picture changed: the highest accumulation of K in stems was found in *H. alchorneoides* (252 kg/ha, figure 10.3), accounting for 58.7% of above-ground tree biomass K. This quantity was followed by *V. hondurensis* with 175 kg/ha, which represents 76.8% of above-ground tree biomass K. Thus, whole-tree harvest of *H. alchorneoides* and *V. hondurensis* may have the greatest effects on the K budget. *S. excelsum* and *V. ferruginea* had 33.6% and 35.4%, respectively, of K in the stem. The recycling of K from leaves and branches could be relatively more important when considering these latter two species.

Phosphorus

V. hondurensis and *H. alchorneoides* had the highest proportions of stemwood phosphorus (P) (72.4% and 62.1%, respectively) (figure 10.4). *S. excelsum* and *V. ferruginea* had relatively lower amounts of stem P, also correlating with proportions of above-ground tree biomass P (43.9% and 48.7% respectively).

Our results confirm earlier reports on the negative effects of the harvest of whole trees on site nutrient pools: for example, Bruijnzel and Wiersum (1985) studied nutrient input/output budgets for *Agathis dammara* plantations in upland Java. Their results, calculated for a thirty-year rotation, showed that total harvesting of the trees would remove the entire inputs of K and Ca, almost half of the Mg input, and twice the input of P. The authors concluded that, in order to avoid shortages of nutrients, especially P, harvesting of whole-tree biomass should not be practiced. Our results also suggest that leaving plant residues in the site could greatly decrease the negative impacts of nutrient removal at harvest, with varying consequences that would depend on the individual species.

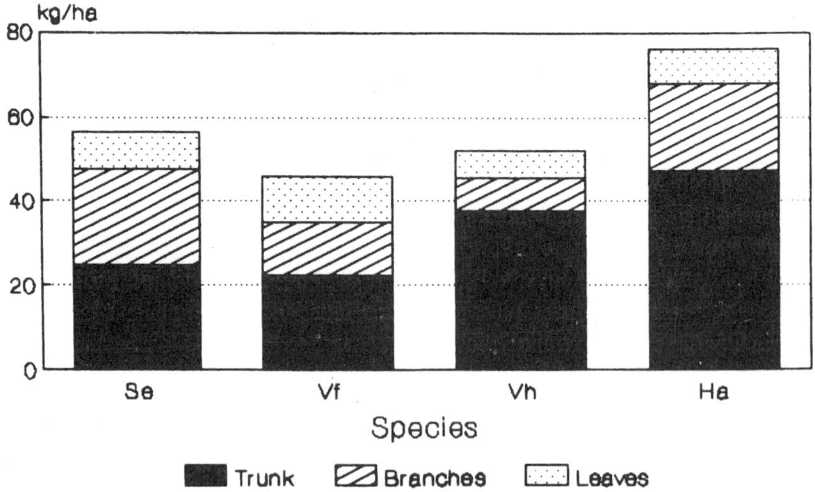

Figure 10.4. Nutrients in Tree Biomass - Phosphorus.

Altering the rate of nutrient removal from the site is probably one of the most important design considerations when planning sustainable plantations (Wang et al. 1991). The tree species as well as the parts of the tree removed will determine the nutrient "cost" of removal. This "cost" can be assessed in advance by nutrient and biomass sampling and estimation. By comparing the quantities of nutrients removed from a variety of species in numerous tropical sites, these authors argue that nutrient removal increases from 13% to 253%, while biomass removal increases less than 50%, when harvesting of whole-tree biomass, rather than harvesting only stems, is practiced. Additionally, they argue that shorter rotations will be more affected by this difference because the fractions of leaves and small branches of the total biomass tend to decrease with tree size. Bruijnzel and Wiersum (1985) found that stemwood P concentrations of 35-year-old *Agathis dammara* were lower than those of younger trees, whereas no such differences were found for other nutrients, and they suggest that the main effects of shorter rotations will be an increase in site disturbance.

Impacts of Trees on Soil Nutrients

Similar to the 1988 findings, higher levels of soil organic matter and N were found in the tree plantation than in grass, with values close to those of secondary forest (table 10.2), but unlike 1988, these differences were not statistically significant ($P<0.05$). P content was higher in grass than in the plantation or in the forest (table 10.2). Within the tree plantation, there were no significant differences among species in soil cation content. However, there was a trend of higher Ca under *S. excelsum* and lower Ca in *H. alchorneoides*. Higher Mg was recorded under the two *Vochysia* species, with lower values again in *H. alchorneoides* (table 10.2).

Results of similar measurements taken in May, 1990, and again in May, 1991, revealed trends like those found in 1988 and 1989. When examining data from 1988 to 1991, no increasing or decreasing trends over time were detected for any individual nutrient (Montagnini and Sancho, unpublished data). Apparently, site improvement was observed in 1988, when the trees were 2.5 years old and had developed closed canopies. Following this initial site improvement, further testing could not detect any additional positive changes in soil chemistry. In fact, as trees approach their maximum-production phase and as the "recyclable" portion (leaves and branches) of the tree decreases in relation to that of the stem, a decrease in soil nutrients can be expected (Lundgren 1980).

The impacts of plantation tree species on soil nutrient reserves will depend on the annual nutrient uptake of the trees in relation to the nutrient-supplying capacity of the soil, nutrient recycling (while the trees are living), and the parts of the tree removed, whether the whole tree or stemwood, and their biomass and nutrient content at the time of harvest. A look at these relationships for *V. hondurensis*, the fastest-growing and seemingly the most

Table 10.2. Organic Matter, Total N, P, pH, Ca, Mg, and K in Soils Under the Four Native Tree Species in Plantation, Grass and Secondary Forest, May 1989[1].

Site	Depth (cm)	OM	N	P	pH	Ca	Mg	K
		(%)		(mg/kg)		(cmol/kg)		
S. exc.	0-15	4.50a	0.278a	2.4a	5.1a	0.68b	0.44a	0.13a
	15-30	3.29a	0.224a	2.1a	5.1ab	0.52bc	0.22bc	0.14a
	30-60	1.88a	0.196a	1.8b	5.1a	0.54a	0.16a	0.14a
V. fer.	0-15	5.06a	0.32a	3.24a	4.98a	0.63bc	0.53bc	0.16a
	15-30	3.66a	0.248a	2.48a	5.03c	0.35d	0.20c	0.10a
	30-60	2.94a	0.200a	2.50b	5.07a	0.33a	0.16a	0.15aa
V. hon.	0-15	4.30a	0.304a	2.30a	5.20a	0.47bc	0.50bc	0.10a
	15-30	3.16a	0.232a	1.82a	5.08ab	0.38cd	0.22bc	0.07a
	30-60	2.42a	0.202a	2.00b	5.13a	0.36a	0.15a	0.06a
H. alc	0-15	5.16a	0.232a	1.5a	5.1a	0.31c	0.21a	0.09a
	15-30	2.77a	0.248a	1.5a	5.1ab	0.45bcd	0.19c	0.10a
	30-60	1.21a	0.158a	1.7b	5.2a	0.46a	0.20a	0.10a
Grass	0-15	3.98a	0.296a	4.1a	5.2a	0.57bc	0.38a	0.22a
	15-30	2.94a	0.236a	3.4a	5.1ab	0.51bcd	0.27bc	0.17a
	30-60	2.46a	0.194a	8.9a	5.2a	0.47a	0.20a	0.13a
Forest	0-15	5.11a	0.288a	2.3a	5.2a	1.16a	0.49a	0.21a
	15-30	3.83a	0.244a	2.0a	5.2a	0.92a	0.45a	0.17a
	30-60	2.48a	0.206a	1.4b	5.2a	0.62a	0.27a	0.12a

[1]. Differences between sites for a given depth and parameter are statistically significant when means are followed by different letters.

nutrient-demanding species of this study, will illustrate this point. Nutrient retention by *V. hondurensis* (calculated by dividing biomass nutrient by plantation age) was an average of 58 kg of N, 181 kg of Ca, 57 kg of K, 22 kg of Mg, and 13 kg of P/ha/yr. The N, Ca, Mg, and K values are all twice as much as those reported by Wadsworth (1983) for teak plantations but the value of P is similar. Although these nutrient values are high, they should be compared with the nutrient-supplying capacity of the soil. For example, Wadsworth (1983) compared data from the rates of annual nutrient uptake of various agricultural crops on upland Ultisols and Oxisols in Puerto Rico (N = 90 - 120 kg/ha/yr, K = 50 - 90, Ca = 86 - 109, Mg = 68 - 98), with the rates of mean annual nutrient retention for fast-growing teak and pine plantations. He concluded that the capacity of the soils to supply nutrients was enough for the needs of the plantations and that plantation trees could be harvested without creating soil deficiencies, except maybe for P. Wang et al. (1991) also reported that the rate of removal of N, P, Ca, Mg and K on an annual basis for *Casuarina* and *Albizia* plantations in Puerto Rico was similar to that of crop systems such as maize and sorghum.

In our analysis, we are not considering "nutrient-supplying capacity" of soils, but rather we are comparing the stocks of above-ground nutrients for each tree species with those in the soil. These quantities were calculated by multiplying the average soil nutrient concentrations for each soil layer under each species to a depth of 60 cm (table 10.2) times the weight of a hectare of the soil for the corresponding layer. Values were corrected by estimating soil bulk density for each type of vegetation (Montagnini and Sancho, unpublished data). If we assume that the nutrient content of the grassy area—where there are no trees—represents the quantity of nutrient stocks before the establishment of the plantation, and if we compare these values with nutrient concentrations found under trees, we can obtain an approximation of the effects of tree uptake on soil reserves.

Comparing soil nutrient budgets (figures 10.1-10.5) with nutrient content of tree biomass (figures 10.6-10.10), we observe that after four years, soil Ca reserves tended to decrease under all the species except for *S. excelsum*. In contrast, soil under the secondary forest appears to have almost twice as much Ca as under *V. hondurensis*, *V. ferruginea*, or *H. alchorneoides* (figure 10.7). Mg and K tended to be lower under *H. alchorneoides* (figures 10.8 and 10.9). However, the most notable overall decrease in soil nutrients occurred for P (figure 10.10), where values under all of the tree species were lower than those under grass and similar to those under the secondary forest. (We do not discuss statistically significant differences here because these measurements are already shown for soil nutrient concentrations, table 10.2). The figures for N are more difficult to compare because soil N was calculated from total N (TKN) values, which tend to be approximately 100 times larger than available N (NO_3 +NH_4), and because N fixation by *S. excelsum* is an additional input that needs to be examined separately. We are not considering other factors affecting nutrient concentration levels such as weathering or rain, the ability of the trees to absorb nutrients from below 60 cm soil depth, or recycling mechanisms (which we will examine next). However, our results tend to agree with the conclusions of Wadsworth (1983) and Bruijnzel and Wiersum (1985) that P, with its potential for depletion, is the most critical element to be considered. Our results also underscore the importance of promoting nutrient-recycling mechanisms through management practices, both in the living stand and at the time of harvest.

Biomass and Nutrient Concentration of Understory Vegetation

Nitrogen accumulation in the total understory biomass was higher under *S. excelsum* plots (14.9 kg/ha), although this amount represented only 3.6% of whole-tree biomass N (table 10.3). For *H. alchorneoides*, understory biomass N was 5.7 kg/ha, or 2.3% of above-ground tree biomass N. For the other nutrients, accumulation in understory biomass under *S. excelsum* ranged from

0.8% to 7.7%, and under *H. alchorneoides*, it ranged from 0.85% to 4.1% of above-ground tree biomass nutrients (table 10.3).

Table 10.3. Biomass and nutrient content of understory vegetation and forest-floor litter[1].

(a) Understory vegetation

	Biomass (kg/ha)	N	Ca	Mg (kg/ha)	K	P
S.exc.	874	14.9(3.6)	2.9(0.8)	3.6(7.7)	8.8(4.2)	1.1(2.0)
H. alc.	425	5.7(2.3)	3.3(1.6)	1.9(2.9)	3.3(0.8)	3.1(4.10

(b) Forest-floor litter[2]

	Biomass (kg/ha)	N	Ca	Mg (kg/ha)	K	P
S. exc.	5612	95.1(23.0)	41.6(11.1)	8.2(17.6)	6.6(3.1)	4.3(7.6)
V. fer.	17215	240.3(137.0)	187.6(47.9)	19.1(35.5)	12.1(7.5)	15.5(33.8)
V. hon.	11084	134.0(57.0)	170.0(23.4)	26.4(24.3)	9.7(4.2)	11.6(22.3)
H. alc.	4238	39.2(15.8)	55.0(27.0)	11.9(17.9)	6.8(1.6)	12.9(16.9)

1. Numbers between parentheses are percent in relation to whole-tree biomass nutrients.
2. Totals, including leaves, fragments, branches.

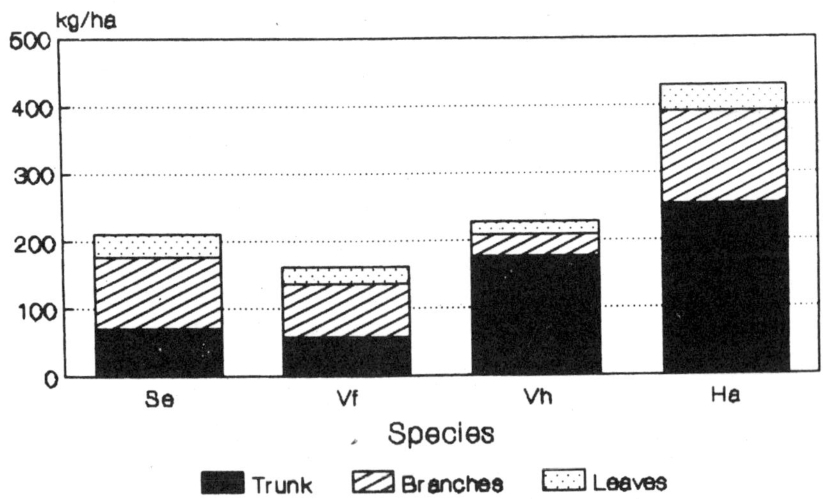

Figure 10.5. Nutrients in Tree Biomass - Potassium.

Since vegetation in the understory apparently accounts for a relatively small proportion of nutrients, as compared to the whole tree, manipulations of the understory should have little effect on nutrient recycling at the site. For example, weeding should have a relatively minor effect on nutrient recycling, unless the understory is removed many times a year. This hypothesis also suggests that intercropping with annual, herbaceous species that attain similar amounts of biomass as that of the understory found under *S. excelsum* or *H. alchorneoides* will not have a substantially negative effect on the balance of nutrients at the site. Since samplings of understory biomass were taken at a time when the quantity of biomass was at its peak, the amounts reported here are considered to approximate those that could be attained by crops. However, this factor deserves further research, since nutrient requirements and parts of plants and trees removed at harvest will vary according to the crops. Our results, however, tend to agree with Bruijnzel and Wiersum (1985), who concluded that the use of controlled intercropping on tree plantations of upland Java, coupled with precautionary measures to minimize soil erosion, was an acceptable way to conserve nutrients. They argue that, in addition to their socioeconomic benefits, the use of such "taungya" practices could also prove advantageous since farmers may be willing to use fertilizers for the crops and the residual effects of these applied nutrients may increase the yield of the trees.

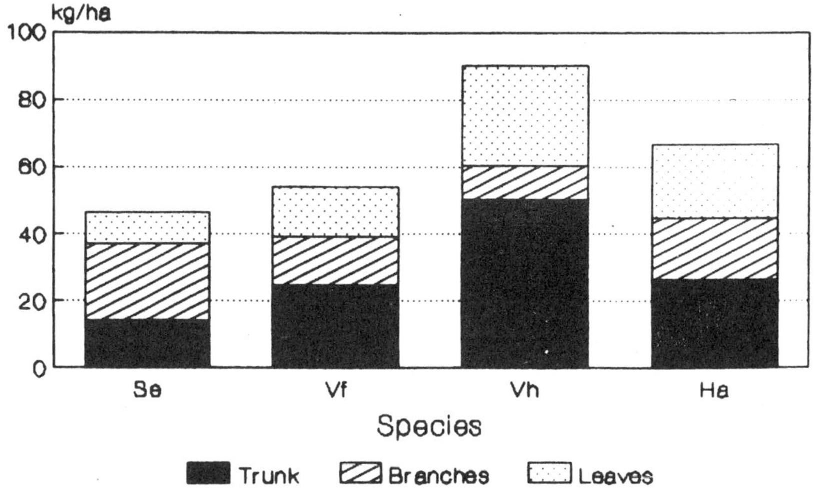

Figure 10.6 Nutrients in Tree Biomass - Magnesium.

Nutrient Accumulation in Forest-floor Litter

The nutrient accumulation and forest-floor biomass were highest under *V. ferruginea*. N in forest-floor litter under *V. ferruginea* was greater than this species' above-ground tree biomass N (table 10.3). As noted above, *V. ferruginea*'s leaf-and-branch biomass accounts for a large proportion of its above-ground tree biomass. This species is self-pruning, a characteristic which enhances shedding of branches and leaves, and occasional prunings may have added even more tree litter to the forest floor. Results of our studies on the rates of litterfall and litter decomposition (Montagnini et al. 1991) suggest that litter decomposes relatively slowly under *V. ferruginea*, a factor explaining the high accumulations noted above. *V. hondurensis*, *H. alchorneoides*, and *S. excelsum* show faster rates of litter decomposition. Forest-floor litter Ca, Mg and P were also considerable under *V. ferruginea* (table 10.3), a finding especially relevant for P, since there is a potential deficiency of this element at the site, as mentioned earlier. Nutrients from forest-floor biomass were also higher under *V. hondurensis*. Again, this result was more significant for N, Ca, Mg and P. Thus, in spite of this species' fast growth rate, nutrient recycling from litter may at least partially compensate for depletion of soil nutrients. While the reverse was true for P, forest-floor litter N was more than double under *S. excelsum* than under *H. alchorneoides*, even though both species had similar forest-floor litter biomass (table 10.3).

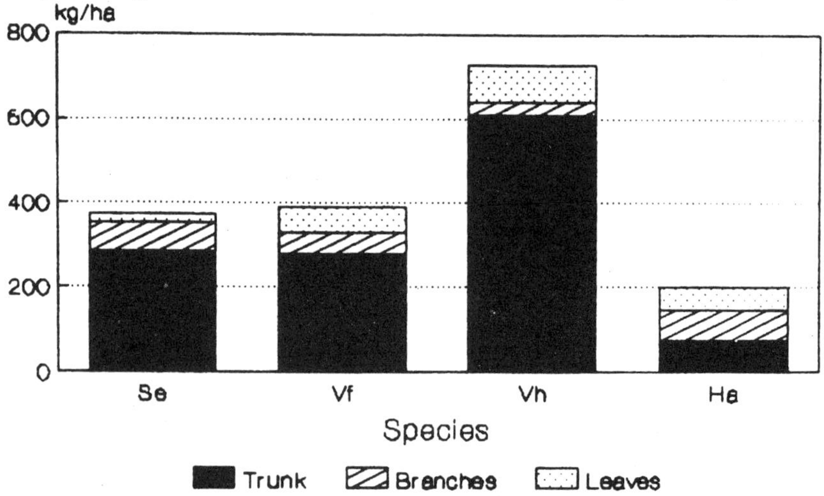

Figure 10.7. Nutrients in Tree Biomass - Calcium.

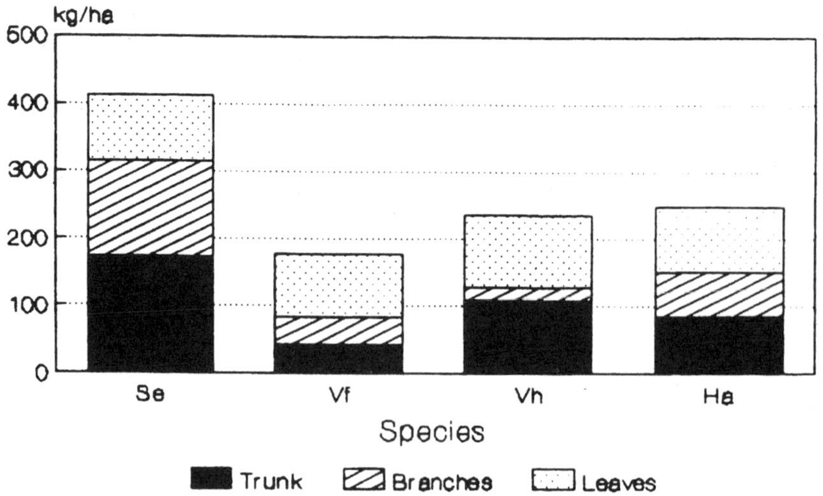

Figure 10.8. Nutrients in Tree Biomass - Nitrogen.

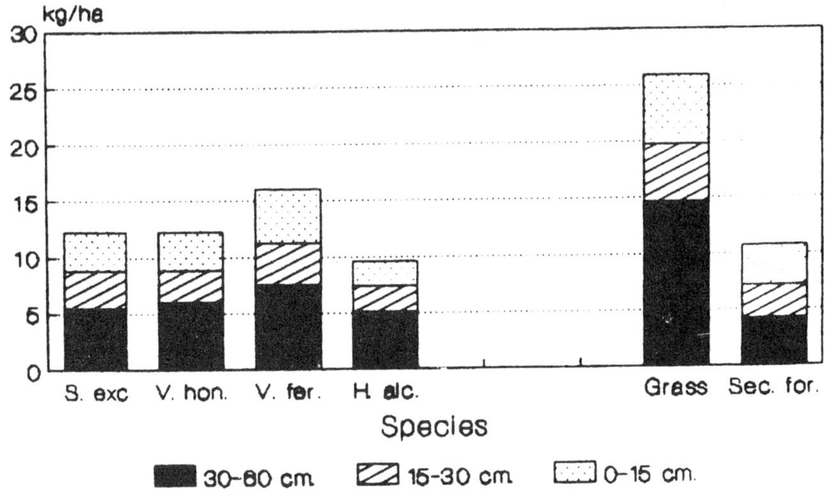

Figure 10.9. Soil Nutrient Balances - Phosphorus.

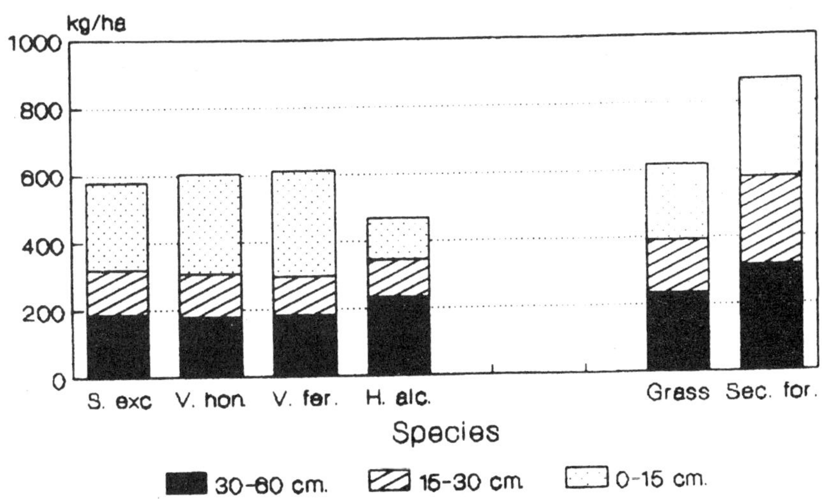

Figure 10.10. Soil Nutrient Balances - Potassium.

These results show that the forest floor is an important place for the accumulation and recycling of nutrients, particularly for N, Ca, Mg and P, but less for K, with marked differences among tree species. If the forest floor is affected by burning or during weeding, a substantial loss of organic matter and nutrients may occur. Wang et al. (1991) also found that with the exception of K, nutrients in forest-floor litter were equivalent to a large proportion (16-50%) of the nutrients contained in the above-ground tree biomass. They concluded that if the litter were left on the floor after harvest, it would represent a substantial reservoir of nutrients for the next rotation.

CONCLUSIONS AND RECOMMENDATIONS

1. Many indigenous tree species with good timber value can grow in open plantations on sites of low fertility and can exhibit fast growth and potential beneficial effects on soil nutrients. Moreover, their effects on soil nutrients may be observed early in the term of rotation, at canopy closure.

2. Tree species differ in their rates of nutrient accumulation in their biomass and in their tissue nutrient content. For a given species, the same trends do not hold for all nutrients: for example, a species may have the greatest effect on site Ca, but its influence on K or N may be minimal; another species may have a more significant influence on site K or P.

3. The establishment of mixed-tree plantations should be an appropriate strategy for combining the nutrient requirements of different tree species with their effects on site nutrients, so that a site may not become depleted of any particular nutrient. However, even with mixtures of trees, site deficiencies are possible for K and P in the long run.

4. Harvest of whole trees will have more negative effects on site nutrients than stem harvest only. The effects will vary according to the species and tree parts removed. Depletion of site nutrients will be greater with shorter rotations because younger trees have a larger proportion of leaf+branch tissue in relation to their stems than older trees, i.e., the potentially "recyclable" portion of the tree is greater in young trees; and harvesting shorter rotations will increase the frequency of removing nutrients from the site as well as the site disturbances associated with harvest operations (soil erosion, compaction, disturbance of the forest-floor layer, etc.).

5. Apparently, growth of understory vegetation and the corresponding accumulation of nutrients plays a relatively small role in affecting the site's nutrient budget. Therefore, agroforestry practices in the understory, such as weeding and intercropping with annual species, may not be that critical for the preservation of nutrients in the site. This situation will vary with the species planted and their management. Intercropping during early stages of tree growth, as long as the crops' nutrient requirements and their management do not provoke other deleterious effects (i.e., soil erosion, excessive nutrient removal with repeated harvests), is an alternative way to accelerate capital returns on the land and thus to stimulate local farmers to plant trees.

6. Forest-floor litter represents a major component of nutrient accumulation and recycling. Practices that affect the forest-floor litter, such as burning for weed control, harvest of litter for fuelwood or mulch, etc., can have serious deleterious effects on site nutrients.

7. The measurement of tree-tissue biomass and nutrient concentrations early in a rotation (for example, when thinning) can give a good indication of the potential impact of management practices on the conservation of site nutrients.

8. References to the rates of nutrient extraction by agricultural crops common in the region can serve as indicators of the "nutrient-supplying capacity" of soils to be compared with the rates of tree species' nutrient uptake in order to assess potential nutrient deficiencies in the site.

ACKNOWLEDGMENTS

This project was funded by the International Foundation for Science (IFS) and by a grant from the A. W. Mellon Foundation to Yale University School of Forestry and Environmental Studies. For their collaboration, we thank: H. and V. Alvarado, La Selva Biological Station; personnel of the Soils Laboratory, Center for Agricultural Research, University of Costa Rica; and H. Asbjornsen and B. Auer (Yale University, School of Forestry and Environmental Studies) for their help with data analysis.

Literature Cited

Allen, O. N. and E. K. Allen. 1981. *The leguminosae: a source book of characteristics, uses, and nodulation*. University of Wisconsin Press, Madison, Wisconsin.

Alpizar, L., H. W. Fassbender, J. Heuveldop, H. Folster, and G. Enriquez. 1986. Modeling agroforestry systems of cacao (*Theobroma cacao*) with laurel (*Cordia alliodora*) and poro (*Erythrina poeppigiana*) in Costa Rica. I. Inventory of organic matter and nutrients. *Agroforestry Systems* 4:175–189.

Bruijnzel, L.A. and K. F. Wiersum. 1985. A nutrient balance sheet for *Agathis dammara* Warb: plantation forest under various management conditions. *Forest Ecology and Management* 10:195–208.

Bruijnzel, L. A. 1991. Nutrient input-output budgets of tropical forest ecosystems: a review. *Journal of Tropical Ecology* 7:1–24.

Chambers, R. and M. Leach. 1990. Trees as savings and security for the rural poor. *Unasylva* 41:39–52.

Chudnoff, M. 1984. *Tropical timbers of the world.* USDA Forest Service, Agricultural Handbook 605, Washington, DC.

Dommergues, Y. R. 1987. The role of biological fixation in agroforestry. In: *Agroforestry: a decade of development* (Steppler, H. A. and P. K. R. Nair, eds.). Nairobi, Kenya: International Council for Research in Agroforestry. pp 245-271.

Espinoza Camacho, M. y Butterfield, R. 1989. Adaptabilidad de 13 especies nativas maderables bajo condiciones de plantacion en las tierras bajas humedas del Atlantico, Costa Rica. pp. 159-172. In R. (ed.). Manejo y provechamiento de plantaciones forestales con especies de uso multiple. Actas Reunion IUFRO, Guatemala, 3-7 de abril 1989. CATIE, Turrialba, Costa Rica.

Evans, J. 1987. Site and species selection - changing perspectives. *Forest Ecology and Management* 21:299–310.

Fassbender, H. W. 1984. Bases edafologicas de los sistemas de produccion agroforestales.

Serie Materiales de Ensenanza No. 21. Centro Agronomico Tropical de Investigacion y Ensenanza (CATIE), Turrialba, Costa Rica.

Gladstone, W. T., and F. T. Ledig. 1990. Reducing pressure on natural forests through high-yield forestry. *Forest Ecology and Management* 35:69–78.

Gonzalez, E. Preliminary seed trials for 36 native tree species. *Revista de Biologia Tropical* (Costa Rica). 39(1):47-51.

Gonzalez, E., R. Butterfield, J. Segleau, y M. Espinoza. 1990. Primer Encuentro Regional sobre Especies Forestales Nativas de la Zona Norte y Atlantica. Memoria. 28-29 Julio 1989. Chilamate, Costa Rica. Instituto Tecnologico de Costa Rica. Cartago, Costa Rica.

Hartshorn, G. 1983. Plants. In: *Costa Rican Natural History* (D. Janzen, ed). University of Chicago Press. pp 118-157.

Hartshorn, G. and B. Hammel. Trees of La Selva. Unpublished manuscript., La Selva Biological Station, OTS, Apdo 676, 2050 San Pedro, Costa Rica.

Holdridge, L. R. y L.J. Poveda. 1975. *Arboles de Costa Rica*. Centro Cientifico Tropical, San Jose, Costa Rica.

Lugo, A. E., S. Brown, and J. Chapman. 1988. An analytical review of production rates and stemwood biomass of tropical forest plantations. *Forest Ecology and Management* 23:179–200.

Lugo, A. E., D. Wang and F. Bormann. 1990. A comparative analysis of biomass production in five tropical tree species. *Forest Ecology and Management* 31:153–166.

Montagnini, F., B.L. Haines, L.R. Boring, and W.T. Swank. 1986. Nitrification potentials in successional black locust and mixed hardwood forest stands in the southern Appalachians, U. S. A. *Biogeochemistry* 2:197–210.

Montagnini, F. and F. Sancho. 1990a. Influencia de seis especies de arboles nativos sobre la fertilidad del suelo en una plantacion experimental en la llanura del Atlantico de Costa Rica. *Yvyrareta* (Argentina) 1(1):29–49.

Montagnini, F. and F. Sancho. 1990b. Impacts of native trees on tropical soils: a study in the Atlantic lowlands of Costa Rica. *Ambio* 19(8):386–390.

Montagnini, F., F. Sancho, K. Ramstad, and E. Stijfhoorn. 1991. Multipurpose trees for soil restoration in the humid lowlands of Costa Rica. In: *Research on multipurpose trees in Asia. Proceedings of an International Symposium, November 19-23, 1990, Los Banos,*

Philippines (D. A. Taylor and K. G.MacDicken eds.). Bangkok: Winrock International Institute for Agricultural Development. pp 41-58.

Nair, P. K. R. 1989. The role of trees in soil productivity and protection. In: *Agroforestry systems in the tropics* (P.K.R. Nair, ed.). Dordrecht, The Netherlands: Kluwer Academic Publishers/International Council for Research in Agroforestry. pp 567-589.

Parrotta, J. A. 1989. The influence of management practices on the biogeochemical stability of tropical fuelwood plantations. *Tropical Ecology* 30:1-12.

Perry, D. A. and J. Maghembe. 1989. Ecosystem concepts and current trends in forest management: time for reappraisal. *Forest Ecology and Management* 26:123-140.

Russell, C. E. 1987. Plantation forestry. The Jari project, Para, Brazil. In: *Amazonian rain forests. Ecosystem disturbance and recovery* (M.G.R. Cannell and J.E. Jackson, eds.). Springer-Verlag, New York: Ecological Studies 60. pp 76-89.

Sanchez, P. A., C. A. Palm, C. B. Davey, L. T. Szott and C. E. Russell. 1985. Tree crops as soil improvers in the humid tropics? In: *Attributes of trees as crop plants*. (Cannell, M .G. R. and J. E. Jackson, eds.). Abbots Ripton, Huntingdon, England: Institute of Terrestrial Ecology, Natural Environmental Research Council. pp 327-350.

Sanchez, P. A. 1987. Soil productivity and sustainability in agroforestry. In: *Agroforestry: a decade of development* (Steppler, H. A., and P. K. R. Nair, eds.). Nairobi, Kenya: International Council for Research in Agroforestry. pp 205-223.

Sancho, F., and R. Mata. 1987. *Estudio detallado de suelos*. Estacion Biologica La Selva. Organizacion para Estudios Tropicales, San Jose, Costa Rica.

Standley, P. C. 1937-38. *Flora of Costa Rica*. Field Museum of Natural History Chicago.

Wadsworth, F. H. 1983. Production of usable wood from tropical forests. In: *Tropical Rain Forest Ecosystems: structure and function. Ecosystems of the World 14A* (F.B. Golley, ed.). New York: Elsevier. pp 279-288.

Wang, D., F. H. Bormann, A. E. Lugo and R. D. Bowden. 1991. Comparison of nutrient-use efficiency and biomass production in five tropical tree taxa. *Forest Ecology and Management* 46:1-21.

Whitmore, J. L., and G. S. Hartshorn. 1969. *Literature review of common tropical trees*. Tropical Forestry Series, Contribution No. 8. Institute of Forest Products. College of Forest Resources. University of Washington, Seattle.

Young, A. 1989. *Agroforestry for soil conservation*. ICRAF, Science and Practice of Agroforestry. C.A.B. International and International Council for research in Agroforestry Wallingford, UK.

CHAPTER 11

THE DEVELOPMENT OF REGENERATION METHODS FOR MIXED-DIPTEROCARP FORESTS: PAST, PRESENT, AND FUTURE

D. L. Grebner
P. M. S. Ashton

INTRODUCTION

Accelerating deforestation has forced us to look for more effective ways to manage natural resources. With current social, economic, and political conditions, exploitation of forests will continue unabated. The slow process of collecting and analyzing biological data cannot keep pace with the rapid rate of deforestation. In Southeast Asia, however, we know enough about the silviculture of mixed-dipterocarp forests to achieve the elusive goal of sustainable management. Past research on these forests has developed a body of knowledge that is unmatched for any other moist, tropical forest region in the world.

Rather than providing a detailed review of the literature, we will present a philosophical examination of silvicultural research and development for mixed-dipterocarp forests. First, we will summarize the knowledge upon which current methods of regeneration are based, especially their dynamics in space and time. We will examine methods of establishing regeneration and comment on their relative successes and failures. Even though these forests could be managed on a sustainable basis, they are now exploited in economically nonsensical and ecologically deplorable ways. Known or tested methods to establish natural regeneration are not applied, or there are misconceptions about the potential economic and ecological benefits of good silviculture. We will summarize by suggesting more biologically flexible and socially acceptable methods to regenerate mixed-dipterocarp forests.

CURRENT KNOWLEDGE OF STAND DYNAMICS

Complex and diverse forest ecosystems, such as mixed-dipterocarp forests, and the climatic conditions associated with them, demand greater understanding of their structural and functional changes. Regions where mixed-dipterocarp forests are located experience heavy annual rainfall of around 3,000 mm. Dry spells, frequently interrupted by convectional rainstorms, usually last no longer than a month. Seasonal variation in temperature is negligible, hovering about 1° to 2° around 27°C. However, daily temperature variations can be as great as ± 6°C.

Understanding Structural Change in Time and Space

Stands within mixed-dipterocarp forests are composed of differently-aged groups of trees ("mini-stands" after Smith 1986) within which species occupy different horizontal and vertical strata over time (Whitmore 1984, Oliver and Larson 1990, Ashton 1992). Trees can be grouped into particular *regeneration guilds*, composed of species with common silvical characteristics, such as successional status, shade tolerance, and growth strategy (Whitmore 1984, Swaine and Whitmore 1988). Figure 11.1 illustrates that *regeneration guilds* are defined by the successional growth phase they dominate and by the stratum their mature crown occupies in the gap, building, and mature stages of stand development (Watt 1924, 1947, Cousens 1974, Whitmore 1984, Ashton 1964, for review see Ashton 1992). Remaining guilds represent species that occupy habitat strata—dominant canopy, nondominant canopy, subcanopy and understory—of the mature phase (modified after Ashton 1992).

In mixed-dipterocarp forests, the differentiation of guilds based on successional status and stratum of crown occupation is more marked than in other types of tropical forests. Early successional guilds, which dominate the canopy during gap and building phases of stand development, rely on seeds that are buried, windblown, or dispersed by animals. The seeds germinate and become established after a site disturbance—such as a windstorm, landslide, or lightning (Fox 1972, Liew and Wong 1973, Ng 1978, 1980, Putz and Appanah 1987). Apart from these early successional species, the majority of guilds within mixed-dipterocarp forests become established and develop through advanced regeneration, that is, their seedlings or saplings are beneath the forest canopy before a disturbance creates or releases growing space in the canopy, or their roots or stumps survive a disturbance to sprout. The ability of mixed-dipterocarp forests to develop strata of trees representing different guilds depends largely on disturbances that assure the perpetuation of different species in advanced regeneration (Whitmore 1984, 1990, Ashton 1992). If the structure and composition of the forest are to recover following a disturbance, all guilds must be present and develop together over a relatively short time (Nicholson 1960, de Zoysa et al. 1990, Ashton 1992). Therefore, the period immediately after a disturbance is critical for the initial development of a new and complete stand.

Canopy trees of the mature phase of a mixed-dipterocarp forest are essential for the future structural and functional integrity of the stand. In the mature phase of stand development, the crowns of these trees modify the effects of radiation and rainfall, creating a buffer against microclimatic

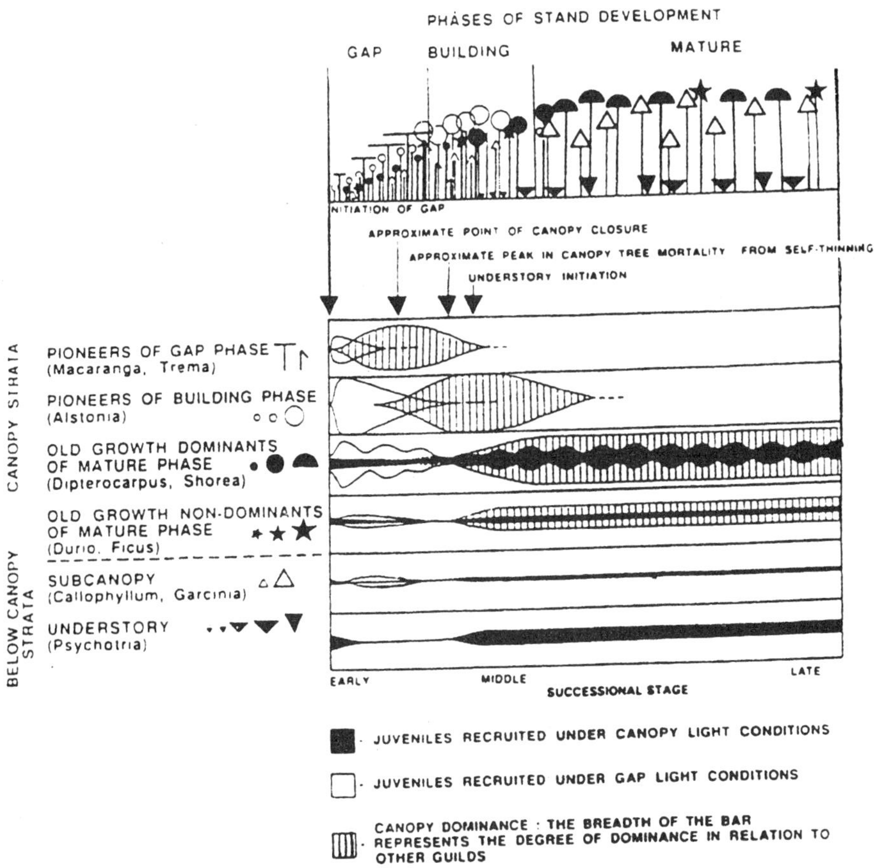

Figure 11.1. A Generalized Diagram Depicting Recruit Frequency and Canopy Dominance of Guilds over Different Successional Stages of Stand Development for Mixed-Dipterocarp Forest. Note the Periodic Recruitment of Seedling for Tree Species Belonging to the Old Growth Dominant Guild. Also, Juveniles Recruited Under Canopy Light Conditions Represent Advanced Regeneration (This diagram has been modified after Ashton 1992a).

extremes and promoting diverse micro-environments on which other plant and animal species depend. The majority of these tree species are similar in reproduction, growth, and development and form the dominant guilds in the mature phase of a stand's development (see Putz & Ashton, in press). Many of these species coexist in the same forest as closely related taxonomic groups —*Dipterocarpus*, *Dryobalanops*, *Hopea*, *Mesua*, *Palaquium*, *Shorea*, *Syzyguim*.

In silviculture, the process of stand development has often been neglected, misunderstood, or oversimplified (Oliver and Larson 1990, Ashton 1992). To be successful, the dominant regeneration guild in the mature phase of mixed-dipterocarp forests must develop with the complete stand, through all of its phases. For example, short-lived pioneers of the gap phase may "nurse" shade-tolerant trees, helping them to establish by shielding their crowns from climatic extremes (Ng 1980, Whitmore 1984, Hallé et al. 1978, Ashton and Ashton 1992, Ashton et al. 1992a). Under certain circumstances, the pioneers may temporarily suppress regeneration of trees. However, in these cases, they often have a stronger deleterious effect on vines and other weeds that need light (Whitmore 1984). The more columnar, crowned trees that occupy the canopy of the building phase live longer than those species that occupy the canopy of the gap phase, but they thin rapidly. They influence the shape of the stem and crown of shade-tolerant trees, originally "nursed" by the pioneers, by providing them with side shade (Ashton 1978, Ng 1979, 1981, Hallé and Ng 1981, Ashton et al. 1992a, 1992b).

The spatial and temporal scale of regeneration depends on the size, frequency, nature, and predictability of disturbance (Anderson 1961, Brunig 1964, Ashton 1992). For instance, the size, frequency, and type of disturbance are predictably different in undulating topography, from ridge to valley, in southwest Sri Lanka (Ashton 1992). Speculations are similar for mixed-dipterocarp forests in Sarawak (Ashton and Hall 1992). Thus, stands within an unmanipulated old-growth forest are like mosaics, composed of shifting mini-stands, all having the same successional processes of development, but of varying sizes and at different rates and stages.

Maintaining diversity of species in forests and developing appropriate methods of regeneration require that we recognize variations among site locations, and the number and types of plants associated with them at the stand level, within mixed-dipterocarp forests. While silvicultural manipulations should essentially follow the general developmental processes across the forest's topography, their intensity and exact nature should be site-specific, reflecting the capacity of a particular stand to respond. Recognizing that management of stands should be site specific is essential for accommodating variations in their rates of successional development, and hence the implications for management for site index and rotation length.

Understanding Functional Change in Time and Space

Mixed-dipterocarp forests are notoriously poor in incorporated organic matter, and almost all soils on which they develop are classified as ultisols or oxisols (Leamy 1966, USDA 1975). On stable landscapes, these soils are highly weathered, deep, and poor in nutrients (Leamy 1966, Sanchez 1976). On steeper slopes, landslips constantly truncate the soil profile and expose little-weathered parent material (Baillie and Ashton 1983). If the forest is removed, surface horizons, particularly those with a sandy texture and weak structure, are highly susceptible to erosion (Daniel and Kulasingam 1974).

Unlike some other forest regions, the reaccumulation of organic soil appears to be a fragile process in mixed-dipterocarp forests (Ashton and Hall 1992). Although nutrients are exchanged throughout this forest ecosystem, the surface layer is the most active portion of the site for nutrient conservation and cycling (Ashton and Hall 1992). This is where fine roots actively absorb nutrients from decomposing litter (Yoda 1978, Proctor et al. 1983), and where the turnover of nutrients is high (Whitmore 1984). These nutrient cycles, changing both in time and space, reflect the same successional processes that influence the structure and composition of stands (Ogawa 1978, Anderson and Swift 1983, Gong and Ong 1983). In mixed-dipterocarp forests, associations of tree species that occur across the landscape are correlated with types of soils and with the level of nutrients in the soils (Ashton and Hall 1992).

Although conditions in many mixed-dipterocarp forests are unfavorable for nitrogen-fixing species, they are suitable for trees with ectotrophic mycorrhizal associations, which include members of the *Leguminosae* (*Caesalpinoideae*), *Dipterocarpaceae*, and *Myrtaceae* (Janos 1988). Species with ectotrophic mycorrhizal associations are among the most soil- and site-specific and are found in some of the most diverse nutrient-impoverished forests in the world (Ashton 1964, Ashton and Hall 1992). The dominance of these tree species in mixed-dipterocarp forests demonstrates that ectotrophic mycorrhizal associations facilitate specialization in uptake of soil nutrients in species combinations that are site specific. It implies that mixed-dipterocarp forests exploit these soils that are poor in nutrients more efficiently than would other crops that could be cultivated.

The relatively closed nutrient cycle on the infertile soils of mixed-dipterocarp forests is sustained by raw humus and sufficient accumulation of organic matter within the soil. Long-term sustainability of the forest depends on how organic matter in the soil is affected by timber harvesting and silviculture. Raw humus accumulates slowly in tropical regions. Consequently, it must be conserved if dipterocarp forests are to be managed (Putz & Ashton, in press).

PAST AND PRESENT METHODS OF REGENERATION

Manipulation of the canopy is the biological basis of silviculture in a natural forest. Controlling the size of gaps in the canopy influences the composition of species established at the groundstory. Foresters have used this empirical principle of silviculture in continental Europe since at least the twelfth century. Silvicultural practice reached forests of colonial Malaysia via experiences in India and Burma and was strongly influenced by teutonic traditions (Brandis 1895).

Silviculture of mixed-dipterocarp forests changed dramatically when these forests were exploited at the turn of the last century by British and Dutch colonial governments. For instance, the Malaysian rainforest was managed originally during the 1890s to 1920s to increase the yield of latex, used for the insulation of telegraph cables, from gutta percha (*Palaquium gutta*, *Sapotaceace*). Later, slower-growing durable woods—heavy meranti—were in demand for construction, and now, there is a demand for faster growing plywoods and timbers—light meranti (*Shorea* spp.).

Much of the early silviculture of mixed-dipterocarp forests originated on the Malaysian peninsula. Seedlings of the principle species were often present in the understory of old-growth forests in the lowlands. These provided successful stocking and grew after the removal of the overstory. The first attempts to create artificial openings were made by girdling selected trees within stands, killing them by removing a band of bark from around their trunks. Addition of herbicide to the girdle hastened this process.

From the 1920s to the 1940s, demand for heavy hardwood for construction encouraged the selection and removal of only a few large stems per hectare. This pattern of harvesting required the application of a silvicultural treatment called *release cutting*. Undesired species were removed to create space within the forest for growth to improve the quality of young stands (Smith 1962). This shelterwood method of regeneration was called the *improvement method* (Sanger-Davies 1932, 1937). By the postwar era, demand increased and prices rose for an increasingly long list of species. As a consequence, all marketable species were removed in a series of harvesting operations at or toward the end of a stand's rotation period. The existing canopy completely disappeared, leaving immature residuals and advance regeneration for the development of a future stand.

At first, repeated release cuttings were done to aid the growth of this advance regeneration, but experimental plots of untreated regeneration showed that the faster-growing commercial species eventually gained dominance unaided (Barnard 1954). Here is the basis for the Malaysian Uniform Shelterwood (MUS) method, the results of 20 years of experiments that were systematized by Wyatt-Smith (1963). Similarly, various shelterwood methods were developed in other regions of tropical Asia for mixed-dipterocarp forests—southwestern India (Kadambi 1954), the Andaman Islands (Chengappa 1944), Sri Lanka (Holmes 1957, 1958).

Several important biological and socioeconomic factors illustrate the deficiencies of past methods of shelterwood regeneration. These methods developed by focusing on manipulating stands for single values that changed over time—gutta percha, heavy hardwoods, light hardwoods. Structurally complex and diverse stands were simplified through various intensive techniques to exclude those species that were perceived to be of no value. Silvicultural systems, requiring intensive management, were designed to produce a single, valued species. Socioeconomic values, especially prices, but also political signals, change faster than any stand can grow. Old silvicultural systems became biologically inflexible, obsolete, incapable of responding to changes in market values, and in that sense, they failed.

By the mid-1970s, economic pressures in the postcolonial period led to the use of polycyclic systems to replace the older shelterwood regeneration methods. These methods of selecting single trees for harvesting specify diameters so that only large trees are removed at each cutting cycle. Selected trees are therefore removed in a continuous series of felling operations so that the interval of time between cutting is less than the time required for any individual tree to mature. This approach was politically attractive because the potential harvest of commercial timber was every 10 to 30 years rather than at least 50 years. The Malaysian Selection System, based on this rationale, was adopted for dipterocarp forests in the hills. The Philippines and Indonesia began to use this system as well.

Under these systems, harvesting stands with intermediate-size classes of commercial tree species that are in the canopy is tantamount to high grading. This results in the deterioration of the stand and should be avoided. Mixed-dipterocarp stands in the mature phase of development contain species that occupy particular habitat strata. When these stratified old-growth stands are managed as "all-aged," using selection methods that create small openings, canopies often close, because the crown of the residual overstory expands or because noncommercial trees in the subcanopy grow. The stature of the forest declines when species in the canopy are removed. Then, species in the subcanopy close to form a new, lower canopy. Regeneration, establishment, and growth of tree species that have a greater need for light—those trees that dominate the forest canopy in its mature phase—are inhibited.

Present selection systems in silviculture fail because they ignore the inherent dynamic capabilities of mixed-dipterocarp forests. Established rules are not enforced, often because of political intervention. Continued, frequent harvesting causes damage to regeneration and the forest floor as well as soil compaction and erosion because skid trails and logging roads are used too often for vegetation to grow.

The postwar, political emphasis on utilizing forest resources has forced us to consider other systems to replace or modify those of the past. Previously, assumptions about the nature of stand development for mixed-dipterocarp forests were made conveniently, and "management" sanctioned the exploitation of forests. In reality, little attention is devoted to creating the conditions necessary for establishing new stands. Managers ignore the biological limitations of stand development, and during the postwar period, substantial degradation of forests has occurred.

FUTURE METHODS OF REGENERATION

Considerations and Requirements

How do we manage mixed-dipterocarp forests sustainably for the future? We are suggesting that the shelterwood methods of the past be combined with the multiple socioeconomic necessities of the present. Research has demonstrated that these methods of regeneration are the most compatible with natural regeneration processes of tree species that form the canopy of mixed-dipterocarp forests. Shelterwoods are more economically cost-effective than selection methods. Commercial timbers are removed in several cutting sessions over a relatively short time span at the end of the rotation. If harvesting is controlled, there is little damage to the growth and development of the new stand. In addition, long-term costs are less for administration and the establishment of a timber-harvesting infrastructure.

The most essential aspect of this plan is that mixed-dipterocarp forests should be managed for their ecological processes and for an array of desired services, not for their products alone. This recognition will promote the ecosystem's viability and its resilience both to capricious changes in value and to biological degradation. Methods of shelterwood regeneration therefore must include the functional and structural limitations of natural forest succession.

By implication, maintaining the integrity of the composition and successional process in the forest ecosystem will also facilitate the production of many other goods and services besides timber. Nontimber products gathered from mixed-dipterocarp forests include fruits, nuts, latex, medicines, and wildlife, all of which produce more or less quantifiable yields. Other services, more difficult to assess, yet beneficial to society, include recreation, aesthetics, carbon sequestration, and, most importantly, genetic conservation.

In the future, maintenance of the composition and structure of mixed-dipterocarp forests must incorporate manipulations within shelterwoods that more closely emulate natural disturbances. This technique would facilitate the perpetuation and differentiation of advanced regeneration guilds, which, in turn, will promote both the temporal and spatial stratification of a stand and the diverse micro-environments upon which plant and animal species depend.

Methods of shelterwood regeneration must focus on the establishment of advanced regeneration of tree species, dominant in the mature phase of stand development, that form the canopy. Failure to maintain the presence of these species will promote degradation of the canopy structure and the usurpation of growing space by unmerchantable subcanopy species. Essential processes of successional development—gap, building, mature—must be maintained and protected until a suitable period during the mature phase of development before significant harvesting can occur. Early re-entry, or premature harvesting, may prevent the establishment of some tree species and perhaps may introduce exotic weeds. Both events could dramatically break the structural processes of the ecosystem. In some cases, inappropriate harvesting could transform a mixed-dipterocarp forest into a grass-fire ecosystem. Functional processes are also at risk because repeated harvesting predisposes the fragile organic surface to destruction and alters or damages closed nutrient cycles.

Shelterwood Variations

There are many variations of the shelterwood method for specific conditions (see Smith 1986). Shelterwoods can facilitate the creation of multi-aged stands in the sense of Oliver and Larson (1990). Structural components of a stand can be left as individuals or groups of residuals for wildlife habitats and food sources (e.g., keystone species). Slower-growing canopy species—heavy meranti—may also be left as residuals to grow for two or more rotations as faster-growing canopy trees—light meranti become established. Thus species that require a longer than usual successional period to reach maturity may be accommodated. Where established, advanced regeneration already exists, one-cut shelterwoods—those methods involving no prior cutting for preparation or establishment of a stand—can remove the overstory to release a new stand, as in the Malaysian Uniform System.

Shelterwoods are also suitable for small land holdings, where periodic, rather than continuous, harvesting of timber from the stand is preferable. Alternatively, continuous timber supplies can be secured by applying the shelterwood system to stands that are arranged in strips within the forest. The use of strip shelterwoods facilitates the spatial supervision and organization of sequential cutting operations, especially for community-held ownerships. The spatial arrangement of strips within the topography can also be used to protect soils that are prone to erosion or that serve as important catchment areas for watersheds.

Shelterwoods, compared to selection methods, are less intrusive over the whole rotation, they are less complex and simpler to supervise. However, the period of regeneration for the shelterwood method requires more rigorous supervision and professional guidance. In addition, when there is a dramatic change in canopy cover, with more light penetrating into the forest, weeds that depend on light grow faster and can become a serious nuisance.

CONCLUSIONS

Shelterwood methods of regeneration need to satisfy specific microclimatic and edaphic conditions for the successful establishment of a new canopy. They must be site specific. The planner should develop variations that reflect the landscape, forming a gradient of differing structures across the topography, to provide a continuing source of benefits. Patterns of shelterwoods must duplicate those of natural disturbances, whenever possible. Silvicultural operations must preserve fragile topsoil so that the future productivity of the site is maintained.

While maintaining structural and functional processes at the scale of an individual stand, a manager must also consider larger-scale linkages over the landscape to provide continuity between stands. Corridors located on edaphically sensitive zones—riparian buffers, steep slopes, seasonally inundated bottomlands—protect those stands in the landscape mosaic that are managed for timber production. These corridors prevent soil erosion, enhance wildlife migration, facilitate the conservation of genetic resources, and control floods.

Literature Cited

Anderson, J.R. 1961. The destruction of *Shorea albida* forest by an unidentified insect. *Commonwealth Forestry Review* 40:19–29.

Anderson, J.H. and M.J. Swift. 1983. Decomposition in tropical rain forests. In *Tropical rain forest ecology and management.* (S.L. Sutton, T.C. Whitmore, and A.C. Chadwick, eds.). Oxford: Blackwell.

Ashton, P.S. 1964. Ecological studies in the mixed dipterocarp forests of Brunei State. Oxford Forestry Memoirs 25.

Ashton, P.S. and P. Hall. 1992. Comparison of structure among mixed-dipterocarp forests of Northwestern Borneo. *Journal of Ecology* 80:459–481.

Ashton, P.M.S. 1992. Establishment and early growth of advance regeneration of canopy trees in moist mixed-species broadleaf forest. In *The ecology and silviculture of mixed-species forest.* (M.D. Kelty, B.C. Larson, and C.D. Oliver, eds.). Dordrecht, The Netherlands: Kluwer Academic Publishing Co. pp 101-125.

Ashton, P.M.S. and P.S. Ashton. 1992. Plant resources for agroforestry systems in tropical Asia. In *Agroforestry in South Asia: problems and applied research perspectives.* (W.R. Bentley, P.K. Khosla, and K. Seckler, eds.). New Delhi: Oxford & IBH Publishing Company, Pvt., Ltd.

Ashton, P.M.S., C.V.S. Gunatilleke., and A.U.N. Gunatilleke. 1992a. A shelterwood method of regeneration for sustained timber production in *Mesna-Shorea* forests of southwest Sri Lanka. In *Ecology and landscape management in Sri Lanka: conflict or compromise?* (W. Erdelen, C. Preu, N. Ishwaran, and C. Santiapillai, eds.). Hamburg: Margraf Scientific.

Ashton, P.M.S., I.A.U.N. Gunatilleke, and C.V.S. Gunatilleke. 1992b. A case for the development and evaluation of mixed-species plantations for the wet zone of Sri Lanka. In *Ecology and landscape management in Sri Lanka: conflict or compromise?* (W. Erdelen, C. Preu, N. Ishwaran, and C. Santiapillai, eds.). Hamburg: Verlag Josef Margraf Scientific.

Baillie, I.C. and P.S. Ashton. 1983. Some soil aspects of the nutrient cycle in the mixed-dipterocarp forest of Sarawak, East Malaysia. In *Tropical rainforest ecology and management.* (S.L. Sutton, T.C. Whitmore, and A.C. Chadwick, eds.). Oxford: Blackwell.

Brandis, Sir Dietrich. 1895. *Forestry in British India.* London: Empire Press.

Brunig, E.F. 1964. A study of damage attributed to lightning in two areas of *Shorea albida* forest in Sarawak. *Commonwealth Forestry Review* 43:134–144.

Chengappa, B.S. 1944. The Andaman forest and their regeneration. *Indian Forester* 70: (9,10,11,12).

Cousens, J.E. 1974. *An Introduction to Woodland Ecology.* Edinburgh: Oliver and Boyd.

Daniel, J. G. and A. Kulasingam. 1974. Problems arising from large-scale forest clearing for agricultural use—the Malaysian experience. *Malaysian Forester* 37:152–160.

Fox, J.E.D. 1972. The natural vegetation of Sabah and natural regeneration of dipterocarp forests. Ph.D. Thesis, University of Wales.

Gong, W.K., and J.E. Ong. 1983. Litter production and decomposition in a coastal hill dipterocarp forest. In *Tropical rainforest ecology and management.* (S.L. Sutton, T.C. Whitmore, and A.C. Chadwick, eds.). Oxford: Blackwell.

Hallé, F., R.A.A. Oldeman, and P.B. Tomlinson. 1978. *Tropical trees and forests: an architectural analysis*. New York: Springer Verlag.

Holmes, C.H. 1957. The natural regeneration of the wet and dry evergreen forests of Ceylon I. *Ceylon Forester* 3:15–44.

Holmes, C.H. 1958. The natural regeneration of the wet and dry evergreen forests of Ceylon II. *Ceylon Forester* 3:111–127.

Janos, D.P. 1988. Mycorrhiza applications in tropical forestry: are temperate-zone approaches appropriate? In *Trees and mycorrhiza*. (F.S.P. Ng, ed.). Kuala Lumpur: Forest Research Institute. pp 133-88.

Leamy, M.L. 1966. Proposals for a technical classification of Malaysian soils. Malaysian Soil Service Report #3.

Liew, T.C. and F.O. Wong. 1973. Density, recruitment, mortality, and growth of dipterocarp seedlings in virgin and logged-over forests in Sabah. *Malaysian Forester*. 36:3–15.

Ng, F.S.P. 1978. Strategies of establishment in Malaysian forest trees. In *Tropical Trees as Living Systems*. (P.B. Tomlinson and M.H. Zimmermann eds.). Cambridge: Cambridge University Press.

Ng, F.S.P. 1979. Growth rhythms in tropical juvenile trees. *Actual. Bot.* 3:139–149.

Ng, F.S.P. 1980. Germination ecology of Malaysian woody plants. *Malaysian Forester* 43:406–437.

Nicholson, D.I. 1960. Light requirements of seedlings of five species of Dipterocarpaceae. *Malaysian Forester* 21:193–196.

Ogawa, H., 1978. Litter production and carbon cycling in Pasoh forest. *Malay Nature Journal* 30:367–373.

Oliver, C.D. and B.C. Larson. 1990. *Stand development in North America*. New York: McGraw Hill.

Proctor, Anderson, J.M., P. Chai, and H.W. Vallack. 1983. Ecological studies in four contrasting rainforests in Gunung Mulu national park, Sarawak. I Forest environment. *Journal of Ecology* 71:237–260.

Putz, F.E. and S. Appanah. 1987. Buried seeds, dispersed seeds, and three dynamics of lowland dipterocarp forest in Malaysia. *Biotropica* 19:326–333.

Putz, F.E. and P.M.S. Ashton. In press. The ecology and silviculture of mixed-dipterocarp forest of Southeast Asia. In *Sustainable tropical forest management*. (R.A. Sedjo and M.C. Douglas, eds.). Washington, D.C.: Island Press

Sanchez, P.A. 1976. *Properties and management of soils in the tropics*. New York: John Wiley & Sons.

Sanger-Davies, A.E. 1932. Final fellings, cleanings, and thinnings. *Malaysian Forester* 1:43–48.

Sanger-Davies, A.E. 1937. The tending of mercanti tembaga (*Shorea leprolula*), its rotation and probable yield. *Malaysian Forester* 6:56–61.

Swaine, M.D. and T.C. Whitmore. 1988. On the definition of ecological species groups in tropical rain forests. *Vegetation* 75:81–86.

USDA. Soil Conservation Service. 1975. Soil taxonomy: a basic system of soil classification for making and interpreting soil surveys. In *Agricultural Handbook #436*. Washington, D.C.: USDA Soil Conservation Service.

Watt, A.S. 1924. On the ecology of British beechwoods with special reference to their regeneration I. *Journal of Ecology* 12:27–73.

Watt, A.S. 1947. Pattern and process in the plant community. *Journal of Ecology* 35:1–22.

Whitmore, T.C. 1984. *Tropical rain forests of the far east*. Second edition. London: Clarendon Press.

Wyatt-Smith, J. 1963. Manual of Malaysian silviculture for inland forests. Vol. I. Malaysian Forest Records #23.

Yoda, K. 1978. Organic carbon, nitrogen, and mineral nutrient stocks in the soils of Pasoh forest. *Malay Nature Journal* 30:229–251.

Zoysa, de, N.D., C.V.S. Gunatilleke, and I.A.U.N. Gunatilleke. 1990. Comparative phytosociology of natural and modified rain forest sites in Sinharaja MAB Reserve, Sri Lanka. In *Rain forest regeneration and management*. (T.C. Whitmore, A. Gomez-Pompa, and M. Hadley, eds.). UNESCO, Biology Special Issue #18. Paris: Parthenon Press.

CHAPTER 12

PERTURBATIONS, FOREST ECOSYSTEM FUNCTION, AND SUSTAINABILITY

P.S. Ramakrishnan

INTRODUCTION

In a forest ecosystem, as in all other types of ecosystems, biotic and abiotic components are so interlinked that the flow of energy through the system forms well-defined trophic structures, biodiversity, and material cycles (Odum 1971). Since ecosystems are increasingly affected by human activity, their integrity is altered in many ways. For example, degradation of a site from some types of harvesting practices may cause a drastic decline in biological diversity, with consequent impacts on functions of ecosystems at the trophic level and disruptions of nutrient-cycling. These disturbances make us realize that ecosystem functions are related to their sustainability.

The concept of *sustainability* suggests that ecological systems are used in a way that satisfies current needs and maintains a diversity of options for the future. This leads us into the realm of time and space, dimensions of sustainability. In the dimension of time, one may consider sustainability in the short-term, 15 to 20 years, or in a longer time frame of 50 to 100 years. Perhaps, we are straining this concept if we consider sustainability in an indefinite time frame! Similarly, we may consider an ecosystem in the dimension of space—a complex set of ecosystems often referred to as a "landscape," or even the biosphere as a whole. In the following discussion, I will consider the ecological consequences of deforestation at the landscape level.

CONSEQUENCES OF DEFORESTATION

The consequences of deforestation are two-dimensional—ecological and social. In an ecological context, its impact on the depletion of biodiversity and the disruption of hydrologic and nutrient cycles is the most obvious. In a social context, deforestation has resulted in a vicious cycle of poverty in the tropics, with economic growth and changes leading to a widening gap between the richer few and the poorer majority (Ramakrishnan 1992a).

Biodiversity

Over 200 years ago, Linnaeus visualized about 50,000 species in all. Today, this estimated number exceeds 1.6 million, but we have only meager information on the vast majority of species. Many more are yet to be discovered and catalogued. Some estimates range from approximately 30 million to nearly 90 million species, and scientists guess that about two-thirds of them are from the tropics. With approximately 500,000 tropical species catalogued thus far, the rate at which species are being lost even before they are catalogued is conjectural.

A significant impact of converting forestland to agriculture is the rate of species extinction. Tropical forests are disappearing at a rate of at least 17,000 km^2/yr (see Gowen et al., this volume), with perhaps as much forest being degraded each year. The implications for biological diversity are alarming. Twenty percent of all species inhabit Latin America, excluding Amazonia, and another 20% are in the Asian and African tropics, excluding the Zaire basin. If present patterns of deforestation continue in these regions, at least a 15% loss in biota could occur by 2000 AD. If we add deforestation in Amazonia, the rate of species extinction could exceed over 70%. Many biologists have predicted much higher rates. Regional rates of species extinction correspond to ecosystem type and disturbance regime. Obviously, human interactions are often critical.

Desertification

Desertification is a term that explains decline in the biological productivity of an ecosystem. While this phenomenon is linked to arid, semiarid, and subhumid ecosystems, its impact could be most dramatic, even in the humid tropics. Impoverishment of terrestrial ecosystems that have been adversely affected by man may become apparent in a variety of ways—accelerated erosion, salinization, and decline of site quality. A major consequence of deforestation is its adverse effect on the ecosystem's hydrology, with related soil and nutrient losses. This consequence is eloquently illustrated by the classical study of the clear-cut, temperate ecosystem that was closely monitored by Hubbard Brook in its phase of forest regrowth (Likens et al. 1977). We conducted a similar study, but one that links ecological sciences with the social, economic, and cultural dimensions of shifting agriculture (*jhum*) in humid, tropical forests. Our study revealed more complex interactions (Ramakrishnan 1992b).

The consequences of deforestation invariably arise from site degradation through erosion (Zacher 1982), with tropical Asia, Africa, and South America having the highest levels. Of all forms of erosion, that caused by precipitation occurs over approximately 50% of the earth's land surface. The magnitude of erosion by water in Asia is illustrated by the silt load of the Yellow River, with an estimated high of 1,850 tons/km^2/yr. Some of its tributaries carry as much as 7,000 tons/km^2/yr. Researchers have recorded very high erosion rates for other rivers in the tropics. Soil degradation and decline in the quality of available water—the ultimate consequence of site desertification—leads to the consequent decline in yields of food, fodder, and fuelwood that is essential for the economic well-being of rural communities.

Social Disruption

Agriculture and animal husbandry are two important economic activities of rural communities. Ill-conceived agricultural activities are often responsible for social disruption. In the humid tropics, for example, forest farmers, who are traditionally involved in shifting agriculture, combine this practice with swine husbandry and raising poultry while living in harmony with their environment. The cycle of shifting agriculture—the length of the fallow period between two successive plantings on the same site—is long. Animal husbandry is based on grazing, browsing, and scavenging of detritus or feed that is unfit for human consumption. Indeed, shifting hill agriculture—with a variety of other, complementary systems, like rice systems and home gardens in the valleys—harmonize with the environment of the forest farmer.

According to mid-1970 estimates, there were approximately 140 million forest farmers on two million km^2 of tropical forests. These farmers are believed to eliminate another estimated 100,000 km^2 of forest each year, of which a substantial part is closed forest. Large-scale timber harvests for industry have depleted forest cover. Pressures from the increased population on areas where shifting agriculture is practiced have led to a shortened agricultural cycle, which causes social tensions (see box 12.1).

In addition, harvesting forests for fuelwood is an important cause of deforestation, which, in turn, directly affects the economy when there are extreme shortages of this commodity. In the Asian tropics, about 600 million people experience an acute shortage of fuelwood. Such deficiencies affect the economy indirectly when cow dung, a valuable fertilizer for crop production, is burned for fuel. Over 400 million tonnes/yr of cow dung are burned in Asia and Africa. In economic terms, this substitution of fertilizer for fuel represents a 50-kg decline in the production of grain for each tonne of cow dung that is burned.

Poverty and deprivation, caused by deforestation, will eventually result in accelerated conversion of forestland to agriculture, with related animal husbandry and domestic activities, in rural, poor areas. At least one group of researchers predicts that clearing land for agriculture would significantly influence levels of cutting and exporting tropical hardwood. Viewed in this context, the future scenario for the developing world is grim unless we take urgent measures to reverse this current trend.

Box 12.1. Shifting Agriculture (*jhum*) and Social Disruption in Northeastern India (Ramakrishnan 1992).

Northeastern India has over 100 linguistically and culturally distinct tribes. Different tribes often live short distances from one another, in some cases only a few kilometers. Shifting agriculture—or *jhum*, as the tribes call it—is the major economic activity. This highly organized agroecosystem is based on empirical knowledge that has accumulated for centuries. It is in harmony with the environment as long as the *jhum* cycle—the length of time that the land is fallow between two successive plantings—is long enough to allow the soil to recover fertility that was lost during the crop phase.

The valley system of wet rice cultivation and home gardens supplements the *jhum* system. This system is sustainable year after year because soil washed down from the hills provides sufficient fertility for rice crops. Home gardens, found throughout the region, have economically valuable trees, shrubs, herbs, and vines. These gardens form a compact multi-storied system of crops—fruits, vegetables, and medicinals—which also provide cash. The valley system, in its structure and function, imitates a natural forest ecosystem. There may be 30 or 40 species in less than a hectare. It represents a highly intensive system of farming that is in harmony with the environment. Animal husbandry, traditionally centered around pigs and poultry, is part of this land use. The animals recycle food from the agroecosystem or other food wastes that are unfit for human consumption.

The *jhum* cycle has been reduced to four to five years or less because of pressure from increased human population and decline in available land from extensive harvesting of timber for industrial use and from *jhum* itself. In urban areas, where population densities are high, a rotational/sedentary system of agriculture, rather than slash-and-burn, is used. Although the rotational/sedentary system is an attempt to maximize production with depleted soil fertility, its yield is often below the level of subsistence. Goat or cattle husbandry, which is inappropriate for the area, could cause rapid site deterioration through indiscriminate grazing, browsing, and removal of fodder. Such deterioration has occurred in the Himalayas. Serious disruption of the social fabric is the result. Consequently, demand for food and an integrated approach to managing the forest come face to face.

Rapid site degradation of once forested areas has several consequences—infertile soils, poor vegetational cover, invasion of exotic weeds—which, in turn, cause simplification of the ecosystem over vast tracts of land (Ramakrishnan 1991). In northeastern India, and in many parts of the tropics in southern and Southeast Asia, exotic weeds, such as *Eupatorium* spp. and

Mikenia micrantha, are replacing vast areas of forest. Elsewhere, in India, we find exotic weeds such as *Lantana* or *Parthenium*. All of these weeds come from the Latin American tropics.

PEOPLE'S INITIATIVE AND SUSTAINABLE MANAGEMENT

Perturbation is an inherent feature of all ecosystems. Natural perturbations are usually minor, although major ones occur infrequently. However, human activities cause more serious disturbances because of their frequency, intensity, or both. Contending with ecosystems that have been severely disturbed by man and managing them sustainably demands the application of knowledge and skills from a variety of disciplines.

Traditional Knowledge

We need knowledge about ecology to make management decisions. Knowledge of areas such as biology and physiology of trees, patch dynamics, ecophysiology of developing forest communities, reproductive biology and nutrient-cycling processes could be integrated into current management practices and future management options.

A case study from northeastern India (see box 12.2) illustrates how an integrated approach to land-use management within the broad framework of a landscape is significant. Agriculture, animal husbandry, and domestic activities of a rural ecosystem are interlinked in the forestry sector in northeastern India where shifting agriculture is traditionally practiced (Ramakrishnan 1992). Here, traditional knowledge plays a key role. For example, even though a number of native species may be important elements for strengthening agroforestry, which is a part of shifting agriculture, the Nepalese alder, *Alnus neplensis*, traditionally conserved by tribal societies, was readily accepted by the local communities. Similarly, some of the important bamboo species are also traditionally valued by the local communities. When the cycle of shifting agriculture is reduced to two to three years or less, many tribal communities plant a leguminous tuber crop, *Fleminga vestita*, in the rotational fallow phase, or even in the sedentary phase. This little known food crop is an important source of protein during the lean period of the year when traditional foods are in short supply. Both *Fleminga vestita* and *Alnus nepalensis* are important species for agroforestry and early successional forest ecosystems because they have nitrogen-fixing characteristics. Similarly, some of the more important bamboo species, such as *Dendrocalmus hamiltonii*, *Bambusa tulda*, and *B. khasiana* play a key role in nitrogen (N), phosphorus (P), and potassium (K) conservation, particularly in early successional environments.

Box 12.2. Shifting Agriculture (*jhum*) and Sustainable Development in Northeastern India (after Ramakrishnan 1992a).

The following strategies suggested by Ramakrishnan and his co-workers for improving the system of land use and resource management in northeastern India are based on a multidisciplinary analysis. Many of these proposals have been implemented.

Transfer technology from one tribal area to another to improve *jhum*, valley-land, and home garden ecosystems. *Jhum* practiced by a hundred tribes in diverse ecological conditions results in a variety of crop and yield patterns. For example, an emphasis on growing potatoes at higher elevations has significantly increased economic yield, despite the low fertility of the more acidic soils at higher elevations.

Maintain a cycle of *jhum* that is a minimum of 10 years—the time needed to achieve sustainability if money, energy, or soil fertility are criteria for evaluation—with greater emphasis on other land-use systems, such as traditional cultivation of home gardens in the valleys.

Accelerate regeneration in the fallow period after *jhum* by introducing fast-growing native shrubs and trees.

Deduce the time needed to regenerate forests and accelerate the restoration of degraded lands through an understanding of the strategies and "architecture" for planting and growing trees and for planting appropriate mixtures of species in time and space.

Improve animal husbandry by using improved breeds of swine and poultry.

Redevelop ecosystems in villages by introducing appropriate technology to relieve drudgery and to increase the efficient use of energy—cooking stoves, agricultural implements, biogas generation, small hydroelectric projects, etc. Promote crafts and products based on leather, bamboo, and other woods.

Strengthen conservation based on traditional knowledge and values with which tribal communities can identify—the revival of the sacred grove, which is steeped in cultural tradition could enable each village to have a protected forest, even though few are now left.

These examples underscore the parallel between ecologically valued and socially significant species that could benefit sustainable development of traditional societies. These species link ecological and social values to ensure that local people who belong to traditional societies will participate in sustainable rural development.

Besides shifting agriculture, there are other systems of land use employed by traditional societies in which concepts of agroforestry based on traditional knowledge and economic and social values are interlinked. Home gardens, planted with economically important tree species as well as shrubs, herbs, and climbers, form another such efficient ecological system. In fact, the ecological processes involved in the organization and maintenance of home gardens correspond to those in the natural forests, where traditional rural societies live (Ramakrishnan 1992b).

Shifting agriculture is part of an agroforestry system whose tradition can be traced back to antiquity. Through agroforestry, the farmer has linked his family and his practice of animal husbandry to the forest ecosystem. Thus, agriculture, animal husbandry and domestic activities of the village are closely linked with the forest ecosystem to provide food, fuelwood, and timber. Such practices have helped to conserve and protect the often limited natural resources of many societies (Nair 1984). The ultimate objective is to optimize the beneficial effects of interactions among woody species with crops and/or animals. In this way, we may diversify production, reduce reliance on external goods and services, and make the farming system sustainable (Gliessman et al. 1981).

Restoration Ecology

Totally desertified graded sites need to be treated initially with herbaceous species and some shrubs before tree species can be introduced. A variety of growth strategies for early successional herbaceous species could be the basis for choosing appropriate species and their locations on a slope. For example, we may choose species for their reproductive characteristics—seed or vegetative reproduction, or r- vs. k- strategy—or for their ability to absorb and use nutrients—species' C_3 or C_4 photosynthetic strategies, which have varying efficiencies of using nutrients. We may take advantage of the "architectural" attributes of shrubs and trees—that they may be planted in a way that forms a coherent whole within the landscape—when they are introduced into the forest ecosystem during its redevelopment (Ramakrishnan 1992).

We are reasonably certain about the long-term ecological implications for the forest if we plant native, rather than exotic, tree species. The temptations to use exotics, such as *Eucalyptus* spp. or *Lucaena leucocephala*, for afforesting degraded lands in the Asian tropics is great. These species are often simpler to use. The potential of native tree species is often overlooked. Mixed plantations have distinct ecological advantages over monoculture forestry. Mixed stands can be as economically productive, or even higher than monocultures, if compatible species are planted. Fast-growing, early successional species that demand light could be compatible with shade-tolerant mid- or late-successional species. In this case, the mixture of species exploits the availability of light at different canopy levels to optimize production per unit of area. Such a mixture would also allow optimum use of nutrients from the soil profile. The early successional species exploit surface soil layers, and the late successionals absorb nutrients from lower levels. A "condensed succession" could revegetate damaged sites through an appropriate mix of various species categories and through a timely introduction of appropriate species into the mixture. More work in this emerging area of study would be most rewarding.

Social forestry programs in developing countries should identify fast-growing native species for plantations, which would meet rural populations' needs for fodder, fuelwood, and timber. For optimal mobilization and use of resources, the planting of early successional species that have a fast growth rate with a rapid turnover of leaves—and consequently, a rapid turnover of nutrients through litterfall—are appropriate. Systems of social forestry that develop village woodlots, either on private or common land, together with agroforestry, would relieve the pressure on natural forests and contribute to conservation. In the final analysis, restoration ecology, if it is to be effective, has to include not only ecological, but social, economic, and cultural considerations.

CONCLUSIONS

Making forestry an integral part of sustainable development in rural communities is essential. In this way, we may adequately provide food, fodder, fuelwood, and timber to meet daily needs and to provide cash income to improve the quality of life of rural poor people. Multipurpose species may play a useful role in sustainable development because they reduce pressure on natural forests. We need to effectively manage tropical forests because the rural poor depend on local forests first to meet their needs. But, effective management entails a combination of disciplines, including ecology, sociology, and forestry. Including forestry as an integral part of sustainable development in rural communities is an essential aspect of our work. While global environmental issues are important, we must not allow them to divert our attention from solving local problems.

Literature Cited

Gliessman, S.R. 1990. Integrating trees into agriculture: Garden agroecosystem as an example of agroforestry in the tropics. In *Agroecology: researching the ecological basis for sustainable agriculture*. (S.R. Gliessman, ed.). New York: Springer-Verlag. pp 160-168.

Likens, G.E., F.H. Bormann, R.S. Pierce, J.S. Eaton, and N.M. Johnson. 1977. *Biogeochemistry of a forested ecosystem*. New York: Springer-Verlag.

Nair, P.K.R. 1984. *Soil productivity aspects of agroforestry*. Nairobi: International Council for Research in Agroforestry (ICRAF).

Odum, E.P. 1971. *Fundamentals of ecology*. 3rd Edition. Philadelphia: W.B. Saunders Co.

Ramakrishnan P.S. (ed.). 1991. *Ecology of biological invasion in the tropics*. New Delhi: National Institute of Ecology.

Ramakrishnan P.S. 1992a. Tropical forests: exploitation, conservation and management. *Impact of science on society*. 166:149-162.

Ramakrishnan P.S. 1992a. *Shifting agriculture and sustainable development: an interdisciplinary study from northeastern India*. MAB Book Series. Carnforth, Lancs, U.K. and UNESCO, Paris: Parthenon Publishing.

Zachar, D. 1982. *Soil erosion*. Amsterdam: Elsevier.

CHAPTER 13

BIOFUELS: DEVELOPMENT ASSET OR ECOLOGICAL THREAT?

James H. Cook

This volume focuses on the potential role of forests and wood-based biomass energy as assets for rural development. Many of the challenges facing developing countries have been mentioned. We have discussed the ongoing energy crisis—the shortages of fuelwood, fossil fuels, and electricity. We have considered the relationships of energy with such crucial issues as poverty, poor nutrition, the lack of adequate health care, and the displacement and disenfranchisement of indigenous peoples. We have recognized that rapid population growth is erasing the progress that we thought we had achieved. And we have acknowledged the important role that women must play in addressing all of these challenges.

LOSS OF ECOLOGICAL BALANCE

We have largely ignored, however, the challenges faced by other species—wild animals, plants and many other organisms that comprise our planet's ecosystems. These other species face rapid and accelerating loss of their habitats, mostly from the increasing demands for land and other resources by human beings. Deforestation, the loss of wetlands, and the destruction of their habitats are rendering these species homeless on a massive scale. We must also consider the potential impacts of global change to the assaults on the local environments of wild animals and plants. Depletion of stratospheric ozone is leading to an increased UV-B flux, and increased concentrations of greenhouse gases may warm the climate and, consequently, increase the sea level. The economic implications of both local and global damage to the environment are profound. For ethical as well as economic reasons, it is crucial to preserve ecosystems, the species that inhabit them, and the fragile balance between the two.

I am not going to discuss all of these issues—that would take too long. Still, I would like you to keep them in mind as the context for my remarks. I am going to consider how biomass could help meet the demands for energy of people in developing countries—from an ecological and, often, nonhuman viewpoint.

MANAGEMENT'S CRUCIAL ROLE IN PRODUCING BIOMASS

First, it is important to recognize that biomass is not the only option that we have for renewable energy. Other possibilities include photovoltaic and solar thermal systems, wind turbines, and, perhaps, small-scale low-head hydroelectric units. It is crucial to remember that all energy sources have impacts of one sort or another on the environment. Given these considerations, the total impact can be reduced in two ways. We can minimize our demands for energy by maximizing efficient use of that energy, and we can minimize the adverse impacts of each unit of energy that we use. The optimal choice will be different for each place, each time, and each scale that we consider.

What would be the source of the biomass to meet demands for energy in developing countries? The logical alternatives are natural forests, plantations of trees and herbs, and other more or less intensively managed ecosystems. With current practices in management, natural forests and plantations are complementary: natural forests provide lots of habitat and little biomass, while plantations provide lots of biomass and little habitat. Given current rates of global deforestation, loss of biodiversity, and population growth, it does not seem appropriate to consider increasing the pressure on the natural forests that remain. Rather, it seems obvious, even to the casual observer, that humans should limit and reduce their demands on natural forests by obtaining the biomass that they need from relatively small areas of intensively managed plantations. Indeed, this strategy is the essence of modern agriculture.

Recall, however, that our goal is to minimize the ecological impact of obtaining a given quantity of biomass. Perhaps forests could be managed for increased production of biomass without significantly decreasing their value as a habitat. Alternatively, plantations could be managed for increased value as a habitat without significantly decreasing their production of biomass. The optimal mix will depend on the relative feasibility of these approaches. Time —and research—will tell.

MANAGE FORESTS FOR MULTIPLE YIELDS

I would like to turn now to the chapter by Donald Grebner and Mark Ashton, "The Development of Regeneration Methods for Mixed-Dipterocarp Forests: the Past, the Present and the Future." After a discussion of the complexities inherent in managing these forests, they note that past and present silvicultural systems developed by man have led to simplification of tree stands and, therefore, that they are unsustainable in the long run. They go on to suggest that harvesting of these forests for sustainability will require the development of less intensive silvicultural systems that mimic natural processes of regeneration and succession of tree stands to yield multiple values and to maintain complexity.

In response to a question about the economic feasibility of harvesting biomass for energy, they predict that biomass will play a subsidiary role to that of high quality timber. Indeed, they argue that it would be economically viable to harvest for energy only when harvesting higher value products. On the other hand, they note that harvesting biomass for energy might actually facilitate management of the forest understory.

In their chapter, "Nutrient Budgets of Young Plantations with Native Trees: Strategies for Sustained Management," Florencia Montagnini and Freddy Sancho maintain that highly productive tropical plantations could provide a variety of benefits. They demonstrate that appropriate native species of trees could play an important role in restoring degraded lands because these species enrich the soil with nitrogen and organic matter. They emphasize the economic value of returning these degraded lands to their former capacity for production. And they argue that plantations could diminish the pressure to cut remaining natural forests.

While the management of natural forests for sustainability is clearly determined by the choice of silvicultural system, the key seems to be low intensity management for multiple yields. The ecological impact of plantations depends on their context in the larger landscape, the nature of the former use of the land, and the approach to their management.

TOWARDS SUSTAINABILITY AND RURAL DEVELOPMENT

Given the ongoing global loss of biodiversity, most conservation scientists agree that it will be crucial to maintain and establish networks of reserved natural areas to preserve ecological systems and the species that inhabit them. It will also be crucial to provide the resources needed for economic development. Low intensity silviculture in natural forests and intensively managed plantations may each play important roles in this endeavor.

Similar designs for such networks of natural areas have been proposed by several workers in the field. Basically, a network would consist of variously-sized reserves and interconnecting corridors, embedded in a landscape largely dominated by human uses. At the center of each reserve would be a core area, managed to provide optimal habitat for wildlife. The core area would be surrounded by a series of annular zones, with each successive zone one moved outward from the center as being more intensely managed and used than the preceding one. This series of zones would buffer the core area from the human-dominated matrix—increasing its effective size—and would also provide a variety of economic and aesthetic benefits to the local populace and give them a stake in the reserve system. Such reserve systems can also include existing low intensity use by indigenous populations in ways that are compatible with ecological goals. Indeed, local participation in restorative forest ecology projects is essential to their success.

The core reserves would not be isolated, for they would generally be too small to be self-supporting for the long-term. They would be interconnected by a series of wide corridors that would make migration possible and increase the overall habitat value of the network. These corridors of natural reserves would themselves be buffered from the human-dominated matrix similarly to the core reserves.

For example, the zone immediately surrounding a core reserve would be a national park, a lightly used recreational area where people could visit, but where no private buildings or cars would be allowed. The succeeding zone would comprise resort areas featuring the outdoors, the landscape, and outdoor sports such as skiing. Next would be areas where the extraction of resources would be the prime activity. Some of these extraction activities, such as forestry, would be renewable. It is in the zone where we would find vacation homes alongside rural community agriculture, forestry, and agroforestry. Such uses would serve as a transition to intensely managed farms and tree plantations. Plantations of native trees established on degraded lands could also play a transitional role in the re-establishment of natural forests on degraded lands—for enlarging the core areas or for creating the natural reserve corridors. Further outward from the core reserves would be suburban areas, and, finally, industrial areas and cities.

Part Four

Global and Tropical Forest Economies

INTRODUCTION

The ecological boundaries of future wood energy systems are conditioned by the social and economic factors that affect sustainable use. Forest industrial development, opportunities for temperate wood plantations, the history of commercial logging and wood energy in selected countries, and the emerging social forestry paradigms all shape the role for wood energy in various parts of the developing world.

Experiences captured are presented from rural economies in transition in the Philippines, Indonesia, and Brazil. These chapters highlight the close socio-economic interactions that impact wood energy use. As traditional and new patterns of forest management are recognized by policy makers—be they industrial rubber plantations in Brazil or the less visible forest uses and ownership rights of villagers in Indonesia's Kalimantan woodlands—these patterns become valued for their important social and economic foundations. In Kalimantan, for example, local villagers have extremely intricate property rights that determine forest management. These relationships are critical to any forest management policy determined at national levels if sustainable forestry is to be continued.

Forest economies that drive markets are shaped by the national policies that distort or support wood use at the local levels. Non-timber forest products and the forest products industries make sense under varying conditions, as discussed in this chapter. No universal rules of thumb can be applied everywhere, as our Brazilian case study warns us. Rather local and international users require proper signals to ensure that the most appropriate local uses are being fostered. In this section, we raise cautionary social and economic flags which must be accounted for when deciding if the wood residues noted in Part One are truly available for long term use. These papers point out the need to develop future wood energy strategies by building resource inventory bases from local frameworks. Local realities will determine the resource management reality for rural people.

CHAPTER 14

COMMERCIAL FORESTRY AND RURAL DEVELOPMENT

Roger A. Sedjo

INTRODUCTION

The link between commercial forestry and economic development—rural or otherwise—has been quite controversial and open to interpretation. A major thinker in this area was the recently deceased Jack Westoby. Westoby (1962) saw a strong potential for forestry and the forest industry to contribute to the economic development of the Third World. The forest would provide a resource base upon which a viable industry could be built. Such economic activity would generate local employment and income, as well as establish linkages to other sectors of the domestic economy, and thereby would contribute to overall economic development. The revenues generated by the forest resource could be used to finance investments to further the development of the region. Later, Westoby (1978) became increasingly disillusioned with his initial view. He eventually concluded that the forest industry had contributed little or nothing to the economic development of Third World countries. Rather than provide revenues for financing the development of a region, he observed that the creation of the forest industry had merely provided a cheap source of raw materials for the foreign market. His "dependency theory" described the situation: in essence, forest industry development had simply become a vehicle to provide developed countries with cheap forest resources by exploiting the local populations and by generating local economic dependency.

Recently, other analysts have reexamined the relationship between commercial forestry and economic and rural development. Although not quite that of the traditional dependency theory, their increasingly common assessment is that commercial forestry is not playing the role in Third World economic development that many had hoped it would. We find a reflection of this general pessimism in the recent World Bank position, stated in its latest "Forest Sector Policy Paper," that as a matter of policy, the Bank will *not* provide financial support for commercial logging in Third World countries.

Of course, this important relationship between commercial forestry and rural and economic development does extend, in fact, beyond mere forestry.

It includes the role that exploitation of natural resources in general plays in the economic development of the Third World. I believe, however, that there are some unique features about forestry that make the problem of its contribution to economic development more complex than that of most natural resources. These features include not only the increasing concern about global environmental problems, such as biodiversity and atmospheric carbon, in which forests play a critical role, but also the vast land area covered by forests, the presence of local peoples within the forests, and the variation in *de facto* and *de jure* land tenure arrangements. This discussion will overlook the important environmental issues and will focus on the implications of tenure systems, *de facto* and *de jure*, and on forest industry development and on that industry's contribution to regional economic development.

My major thesis is that many of the difficulties in linking forest industry development to regional economics pertain to its distributional consequences. These consequences, in turn, are related to tenure, i.e., property rights, issues, which are both explicit and implicit, and how these issues often are misunderstood. Although this argument seems almost self evident, one needs only to review the literature to see that the role of tenure and rights—both formal and, especially, informal—in development is not well understood. In fact, we rarely address the question of whether or not the forest industry is successfully contributing to sustainable regional and rural development.

We begin this chapter by discussing some current views on this issue of the relationship of forest industry development to economic growth that have been published in recent literature. We cite examples of where the forest industry has promoted economic development effectively, and where it has not. We consider a particularly difficult case situation—evaluating the role of the forest industry on economic development in Indonesia. Finally, we conclude our analysis by relating these examples to the distributional consequences of forest industry development in many tropical regions.

FOREST INDUSTRIAL DEVELOPMENT: SUCCESSES AND FAILURES

I recently reviewed two books in the journal, *Economic Development and Cultural Change* (Sedjo 1990). The first, entitled *The Hour of the Fox: Tropical Forests, the World Bank and Indigenous People in Central India* (Anderson and Huber 1988), examines a project of the late 1970s that was financed by the World Bank. In this project, in the region of Bastar, India, a tropical forest area of Madhya Pradesh was converted to a forest plantation and a large pulpmill was established in the area. Since the project had largely negative effects upon the local people, the authors are highly critical of both the World Bank and the project. We immediately ask why a project, designed to promote the economic development of a region, should be regarded with such hostility by the local people and also have such severe negative consequences for their welfare.

The second book that I reviewed, *The Forest for the Trees? Government Policies and the Misuse of Forest Resources* (Repetto 1988), is a summary report of a larger volume edited by Robert Repetto and Malcolm Gillis (1988) entitled *Public Policies and the Misuse of Forest Resources*. It is highly critical of commercial forestry's role in development, at least as it is currently practiced in a number of tropical countries. Here, the criticism is directed neither at the forest industry *per se* nor at the development banks, but rather it focuses on the policies that are created largely by the local governments where commercial logging occurs. The authors are primarily concerned about the mechanisms designed to capture the forest resource rents (stumpage fees), especially in the tropics. They ask whether the governments or private interests are capturing the resource rents or stumpage fees, and what the effects of these incentive policies are on both forest management and forest exploitation. Repetto and Gillis argue that far too small a portion of the rents are being captured by the governments. They also argue that the incentive system that was created encourages excessive rates of harvest.

In our recent book, *Global Forests* (Laarman and Sedjo 1991), Jan Laarman and I argue that examples can be found where commercial forestry has contributed to economic development and where large-scale logging has left few significant developmental benefits. Finland, Sweden, and the U.S. South provide examples of regions where forestry has offered positive long-term contributions to economic development. Less successful examples are found in Tasmania, Australia, and in Brazil, a primary exporter of mahogany logs. We submit that, in both of these latter cases, the forest industry has done little to promote the sustainable economic development of the region, but rather it has tended instead to encourage dependency. For other regions, "the jury may still be out" on the question of whether they are successful cases of economic growth or whether they are similar to our exploitative examples. An interesting situation, with which I have some familiarity, is the development of the commercial forest-products industry in Indonesia. Over the past 25 years, the industry has experienced rapid growth, first as an exporter of tropical logs, and then more recently as the world's leading exporter of tropical hardwood plywood.

Today, we can make a case for both success and failure from the Indonesian experience. From the late 1960s to the 1980s, Indonesia was almost exclusively an exporter of wood. In the early 1980s, the Indonesian government gradually imposed a prohibition on the exportation of logs and encouraged the establishment of major plywood facilities. Today, Indonesia is the world's single producer and exporter of tropical plywood and has made the transition from a log-exporting country.

Was this effort a success? It depends on our criteria. The case for success is that thousands of jobs have been created, and millions of payroll dollars have been generated in a host of different locations around Indonesia. In addition, hundreds of millions in dollars of foreign exchange have been generated by the timber industry, first from exporting logs and more recently from exporting plywood.

The negative side of the equation is spelled out by Gillis (1988) and others. The resource rents accrued primarily to the private logging concessionaires. The government, Gillis argues, has not collected its fair share. In addition, Boulter (1990) and others have maintained that the domestic wood-processing industry is highly inefficient. Thus the available rents have been reduced and net foreign exchange earnings probably have fallen also. In summary, Indonesia, it is claimed, has been subsidizing its domestic wood-processing industry with its timber resource rents. Furthermore, Indonesia could have fared far better in generating foreign exchange income by simply continuing to export logs.

Nevertheless, it is clear that employment has been created through domestic wood processing, including employment in mills that are located in outlying regions. However, for this employment to be sustainable in the long term Indonesia will need almost certainly to improve the efficiency of her processing operations (table 14.1).

Table 14.1. Indonesian Industrial Wood Sector's Contribution to Economic Development

Phase I: Log Exports (mid-1960s to early 1980s)	
Goal	**Accomplishment**
Employment	modest
Sustainable Regional Employment	little
Net Foreign Exchange Earnings	substantial
Domestic Linkages	modest
Government Collection of Rents	small
Phase II: Plywood Exports (early 1980s to the Present)	
Goal	**Accomplishment**
Employment	modest
Sustainable Regional Employment	significant
Net Foreign Exchange Earnings	probably reduced from log-export phase
Domestic Linkages	increased from log-export phase
Government Collection of Rents	small

EFFICIENCY, DISTRIBUTION AND LAND TENURE

Economists focus on two types of problems: resource efficiency and income distribution. Under *efficiency*, we usually address the issue of how resources are used to produce the social pie; that is, the sum total of goods and services. When we consider *distribution*, we ask how the social pie is allocated; that is, who pays, who benefits, and how new net income is distributed.

Income distribution depends in part on the ownership of assets and is influenced by issues of land tenure. It has always surprised me that an element of the exploitation of natural resources in rural and economic development that tenure and ownership of resources and the relationship of these ownership patterns to regional development is rarely addressed, at least in the forestry literature. Since forests are a valuable asset, their ownership represents a claim to the potential revenues that are derived from them. The two successful cases that are cited in our book, *Global Forests*, are both examples from regions in which the forest asset is broadly held. Both Finland and the U.S. South have a large number of relatively small-scale forest owners. As owners of assets, these proprietors can participate in rural development in several ways. First, as individuals in the rural sector, they can contribute by selling their labor services (or some of the family's labor services) to the forest-products industry. Second, as owners of assets, they can participate by converting their forest asset, timber, into current income. Furthermore, having land tenure gives these local people the option of investing some of their proceeds into forest management for the long term. They can maintain a sustainable and economically profitable forest activity on their lands. In short, the tenure patterns in the forests of Finland and the U.S. South allow for greater linkages between the development of a forest industry and the local people and their economy.

By contrast, in the unsuccessful examples, local ownership is not present. In Tasmania, the forests are largely owned by the state, and it uses tax revenues to subsidize exploitation of the forests. The participation of local people is largely limited to the labor services. In Brazil, the situation could be characterized as the exploitation of mahogany trees, which are a common-property resource. Lack of tenure has encouraged a "cut-and-run" philosophy because no secure basis is provided for long-term resource development. This outcome is likely even if the short-term problems of market control had been adequately addressed.

The case of Indonesia is one in which the government has ownership of all the forest lands and provides large business interests with the rights to concession logging and to forest development. By not charging a market price for these concession rights, the government allows most of the rents to be captured by the concessionaire. The local people, having no rights to the timber, have no legal claim to the income (rents) from the forest assets when the timber is sold. Thus, local people can participate only minimally in the forest industry by selling their labor services, either for logging or for wood processing. Furthermore, although local people do have traditional rights to some of the nontimber products of the forest, these rights are subservient to the concessionaire's logging rights. Thus, the values of nontimber product from the forest are often lost during logging, since the logger has no claim to any revenue from these products and, thus, has no interest in trying to protect them during the harvest. These factors suggest that local involvement in the development of the forest industry will be modest because most ownership and, therefore, control of the forests are external to the region. Such ownership patterns also suggest that local people in Indonesia will not enthusiastically embrace forest industry development. The local people are losing their traditional rights to some of the forest's products, and, in this sense, support some of the costs of this development.

My description of the Indonesian dilemma brings my argument back to the situation in the Bastar region of India. Why do the people oppose the project? For the same reason as they do in Indonesia: because such a project represents the wholesale transfer of the *de facto* ownership of the forest from the locals to the forest industry company. Since the land is legally owned by the government, then the government can dispose of it in any way that it chooses. However, the traditional use of the land has been the province of the local people, and they are deprived of this use, this implicit property right.

Nevertheless, as table 14.1 suggests, the forest industry's impact on Indonesia's economy is ambiguous. The good news for Indonesia is that, unlike some other countries, Indonesian business interests usually invest the proceeds from their exploitation of resources heavily back into Indonesia. Thus, earnings from forestry do not go to a Swiss bank or to a Manhattan office. Rather they return into the Indonesian economy, although rarely into forest management. The bad news is that the ownership of assets is highly weighted towards the few, and the situation is getting worse. Forest industry and resource development have relatively weak linkages with much of the local economy in the forested regions.

FOREST RESOURCES GENERATE SERIOUS PROBLEMS

Although conflicts of *de facto* and *de jure* property rights can be found with many natural resources, they are probably most common, and most difficult, with forest resources. In forests, more than in most other resources, development often can involve large-scale redistribution of economic benefits from local inhabitants to others. For many other resources, the effects of this redistribution on development are probably nearly negligible. Petroleum, for example, is underground or underwater. Local people rarely experience severe losses of their rights when petroleum resources are developed. Mineral resources typically involve relatively small areas, even when they are above ground (e.g., open-pit mining). Again, local rights are rarely in conflict with development. By contrast, forests involve large areas of land. In many cases, this land is the property of the national government, even though the local population is allowed to use it extensively and has important traditional rights to its resources, particularly, in this case, to the nontimber products.

CONCLUSIONS

A serious effort to develop timber resources or forest products can involve a major redistribution of rights to use the forest resources and assets. Rarely do the claims of local users allow them to capture significant portions of the forest rents. Hence, the conflicts largely involve issues of income distribution. The local people usually lose assets, while the forest industry, often controlled by powerful interests, typically captures them. Ultimately, economic efficicncy—the size of the pie—is a less significant aspect of the conflicts involving the forests than is the distribution of that pie.

Literature Cited

Anderson, Robert S. and Walter Huber. 1988. *The hour of the gox: tropical forests, the World Bank and indigenous people in central India.* University of Washington Press, Seattle.

Boulter, David. 1990. Final Report, Forest Planning Assistance Project, draft. FAO Project INS9. Jakarta.

Gillis, Malcolm. 1988. Indonesia: public policies, resource management, and tropical forests. In: *Public Policy and the Misuse of Forest Resources* (Robert Repetto and Malcolm Gillis, eds.). Cambridge: University Press.

Laarman, Jan G. and Roger A. Sedjo. 1991. *Global forests: issues for six billion people.* McGraw Hill, New York.

Repetto, Robert. 1988. *The forest for the trees? Government oolicies and the misuse of forest resources.* World Resources Institute, Washington, D.C.

Sedjo, Roger A. 1990. Book review in *Economic Development and Cultural Change.* 39 (1):215-201.

Westoby, Jack C. 1962. Forest industries in the attack on economic underdevelopment. *Unasylva* 16(4):168-201.

Westoby, Jack C. 1978. Forest industries for socioeconomic development. *Commonwealth Forestry Review*, 58(2):107-116.

CHAPTER 15

INCREASING GLOBAL TIMBER SUPPLY TO MEET THE WORLD'S GROWING DEMAND FOR WOOD

W.R.J. Sutton

A SOLUTION FOR THE GLOBAL DILEMMA

As forest managers, we are often accused of being too concerned about the production of timber as a raw material for wood and fiber. Critics tell us that we should pay more attention to nontimber values—protection of fauna and flora, the quantity and quality of water, the soil—especially to prevent soil degradation—the carbon cycle, the global climate, and the rights of indigenous peoples. In addition, we need to promote biodiversity and recreation. In this discussion, we will review future global demand for wood and its implications, both for long-term investments in growing timber and for nontimber values.

Rapid growth of the world's population is increasing total demand for wood. We are quickly approaching the limits of timber supply from the world's remaining natural forests and may have already exceeded the levels of supply from their sustainable management. As overall standards of living, including health and levels of income, improve, population growth in poorer nations will probably stabilize. However, an increased standard of living almost certainly means an increased per-capita use of wood. In addition, a wealthier, more educated society will also demand that more forest land be set aside for biological reserves, parks, and other scientific or recreational uses, all of which will further reduce the short-term global supplies of timber.

Dramatic increases in the supply of timber are possible by investing in forest plantations. Plantations can be successful, but it will take a long time before investments can be made with confidence. New Zealand has four to five rotations, each of 25 to 30 years' duration, as a foundation of experience in plantation forestry. For other countries, however, investments in plantations now may only yield returns well after shortages of wood affect timber prices. Using substitutes for wood may alleviate the shortage, but these materials—metal, plastics, concrete—require large amounts of energy to manufacture. We must quickly develop cheap, renewable, and environmentally friendly alternative sources of energy.

The global dilemma is obvious. The solutions are not. We will explore the possibility of a massive investment in plantation forestry, with fast-growing species, using the "lessons learned" from New Zealand's experience as a solution.

GLOBAL CONSUMPTION OF WOOD

Wood is an astonishingly versatile raw material because of its unique combination of properties, especially its strength per unit of weight. Most industrial wood is still used in solid form for sawtimber, plywood, poles, or pilings. Annual worldwide consumption of wood is phenomenal—a total of about 3.5 billion m^3. On a green-weight basis, we consume as much wood as we do food.

The world's population has increased from 2.5 billion in 1950 to over 5.4 billion in 1989, at an annual rate of 1.8%. FAO statistics show that the global harvest of wood has increased during the same time from less than 1.5 billion m^3 to nearly 3.5 billion m^3, an annual increase of 2.3%. Since 1950, about half of the global consumption of wood has been for fuel and half for industry —construction, paper, wood products. Annual per-capita consumption of wood has been rising as standards of living have improved, from less than 0.6 m^3/capita in 1950 to about 0.67 m^3/capita in 1989. Even though more substitutes for wood have been used, wood is still consumed both for traditional and new wood products.

Consumption of wood will increase even more in the future. Table 15.1 shows the FAO (1991a) projections for the demand for wood by major end uses and the United Nation's projections of world population through the year 2010. From 1989 to 2010, the annual harvest of wood will increase by 1.6 billion m^3, while the global population should increase by 2.1 billion. At such levels, the annual per-capita consumption of wood will increase to over 0.7 m^3 by the year 2010. An increase of 1.6 billion m^3 over 21 years represents an annual increase of 77 million m^3. The world must add the equivalent of another British Columbia or six New Zealands to the global harvest each year in order to meet such annual increases of demand.

DIMINISHING GLOBAL FOREST RESOURCES

Global supplies of wood come primarily from our remaining natural forest resources, some of which are being currently harvested for the first time. Some previously harvested forests have been regenerated, often many times, after each harvest, while some land has been returned to forest cover after many years' use for crops or pastures. Usually, shortages in wood supply have been regional, not global, in nature. We thought that enough forests remained for a smaller global population that was not expected to increase so sharply. International trade solved most local imbalances between supply and demand, except for fuelwood and other low-value materials.

Table 15.1. Projection of Wood Demand and Population to 2010

	1989 Production	2010 Projected Production	Increase
Industrial Wood (billion m^3)	1.68	2.68	1
Fuelwood (billion m^3)	1.78	2.39	0.61
Total (billion m^3)	3.46	5.07	1.61
Population billions)	5.1	7.2	2.1

Source: FA0 1991a

The remaining undisturbed, natural forests of the world are under increasing pressures from the demand for wood, caused by the population explosion and rising real incomes. At the same time, the number and quality of natural habitats, and biodiversity, are shrinking. There is increased pressure to establish forest reserves for biological conservation. The conflict between these two competing, and possibly incompatible, demands will certainly intensify.

A review of the world's forest resources shows that we are nearing the limits of forest exploitation. The United States, Canada, and the Newly Independent Republics of the former Soviet Union, countries where wood is harvested from virgin or managed natural forests, have little potential to increase their harvest levels in the near-term. Europe has some potential for a small increase in its harvest, mostly in Sweden and France, at most only an extra 100 million m^3/yr by about 2010. The present level of supply or annual harvest cannot be sustained, let alone increased, from the remaining natural forests in Africa, Asia, Central America, South America, and Oceania.

Concern about the environment has raised our awareness of the high rate at which the tropical forest is being cleared. Clearing the land for crop and animal production, rather than logging, is the primary cause of deforestation. Harvesting tropical sawlogs has been vigorously discouraged by a variety of political and economic sanctions. The prevailing assumption has been that lowering the consumption of sawlogs will slow the loss of tropical forests.

According to FAO production statistics for 1989 (FAO, 1991b), the world's total sawlog harvest was 1,006 million m^3 (figure 15.1). Of this total volume, 29%, or 291 million m^3, of the total harvest, were hardwoods, and just under half of that amount—134 million m^3, about 13%, came from tropical countries (table 15.2 and figure 15.1). The largest producers of tropical hardwood logs in 1989 were Malaysia, with 41 million m^3; Indonesia, with 36 million m^3; and Brazil, with 19 million m^3. Tropical hardwood sawlogs are desirable products, but one alternative is to use wood from plantations to reduce the harvest from natural, tropical forests.

Table 15.2. 1989 Tropical hardwood & World Sawlog Harvest (in million m^3)

Tropical Regions		Volume	%
Africa	**Total**	16	12
Latin America	Brazil	16	12
	Other	15	11
	Total	31	23
Asia	Malaysia	41	31
	Indonesia	36	27
	Others	10	7
	Total	87	65
All Totals (above)		134	100

Major World Groupings	Volume	%
Total tropical hardwoods	134	13
Other hardwoods	158	16
Total hardwoods	292	29
Total conifers	714	71
Total all sawlogs	1,006	100

Source: FAO 1991b

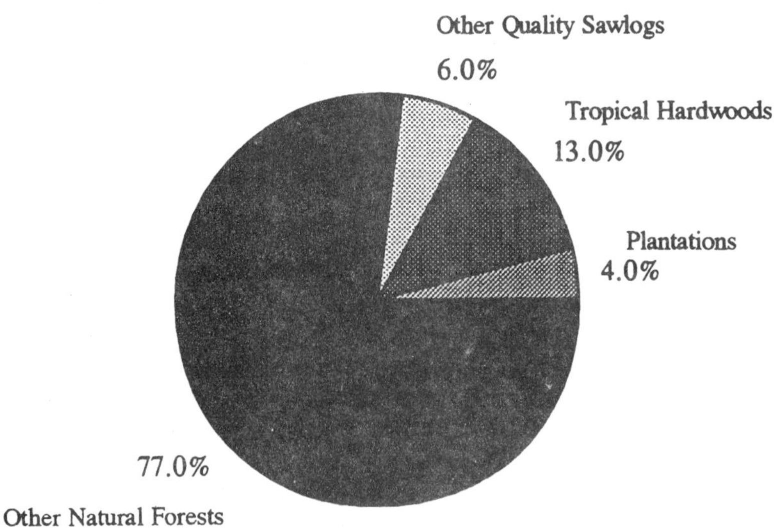

Figure 15.1. 1989 Global Sawlog Harvest (1006 million m^3).

NEW ZEALAND'S EXPERIENCE WITH PLANTATION RADIATA PINE

Only three countries—New Zealand, Chile, and South Africa—produce almost all the industrial wood they need from plantations. Prior to human settlement, 75% of New Zealand's 66 million acres was forested (figure 15.2). While some areas have been deforested recently by volcanic eruptions, most of the remaining unforested area is either too high, too dry, or too wet to support the growth of trees. The Maori began to migrate to New Zealand about 1200 AD. As the native population grew, indigenous forest cover slowly decreased. By the time the major European migration began, around 1840, 53% of New Zealand was forested.

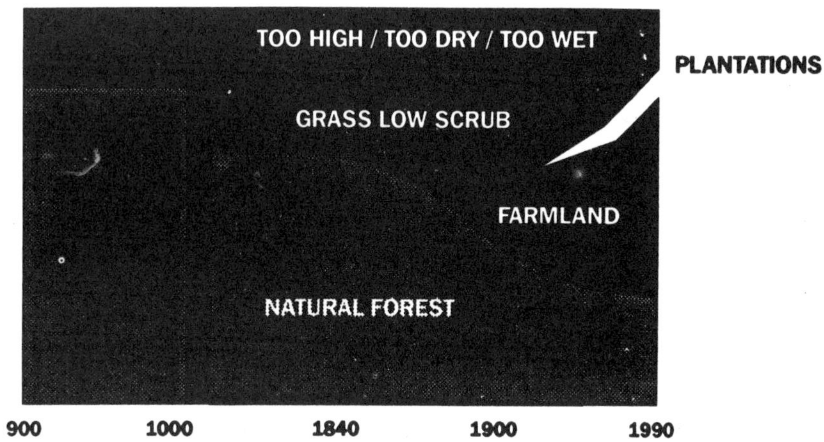

Figure 15.2. New Zealand Vegetative Cover 900-1990.

Forested area in New Zealand was reduced to 28% by 1990, of which 75% is natural forest and almost 25%, about 5.3 million acres, is plantations, primarily of Monterey or radiata pine (*Pinus radiata*) that was introduced from coastal California. Only about 1% of the indigenous natural forest, no more than 350,000 acres, was converted directly to plantations, but this natural forest had been logged at least once. Virtually all plantations are located on areas that were initially cleared for agriculture, since about 99% of originally forested land had been used primarily for sheep farming.

New Zealand's Domestic Supply and Demand: 1910-1990

In 1910, New Zealand's total consumption of domestic wood was about 1.6 million m^3. At that time, some plantations, established in the 1870-80s with tree species introduced from abroad, provided an estimated 3% of the total wood supply, while indigenous natural forests supplied the remainder. Forests were cleared primarily for pasture land, with a great surplus of timber as a result. Stumpage prices were very low. Since New Zealand was covered with so much natural forest of high-quality, old-growth trees, high-quality, large logs dominated its supply of wood. All sawtimber used, except for the small amount that came from plantations, was clear, free of defects, and available for little more than the cost of extraction and sawing. Quality wood was used for everything, whether it was necessary or not.

Even though New Zealand was wealthy in timber, the nation became aware that its forests could not be an inexhaustible source of wood forever. The Royal Commission on the future of forestry had been established by 1913. It predicted that ***"...it is not safe to conclude that there will be any supply of moment at the expiration of 30 years [i.e. by 1943]"***. Indigenous tree species were slow growing and difficult to manage. However, several of the species that had been introduced from abroad had demonstrated high rates of growth. Radiata pine, the outstanding example, exhibited growth rates of around 20 m^3/ha/yr, compared with only 1 m^3/ha/yr for most of the native species.

To ensure that New Zealand would have a future supply of wood, the Commission recommended a major effort to establish plantations. Between 1925 and 1935, enough plantations of sufficient size, mainly of radiata pine, had been established to ensure self-sufficiency in timber for at least the next century. By 1950, 335,000 hectares of plantations had been established, 55% by the State and the remainder by the private sector (figure 15.3). The total timber harvest in 1950 was 2.8 million m^3, 60% from natural forests and 40% from plantations.

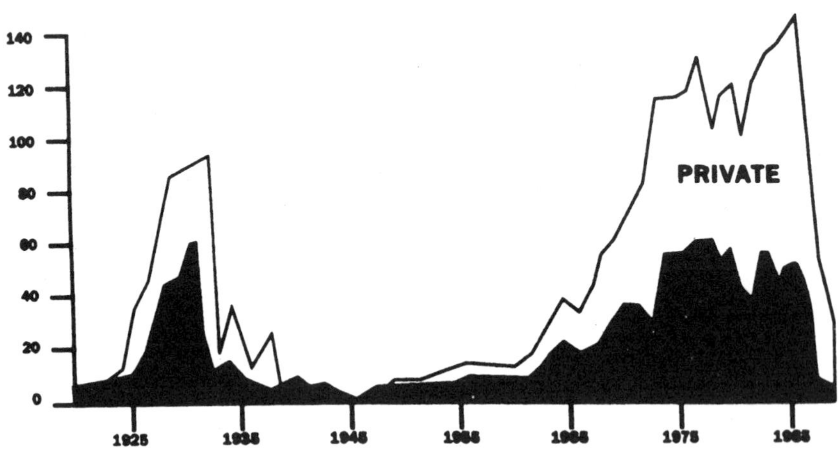

Figure 15.3. Annual New Planting - New Zealand (000's acres).

New Zealand has always consumed large amounts of wood, mostly for housing construction. A political mandate of successive governments has been that timber prices should be maintained at a minimum level so that housing costs could be low. Consequently, trees with high-quality logs in its natural forests were sold until recently for artificially low stumpage prices.

By the 1950s, it became apparent that supplies of high-quality wood from indigenous natural forests were becoming depleted. However, intensely managed plantations, with timely thinning and pruning, could yield the same results. In addition, radiata pine from well-tended plantations had excellent potential for export. J.S. Reid (1953) asked, ***"Can we afford not to prune and thin the stands when the resultant products of such treatment appreciate in value to the extent that may be expected with the species?"*** The era of plantations had begun.

New Zealand's plantations were based on focused research on a large scale. Over the past 30 years, from 8,000 to 10,000 years of research have been invested in plantation forestry by scientists and technicians. Indeed, radiata pine is the most studied plantation species in the world. This research culminated in a second planting boom between 1965 and 1985 to produce wood for export, giving New Zealand an alternative to exporting lamb, wool, and dairy products as a means for obtaining foreign exchange. In 1990, New Zealand harvested 12 million m^3 of wood, with only 3% of this total from natural forests—the exact opposite of the situation in 1910 (figure 15.4). With pressures from environmental organizations, the New Zealand forest products industry has agreed not to harvest wood from natural forests in return for support from environmentalists for large-scale plantations. As of 1991, New Zealand's total area of plantations was 1.2 million hectares.

The quality of wood harvested from plantations has improved gradually. During the last 30 years, from 70% to 80% of plantations in New Zealand have been intensively managed, with such techniques as early pruning to ensure a high proportion of defect-free sawtimber or plywood. However, high-quality, clear wood has only been available since 1990. With increased availability of quality wood and relatively static demand for wood at home, New Zealand now exports about 50% of the wood that it currently produces, earning US$800 million/yr (figure 15.5).

Lessons from New Zealand's Experience with Plantation Forestry

In the 5,000 years that agriculture has evolved, few of the initially promising species of plants, animals, and birds have proved suitable for large-scale commercial use. Agricultural crops are almost entirely even-aged monocultures. Almost all the attempts at mixtures, or uneven-aged management systems, have either failed or have not proved economically viable. The lessons from New Zealand's experience suggests that the future of plantation forestry will be similar. Just as in agriculture, even-aged monocultures have proved to be the only viable management option for commercial plantations.

New Zealand is now almost totally dependent upon radiata pine, but few realize that it was one among hundreds of species tested in plantations and emerged relatively recently as the most successful. L. M. Ellis (1923) noted that ***"200 species have been tested in several hundred acres of plantations."*** Since 1920, when less than 1% of plantings were radiata pine, almost 500 tree species have been tested. Almost all have either failed or proved only partially successful. Although a few other potentially promising species are still being investigated, only radiata pine has proven itself to be a consistently sound commercial investment.

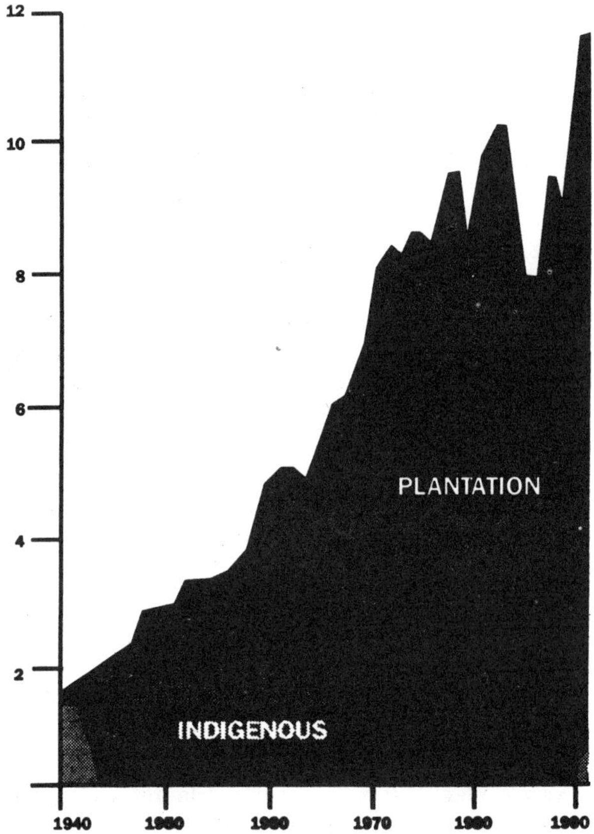

Figure 15.4. New Zealand Total Harvest by Indigenous and Plantation Source (million m^3/yr).

The lessons from New Zealand should be considered and evaluated carefully by any nation formulating a program of plantation forestry. Initial attempts with most species and silvicultural regimes do not achieve their intended objectives. A long period of experimentation is required to determine exactly how species and regimes will perform. New Zealand needed four rotations of sawlog regimes before it could begin to produce high-grade sawlogs with confidence. Other countries may be able to reduce the time required for experimentation, but it is unlikely that they will gather enough information in less than one or two rotations. The period of experimentation also includes investigation about the types of wood products made from the "new" tree species and their marketability. It may take decades and considerable marketing experience before consumers fully accept a species that has not been used before.

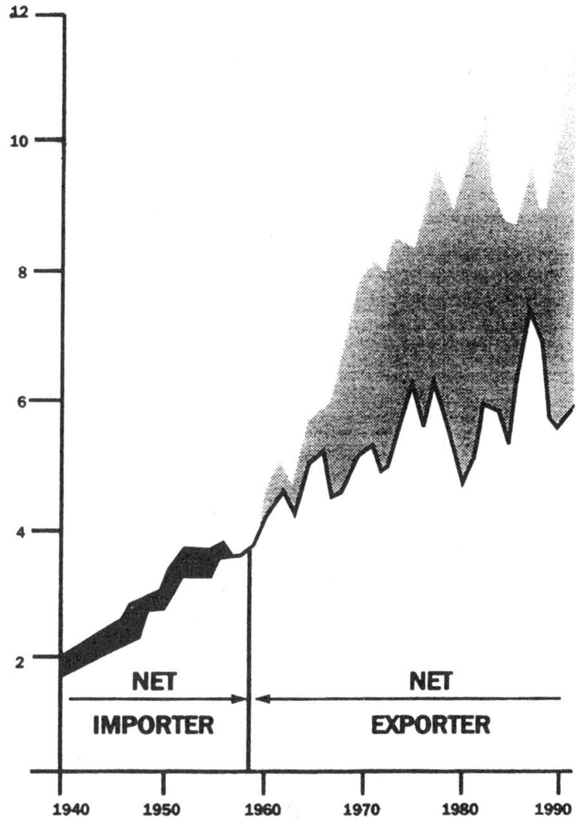

Figure 15.5. New Zealand Total Harvest and Domestic Demand (million m^3/yr).

Long-term vision is required to invest in plantations, especially in the production of high-quality sawlogs, because the market rarely expects future scarcity. For example, sawmillers consistently told owners of radiata pine plantations that high-quality logs of radiata pine would not command a premium price. At that time, before these owners had entered the timber market, it was still possible to purchase large, high-quality sawlogs of indigenous species for very low stumpage prices. The plantation growers, however, were confident of their product because they knew that supplies of quality logs from natural forests were diminishing and that tests of pruned radiata pine from plantations had demonstrated its quality.

The predictions of the plantation owners in New Zealand proved correct. As their logs of radiata pine entered the market in the late 1980s, timber supplies from New Zealand's natural forests dropped to minimal levels, and stumpage prices increased. As the plantation owners had foreseen, radiata pine proved to be an ideal substitute for high-quality wood from the natural forest. Current prices for the pruned logs fully justify the early investment in tending the trees. There was no indication of the long-term profitability of this investment when the trees were first pruned in the late 1950s and early 1960s.

We should note that in North America, where dependence on timber supplies has shifted more gradually from natural forests to plantations of native species, e.g., Douglas fir and southern pine, we may expect higher future prices (e.g., Berck 1978, Berck and Bentley 1992). Perhaps forest farmers in New Zealand had similar expectations, which were obscured by the artificially low prices charged for public stumpage.

PLANTATIONS AS A SUBSTITUTE FOR TROPICAL HARDWOODS

The harvest of tropical hardwoods for sawlogs is large—more than 10 times the current total harvest of all timber in New Zealand, or 10 million m^3 more than the current total harvest of sawlogs from either Canada or all of Western Europe. If a supply of this magnitude were eliminated, the world economy would be jolted. Knowledge of alternative sources of suitable, quality wood would reduce the disruption.

It is unlikely that the deficit of hardwood sawlogs from the tropics will be met from harvesting the natural forests, either virgin or managed, that still remain. Few natural forests are left in the tropics, and all of them are under environmental pressure. Those in other regions are now or soon will be under similar conservation and environmental pressures. At best, natural forests outside of the tropics can only sustain their current levels of production, but it is more likely that these levels will decrease.

Plantations, especially of fast-growing species, have been proposed as an alternative source of timber supply. In tropical countries, harvests from such plantations could replace hardwood sawlogs from natural forests. However, plantations in the tropics will not provide this hoped-for alternative source of wood supply, because these trees must be intensively managed to produce large, high-quality sawlogs. Few reliable figures are available, but it is doubtful that many of the plantations now being established in the tropics will be managed for this product. Teak is the one obvious exception. Most of the plantations in the tropics have been planned to produce short-rotation pulpwood crops.

If carbon sequestering is really a global environmental goal, we need to consider fast-growing plantations in temperate zones. These plantations have a critical advantage over either natural or plantation forests in the tropics because they are significant carbon sinks. While C^4 crops such as maize or sugar cane reach peak levels of photosynthesis at somewhat higher temperatures, rates of photosynthesis generally increase up to 28°C. However, net carbon fixing is not necessarily maximized in tropical forests because metabolic rates also rise with ambient air temperatures, which remain high at night. Radiata pine produces greater volume and stores more carbon in south temperate climates, as in New Zealand, because night temperatures are lower. The trees do not metabolize the day's gross photosynthate at night.

Estimates of total areas in plantations are not readily available. My research suggests that there are around 100 million ha worldwide, about 14 million of which are planted with fast-growing species that have a mean annual increment of at least 14m^3/ha/yr. Table 15.3 and figure 15.2 illustrate the distribution of fast-growing plantations by region, species—hardwood or softwood—and intention of management. I have included areas of plantations being pruned because this technique is the only means of achieving a high-quality sawlog that is equivalent to the hardwood sawlogs currently harvested in the tropics. Although a maximum of 2 million ha of plantations has been pruned, only about half has been specifically managed to produce clear wood.

Table 15.3. 1990 Estimates of Fast Growing* Plantations by Regions, Conifer/Hardwood, and Management (in million ha)

Region	Conifers Pulp**	Conifers Sawlogs	Conifers Total	Hardwood Pulp	Totals
Africa	0.4	0.6(0.5)	1.0	0.9	1.9
Asia	0.4		0.4	0.25	0.65
Central & South America	2.0	1.9(0.3)	3.9	3.9	7.8
Europe	0.05	0.2	0.25	0.8	1.05
Oceania	0.2	1.9(1.2)	2.1	0.1	2.2
Totals	3.05	4.6(2.0)	7.65	5.95	13.6

* Fast growing is defined as having mean annual increment greater than 14m^3/ha/hr.
** Pulp includes all non-sawlog objectives other than fuel.
( ) Volumes in bracket are estimates of areas pruned.
Source: Developed by author from various national statistics.

Let us put the production of hardwood sawlogs from the tropics in perspective by calculating the plantation area of pruned radiata pine that would be needed to replace it. The volume of pruned radiata sawlogs represents about a third of the radiata crop. Radiata averages around 20m^3/ha/yr, so the growth of pruned logs represents about 7m^3/ha/yr. To achieve an annual yield of 134 million m^3, we would need the sustained yield of about 19 million ha (134 ÷ 7) from a plantation. That area is **18 times** the current area of radiata pine that is pruned around the world—mostly in New Zealand—and is 75% of the total land area of New Zealand. In other words we need a managed plantation that is equivalent in size to the total area of New Zealand that is capable of sustaining forest growth in order to harvest enough quality logs equal in volume to the current harvest of tropical hardwood sawlogs.

Plantations are Only a Partial Solution

Table 15.4 presents current characteristics, experience, and production of plantations around the world. Plantations now cover about 100 million ha and currently supply about 10% of the world's industrial wood, about 120 million m^3/yr. However, 85% of planted forests are slow growing, at a rate less than a third as fast as plantation radiata pine in New Zealand. Two-thirds of the fast-growing species that have been planted are pulpwood crops. Consequently, it is not surprising that plantations currently supply less than 5% of the world's sawlogs and peeler logs, about 45 million m^3/yr.

Table 15.4. Global Plantations: Characteristics, Experience, and Production

Characteristics	Experience	Production
Mostly slow growing	Mostly under 30 years old	Less than 10% of the world's wood
Most fast-growing species grown for pulp	Rarely more than one rotation of experience	Less than 5% of the world's sawlogs
Very small areas grown for high-quality sawlog production	Very limited experience with use of wood	Less than 0.5% of the world's high-quality sawlogs

Compared to agriculture, experience in plantation forestry is limited. Like agriculture, however, few tree species which demonstrate potential to respond to management in plantations, either in natural forests or in limited stand management, prove suitable for large-scale plantation forestry. To date, a few species of pines and eucalyptus are the only economic successes, although other conifers and hardwoods, such as Douglas fir and several acacias, appear to be likely candidates. Some environmental groups vigorously oppose the use of most of these successful species, especially the pines and eucalyptus, because they are perceived as unnatural and environmentally unacceptable. But without suitable forest plantations for production of wood, how else will the world be able to obtain its wood supplies?

Plantations Require Both Time and Money

Planted forests are not a short-term solution to the world's decreasing wood supply. Radiata pine is a fast-growing species, one of the few in the world that are capable of sustaining a growth rate of 20m^3/ha/yr. Even so, it requires **25 to 30 years for radiata pine to grow into sawlogs on intensely managed plantations.** There are many other tree species that seem promising for plantations, but these require at least twice as long as radiata pine to reach financial maturity. If we were to make a major effort to establish plantations now with fast-growing species, they would not yield a harvest of sawlogs until 2020 or later. In addition, plantation forests also require not only a huge investment in capital, much in the form of labor, but also they yield no real return on the investment until the moment of harvest. In practice, only countries rich in capital can afford this capital-intensive investment in plantation forestry.

Plantations require not only vision, but confidence. Investors must be able to predict the market at the time of harvest with reasonable accuracy. Conditions of supply and demand operate in complex ways over the 30 years between the decision to plant and the time of harvest. Investors in tree planting—be they individuals or national governments—need assurances that their ventures have a chance to succeed. Investment in plantations to meet the future demand for wood will only be forthcoming if investors' perceived risks and uncertainty are low.

To understand the magnitude of the global problem of wood supply to meet future demand just for industry, ignoring fuelwood, let us assume that world population will be 8.5 billion, with a per-capita consumption of wood of 0.7m^3/yr—a total of 5.95 billion m^3—in 2020. Industrial demand for wood is 3 billion m^3—0.5 x 5.95. Further, let us assume that all wood will come from plantations, which are fast growing, 20 m^3/ha/yr, and that the demand is for solid wood grown in rotations of 25 to 30 years. The supply of wood from existing plantations around the world in 1990—say 800 million m^3—leaves 2.2 (3.0 - 0.8) billion m^3 to come from new plantations. Finally, we will assume that the capital requirements, ignoring interest, are NZ$3,000/ha—about US$1,600. Given these assumptions, we would need 110 million ha of new, fast-growing species (2200 ÷ 20), or new plantations covering over 4 times the size of New Zealand just to meet the industrial requirements for wood in 2020. The requirements for capital would be NZ$330 billion (110 million x $3000) before interest—with no return on this investment for 30 years.

This scenario is possible, but its achievement requires urgent and immediate action on a massive, global scale. The cost and effort for the production of wood from plantations would be less if harvests from the natural forests continue. But the cost and scale of the effort would be much greater if large areas of plantations with slower-growing species were used.

ARE THERE ALTERNATIVES TO THE USE OF WOOD?

Substitutes are not a Viable Alternative

Increasing the use of nonwood products is unlikely to be a practical solution to the growing global scarcity of wood. At best, in cases where such substitutes are suitable, their use would be only a short-term alternative. The manufacture of plastics, metals, and concrete requires vast amounts of capital for plants and heavy equipment. While their use would certainly reduce pressure on the world's forests, it would have a greater adverse impact on the global environment. Considerably more energy would be required for their production than is needed for the wood products that they replace. For example, a product in aluminum requires 50 times as much energy as a wood product having the same strength-to-weight ratio. In the short-term, the only way to increase the supply of energy is by a greater consumption of fossil fuels, with higher levels of atmospheric carbon dioxide and an increase in global warming. Such a scenario would also hasten the depletion of economically viable sources of fossil fuels.

A large-scale shift away from the use of wood would increase demand for energy to manufacture nonwood products and dependence on the availability of cheap, renewable, and environmentally friendly sources of energy. The environmental consequences of greater use of metal, concrete, or plastics would also mean greater use of energy and chemicals, increased mining, and greater disposal problems.

Reconstituted wood made from chips of fast-growing trees on short rotations is another proposed substitute for products of solid wood. However, this is rarely an economic or environmentally acceptable solution because large amounts of energy, chemicals, capital, and technology are required for the manufacturing process. In addition, compared with the strength of solid wood, reconstituted wood has less strength per unit of weight. Large volumes of cheap, chipped wood are needed to manufacture this product, making growing trees for reconstituted wood an economically unattractive option for any plantation grower. In addition, greater use of energy and chemicals makes this option less environmentally acceptable than the use of solid wood.

Recycling, especially of wood fiber, has also been proposed as an alternative to solid wood. However, there are only a limited number of wood-based products that can be recycled. Paper is currently the focus of attention for recycling, and without doubt, recycling of paper will become more common. While it will reduce the demand for fresh supplies of wood, a certain amount of fresh wood will still be required, and there is a limit to the number of times that wood fibers can be effectively recycled. In addition, recycling may not be as cost-effective, energy-effective or pollution-effective as it appears. An alternative may be to use waste wood products either directly for energy production or for the manufacture of alternative products, such as insulation.

Is Stabilization of the World's Population a Solution?

Scarcity of wood is becoming a worse problem as the growth rate of the human population increases. The outlook for reducing or even stabilizing the world's population is not optimistic and may not solve the problem, but it might reduce future adverse impacts on the world's forests. Or it might not. It has been claimed that the most democratic and most promising way of reducing, or even reversing, the population growth is to improve standards of living, which will increase demands for such amenities as more and better housing, more and better schools and hospitals, more books and newspapers, a greater amount of packaging, and higher levels of consumption of hygienic products. In other words, improved standards of living will increase per-capita consumption of wood products derived from both natural and plantation forests.

CONCLUSIONS

The dilemmas are obvious. Existing natural forests are not capable of meeting a greater demand for wood caused by improved standards of living of the growing global population without compromising global biodiversity. Plantations are too small in number, too young, and usually grow too slowly to provide anything more than minimal supplies of wood, especially to meet demands for sawlogs and peeler logs. Some environmental groups are opposed to the few successful plantations of pine or eucalyptus. Substitution of wood with nonwood products on a large scale is not an option, unless there is a major environmentally acceptable breakthrough in their production from resources other than fossil fuels.

Although it is probably already too late, plantations as a solution to diminishing global supplies of wood will require an urgent and massive investment worldwide. It will also require that influential environmental groups, which oppose plantations of pine and eucalyptus, reconsider their fundamental principles and consequent recommendations for the future. Most of the solutions to worldwide shortages of wood involve some environmental compromise. Immediate financing of plantations on a large scale by northern, wealthier countries in suitable, poorer countries of the southern hemisphere is probably the only feasible solution.

New Zealand's experience offers a lesson to the world by suggesting that managed plantations are an acceptable alternative to natural forests as a source of wood. However, an investment in plantation forestry requires at least one 25-to-30-year rotation period—maybe as many as three—to learn how to select appropriate species, to decide how to manage stands, and to determine the use of the wood to be harvested. Only then can any large-scale investor gain enough confidence in the project. Even the most promising, but still untested, tree species or management regime could either fail or only be partially successful in plantations. Additionally, current markets in timber may not indicate future preferences for wood or its quality and price differentials. In other words, any large-scale effort to establish plantations must be based on well-funded, long-term, and concentrated research.

We must stop harvesting our remaining tropical, natural forests. However, there are no significant alternative sources or substitutes for this wood on the large scale that will be necessary in the future. As forest managers, we must be aware of *all* demands on our forests, not only the production of wood. If anything, we have become too focused on the production of wood from our natural forests and have neglected an even more urgent, desperate need—to find a suitable substitute for our dwindling supply of hardwood sawlogs from the tropics.

Literature Cited

Berck, P. and W. R. Bentley. 1992. Hotelling's theory, enhancement and the taking of the Redwood National Park. Submitted to the *Am. Econ. Rev.*

Berck, P. 1978. The economics of timber: a renewable resource in the long run. *Bell Journal of Economics* 9(3): 447–462.

FAO. 1991a. 1961-1989: 2010 Wood & Wood Products-FAO.

FAO. 1991b. FAO Yearbook: Forest Products; FAO Forestry Services 24.

Ellis. L.M. 1923. Comment in second British Empire Forestry Conference 1923. Proceedings and Resolutions.

Reid J.S. 1953. Grade in sawn timber and round products of Pinus radiata. *NZ Journal of Forestry* 6(5):394–400.

Sutton W.R.J. 1991. Are we too concerned about wood production? *NZ Forestry* 36(3):25–28.

CHAPTER 16

COMMERCIAL FORESTRY AND RURAL DEVELOPMENT IN THE PHILIPPINES

Romulo A. del Castillo

INTRODUCTION

Roger Sedjo, in his chapter "Commercial Forestry and Rural Development," recognizes two general issues when he analyzes commercial forestry's impact on rural development—resource efficiency and income distribution. He hypothesizes that "... many of the difficulties in relating forest industry development to regional economics pertain to the distributional consequences of forest industry development." Documented experiences illustrating "successful" and "unsuccessful" cases of the forest industry's contribution to rural development are discussed.

Successful cases have occurred in Finland and the U.S. South, where private owners of forest assets are able to participate in rural development in a variety of ways. On the other hand, the common characteristic of such unsuccessful cases, as those reported in Tasmania, Australia, Brazil, and Indonesia appear to be the state ownership of forests. In these examples, it is a common practice for the state governments to use royalties and other revenues from the forests to subsidize exploitation. As a consequence, the only effect that the commercial forest industry has on rural development is its employment of local people for their labor services in its logging and wood-processing operations, often in low-paying jobs.

A BRIEF HISTORY OF COMMERCIAL FORESTRY IN THE PHILIPPINES

I would argue that the forest situation in the Philippines, where the state owns most of the forest lands, resembles one of Sedjo's unsuccessful cases. The effects of increasing mismanagement of Philippine natural resources, especially in the so-called "development decades" of the 1960s and 1970s, illustrate my point.

Immediately following the Second World War, international demand for Philippine mahogany logs surged and rose steadily to dangerous levels over the next 30 years. Since the country depended on agriculture and natural resources as major sources of foreign exchange earnings, the Philippine government took advantage of the growing foreign markets. Vast areas of forests were given out to private concessionaires under various forms of license agreements covering periods of up to 25 years. Each licensee was required to prepare a forest management plan in which the annual allowable cut was determined on the basis of current volume and growth. Selective logging methods following the principle of sustained yield management were implemented, especially in long-term concession areas. Despite such precautions, however, indiscriminate logging became rampant in practically all accessible and operable areas in the country. Indeed, as early as the 1960s, observers remarked that the Philippines was cutting down its old-growth forests "faster than any country in the world." Those aware of the consequences of such wanton destruction of the forests were so alarmed that they initiated a nationwide program of public education and information on forest conservation.

Finally, in the 1970s, the government began to restrict the export of wood products in round form. Between restrictions on exports and dwindling timber inventories, the volume of log exports declined steadily over the next 20 years, from 2.332 million cubic meters valued at $135.33 million in 1970 to 0.051 million cubic meters valued at $2.86 million in 1990. The Philippine Department of Environment and Natural Resources (DENR), citing the 1990 Philippine Forestry Statistics, has published the following comparative figures which illustrate the result of the government's restrictive policy on log exports.

Table 16.1. Indicators

Indicator	1976	1990
Number of Timber License Agreements	471.0	96.0
Area covered in million hectares	10.32	3.76
Annual Allowable Cut in million cubic meters	21.89	5.05
Total log production in million cubic meters	8.73	2.6
Total log export in million cubic meters	2.332	0.051
*sawlog/veneer in million cubic meters	(2.331)	(0.004)
*pulpwood in million cubic meters	(0.001)	(0.047)
Total value of log export in million dollars	135.33	2.86

The government's policy of imposing a partial ban on log exportation was intended to arrest the destruction of the forests as well as to promote the development of the forest industry sector. In fact, the commercial forest industry prospered, and exploitation of forests continued. The rate of destruction of the forests remained the same, but most of the commercially harvested old-growth forests had vanished by 1990. Recent landsat photos reveal that less than one million hectares of virgin forest remain. Meanwhile, the adverse effect of nearly 40 years of destruction and neglect of forests is obvious in many parts of the archipelago.

Environmentalist groups now advocate a complete ban on logging operations in all parts of the country. Whether or not to ban logging completely or to adopt selective harvesting methods has become one of the hottest issues debated in the Philippine Congress. As long as there remains exploitable forested areas in the uplands and as long as the wealthy and powerful private forest industry continues to profit by exploitation, many fear that the debate might not be resolved.

NEW DIRECTIONS FOR FOREST DEVELOPMENT AND MANAGEMENT

Let me discuss some measures that the government has adopted to ameliorate the deteriorating situation of Philippine forests. Most recently, DENR has initiated fresh approaches to the development and management of forests in the hopes of securing meaningful participation of the local people in the management of their upland forest resources. Examples of these community-based forest management programs include the Integrated Social Forestry Program (ISFP); the National Forestation Program (NFP); the Forest Land Management Agreement (FLMA); and the Community Forestry Program (CFP).

The Integrated Social Forestry Program (ISFP)

Established in 1982, ISFP was conceived to address the twin problems of rural poverty and ecological instability in inhabited forests. Through ISFP, inhabitants of forests are granted secure access to the land they farm. The program provides technical and material aids designed both to increase productivity of the land and to promote sustainability at moderate levels.

Secure tenure of land may be acquired through either the Certificate of Stewardship Contract (CSC) for individual applicants, or through the Community Forest Stewardship Agreement (CFSA) for community organizations. In both cases, farm families are granted renewable 25-year leases on the public land that they occupy and cultivate. In the first few years of the contract, farmers receive technical assistance to apply productive and sustainable farming methods and to develop self-reliance and self-sufficiency.

ISFP's three main areas of assistance are training in agroforestry, land tenure, and community organization. The latter mobilizes groups to acquire stewardship contracts, to promote agroforestry and soil/water conservation activities, and to strengthen local institutions. The program gives greater emphasis to improving existing technologies and practices rather than introducing new ones, except in highly justifiable cases where the technology might provide greater benefits to the farmer and to the environment. Strategies designed to increase the participation of local inhabitants are employed—gathering data, diagnosing field situations, and monitoring technical problems and activities. Cross-farm visits and training courses develop skills both in agroforestry and in strengthening community organizations. Through their active participation in implementing this program, community leaders are trained and prepared to assume the responsibilities of sustaining local development beyond the life of ISFP, which is tentatively scheduled to end in five years.

ISFP's operations are being developed through DENR's Upland Development Program (UDP), which was established in 1980. These agencies, with their multidisciplinary approach, act as DENR's primary research support in its implementation of people-oriented community-based projects in the Philippine uplands. The Upland Development Working Group (UDWG) provides direction for the program and closely monitors UDP's performance. Through the work of its Agency Technical Committee and various task forces, including the Regional Upland Development Committees, the Working Group decides on programs to be implemented and leads in identifying the experiences that might be taught to improve management policies in the uplands and to refine work in the field.

The Working Group is chaired by the DENR Undersecretary for Field Operations and is composed of members from pertinent DENR offices as well as from many other organizations that are concerned with natural resources, including both NGOs and academic institutions. Among the NGOs are the Philippine Association for Intercultural Development (PAFID) and the Philippine Business for Social Progress (PBSP). PAFID's program of assistance seeks to organize cultural communities and to help them secure access to land resources through CFSA or CSC. PBSP, on the other hand, assists in the development and establishment of community organization within DENR. NGOs in academia provide technical support to document (and now monitor) procedures, to develop monitoring and evaluation systems for projects, to improve farming methods in agroforestry, and to conduct research.

The National Forestation Program (NFP)

Initiated in 1988, NFP mobilizes all sectors of society in an accelerated and integrated program of reforestation, watershed rehabilitation and timber stand improvement. In the context of this program, "reforestation" means the replanting of severely or partially denuded forest lands with either indigenous or introduced species, including fruit trees, bamboo, and other trees profitable to industry. In reforestation areas dominated by cogon grass (*Imperata* spp), but where a number of broad-leaf species are found growing naturally, a strategy known as Assisted Natural Regeneration (ANR) may be employed. ANR is a technique used to liberate these broad-leaf trees from competing vegetation with the application of certain silvicultural techniques, such as augmentation planting of a mixture of climax species. The result is a multi-story, multi-species forest stand of climax species produced at a much lower cost.

Watershed rehabilitation" involves the planting of infrastructural and vegetative elements to prevent or minimize soil erosion and to restore an area's ecological balance. Timber stand improvement and enrichment planting are applied in second-growth forests to enhance the growth of future crop trees. These methods are accomplished by removing inferior trees and other superfluous vegetation, and, in some cases, by planting desirable tree species so that the area's stand stocking level may be raised.

A wide variety of entities, such as families of upland settlers, community and civic/religious organizations, entrepreneurs, local governments, other governmental agencies, and NGOs may participate in NFP activities. Any of these may join with DENR in contract reforestation, watershed rehabilitation or timber stand improvement in areas that have been identified by DENR. A contract may cover any one or a combination of the following phases of work: the implementation of the project, the organization and training of the community, or the monitoring and evaluation of the work. In the case of contract reforestation, the project's implementation phase is usually divided into sub-phases, such as production of seedlings, establishment of plantations, development of the infrastructure, and maintenance and protection.

The Forest Land Management Agreement (FLMA)

FLMA provides a way for the government to share with the inhabitants of the forest the responsibility of maintaining and protecting trees in newly reforested areas. This scheme gives participant farmers the privilege of utilizing the area for their own purposes, provided that their farming practices are consistent with sound ecological principles. It is a way of democratizing the access to the land and its benefits for the people who planted and tended the trees so that the NFP and the country may benefit.

A venture that shares production, FLMA provides a long-term contract given by the government over forest areas replanted under NFP. When the contract for reforestation terminates at the end of three years, the community organization which holds the contract in the area may apply for a FLMA. At least 80% of the trees planted must be surviving and properly maintained for the application to be viable. Individual families that are contractors for reforestation must form an association or cooperative to bring the total land area covered to at least 100 hectares in order to qualify. DENR works with local NGOs to assist in organizing the community and teaching various members about forest resource management.

FLMA contracts are for 25 years and are renewable for another 25 years. They are issued to groups who pledge to protect and maintain trees that have been already planted. A manager assumes responsibility for deciding on the use of the forest area during the entire duration of the contract. He may decide to produce desired goods and services, such as growing minor forest products or planting cash crops, fruit trees, or other agricultural crops among the existing trees. He may harvest, process, and sell timber when the planted trees become mature, but such harvesting should be in accordance with the principle of sustained-yield forest management. In return for this freedom to use the land as he chooses, the manager must replant the harvested area with the same tree species, and when he sells his forest products, he must share with DENR a portion of his profits equal to the cost of reforesting another hectare of denuded uplands.

The Community Forestry Program (CFP)

When residual forests are abandoned by a timber licensee, either by choice or by necessity, they are usually treated as common property resources by the local people. This treatment eventually leads to misuse and subsequent degradation. To address the problem, DENR has established a program called the Community Forestry Program (CFP).

CFP grants the dwellers of uplands institutional ownership or stewardship of residual forest areas and thus allows them access to these areas' resources. The contracting community is awarded a 25-year Community Forestry Management Agreement (CFMA), renewable for another 25 years. When there is a mutual agreement between it and DENR, the community organization then has the privilege to harvest, process and sell forest products grown in the area, but it is required to submit a management plan for DENR approval, to comply with prescribed rules and regulations, and to adhere to the principles of sustained-yield management.

Under the terms of CFMA, DENR assists the contractor organization to implement CFP, in particular to establish and strengthen community organization. This assistance, through the services of qualified NGOs, comprises on-the-job training in resource inventory, preparation of forest management and conservation plans, and development of forest-based and other opportunities for achieving economic self-sufficiency, including some that do not depend on forest products.

SUMMARY

The major initiatives in forest management of the 1980s constitute DENR's continuing efforts to find effective approaches to sustainable forestry methods. While the trend in past management policies had been towards resource efficiency, the trend in the programs which I have discussed above is admittedly towards a more equitable distribution of the benefits of forest resources in the Philippines.

Will these initiatives succeed? Or will they, like the forest management approaches of the past, fail? The answer will depend obviously upon the sincerity and commitment of the key participants in the current management of the forest resources: the local people themselves; the intermediary organizations, facilitating POs and NGOs; the implementing government agencies; and the enabling institutions—banks and their lending policies.

If the paramount objective is the relatively efficient management of forest resources, then these initiatives will not have much chance for success. If, however, the primary goals are equity and sustainability, then these new approaches to forest management will have a better chance of success, if the key participants in this scenario have the proper motivation and attitude.

CHAPTER 17

NONTIMBER FOREST PRODUCTS MARKET DEVELOPMENT AS A MEANS OF TROPICAL FOREST PROTECTION

David H. Newman
A. L. Hammett, III

INTRODUCTION

Researchers have generated a substantial body of literature over the past decade that examines nontimber forest product (NTFP) markets and their implication for the protection of tropical forests. Initially, the literature focused on descriptions of existing markets for such products as rattan, resins, and fruits. Recently, work has centered on the potential for expanding NTFP markets and on the political and structural factors needed to accomplish meaningful market development. Economists have regarded extractive reserves, nature-based tourism, and the marketing of genetic resources as necessary and logical extensions of existing NTFP markets. These new markets seem attractive because they rely on *laissez-faire* mechanisms to achieve the broader objective of forest protection. The fact that social, cultural, political, physical, and economic factors may reduce the effectiveness of NTFPs as solutions to forest protection has not limited their intrinsic appeal. In particular, economic theorists often overlook the needs and activities of the new NTFP entrepreneurs with respect to the forests that everyone assumes they are protecting. Development of new NTFP markets may require infrastructure and reasonable economies of scale that may conflict with forest preservation. As NTFP markets develop, efforts must be made to insure that we find and adopt methods of sustainable management in order to insure the long-term viability of both the marketed resources and the tropical forests that produce them.

NONTIMBER FOREST PRODUCTS: A WAY TO PROTECT AND DEVELOP FORESTS

In the past decade, as the world has become increasingly concerned about tropical deforestation and its consequences, conservationists have promoted a number of nontimber forest products (NTFPs) as a way to curb indiscriminate harvesting of timber. A profusion of books, proceedings, and articles has detailed the full array of existing new products and the extent of their markets. This literature has also examined the potentially positive impacts of NTFPs on the environment and the economy and the possibilities for expanding NTFP production while protecting the forest (for example, Plotkin and Famolare 1992, Wickens 1991, Falconer 1990, de Beer and McDermott 1989, Peters et al. 1989, Myers 1984). Our attention focuses increasingly on the fact that forests contain more products than just timber, and we now realize that the loss of natural forests may impose greater costs on society than simply higher prices for timber.

As a fundamental premise of our research, we are convinced that the production of NTFPs may provide a viable way to foster rural development while protecting forests in tropical developing countries. For example, some have proposed the establishment of *extractive reserves* in developing countries, especially in the Amazon region, to preserve indigenous cultures, protect large areas of natural forests, and develop some of the region's wealth (Fearnside 1989, Allegretti 1990, Allegretti and Schwartzman 1989). While a number of concerns have been raised about the basic problem of combining protection of forests with their development (Browder 1990, 1992), there have been continued efforts to employ the concept of *reserves* as a means to facilitate forest management and development. Other marketing schemes for NTFPs—ranging from rubber to rattan to *Rainforest Crunch* ice cream—have also been proposed to protect the tropical rainforest. However, a number of social, political, physical, and economic factors influence the production of these goods. Such factors must be taken into account before we can consider NTFPs to be effective substitutes for timber extraction in the tropics.

In this chapter, we shall examine both existing and potential NTFP markets as well as some issues of economics and policy that are associated with the development of NTFP products. We are particularly concerned that the actions and policies integral to the development of new NTFP markets—such as the development of transportation and other infrastructure development and the achievement of reasonable economies of scale—may conflict with the broader objective of forest preservation. We will first describe the range of NTFP markets, including an assessment of recently emerging, perhaps nontraditional, NTFPs—ecotourism and the marketing of genetic resources—and we will categorize the types of products within these markets. We will then discuss some institutional, economic, and policy issues integral to the expansion of NTFP markets.

MARKETS FOR NONTIMBER FOREST PRODUCTS

The initial difficulty in any discussion of NTFPs lies in their definition. The meaning of *NTFP* has expanded as researchers have identified an increasing variety of goods that are derived from the forest. De Beer and McDermott

(1989), conforming to traditional views of NTFPs, used the following definition:

> The term *nontimber forest product* encompasses all biological materials other than timber which are extracted from natural forests for human use. These include foods, medicines, ornamental plants, wildlife, fuelwood, and raw materials, notably rattan, bamboo, smallwood, and fibers.

However, Wickens (1991) had a much wider perspective in defining the term:

> *Nonwood forest products*—refer to market or subsistence goods and services for human or industrial consumption [that are] derived from renewable forest resources and biomes, bearing promise for augmenting real household incomes and employment.

Thus, the concept of *NTFPs* has been extended beyond specific products that are extracted from the forest to potentially encompass all goods and services from the expanded forest resource biome. This broader definition acknowledges that tastes and technology have modified to include some NTFPs, such as water production and ecotourism, that are not physically extracted from nature. In addition, NTFPs are now regarded as goods that may also have an expanded potential for development. Now, we think of NTFPs as goods that can fit into plans for rural development while providing environmental protection.

We should bear in mind that most NTFPs are not really new, "emerging" products. Historically, NTFPs such as rubber, dyes, spices, medicines, and foodstuffs—rather than fine-quality wood, the primary timber product exported from the tropics—dominated international trade from tropical forest-based economies. Only since the end of World War II has trade in timber become a major component of international trade for tropical developing countries. Although the data are incomplete, exports of tropical hardwood prior to World War II remained fairly steady, at below three million cm^3/yr. Since 1946, however, exports of timber from the tropics had increased more than 20-fold, to over 66 million cm^3 in 1980 (Laarman 1988). To our knowledge, no aggregate data exist on world trade in NTFPs, but given the broad range of goods, including the phenomenal growth of ecotourism, it seems safe to assume that the market values from these products have grown to at least match those of timber.

This long history of use, plus tantalizing reports of new products, have heightened the perception that both national and international markets for NTFPs can provide continued economic growth as well as benefits that have expanded in scope to include environmental protection of forests in tropical developing countries. The historical development of markets for NTFPs, however, makes us regard this hope for expanded economic growth and environmental protection with skepticism. Markets for NTFPs have always responded to economic forces in the same manner as do other traded goods. Increased demand has led to rising prices and attempts to increase output. However, since NTFPs often exist with little or no managerial inputs, the only way to increase their production has been to exhaust existing production sites or to expand the operation into new areas. In addition, producers and entrepreneurs have made little effort to develop sustainable management systems or to determine sustained-yield levels of harvest because many of these products are found in the open and are easily accessible. No managerial control exists. Thus, in many cases, the development of NTFP markets has caused either unsustainable levels of production and degraded forest resources or rapid development of synthetic substitutes that have decimated the market. Substitution of synthetics for chicle from the sapodilla trees in Central America and the intense production of rubber in Amazonia and rattan in Southeast Asia are among the numerous examples of product substitution and resource degradation as markets for such products rapidly develop (Caldecott 1988, Ryan 1991). We may expect that similar depletions of forest resources will occur as existing markets grow and new ones develop.

We hope that our grim outlook will not apply to all NTFPs. Pendelton (1992) lists a number of prerequisites for a successful NTFP venture. It is apparent that economic issues that address the basic production of the goods, in addition to those of biological and market sustainability, are critical. Of paramount importance for any NTFP venture to be a marketing success is that the profit from an activity that degrades the forest resource through overly rapid use or through conversion of the site to other uses should be less than the market values of NTFPs in both the short- and long-run. For this to occur, an NTFP must not be only a profitable enterprise. Property rights which allow for the production of the NTFP must be clearly specified and enforced. Successful marketing of NTFPs might require the balance of a difficult combination of prerequisites—the diffuse production pattern that many NTFPs have displayed, the generally low return to labor from their production, and the difficulty of maintaining social controls on the use of the forest resource in an expanding local economy or remote area.

The current mélange of NTFPs may be sorted into two broad categories by use: **consumptive** and **nonconsumptive**. Some of these two uses are more conducive to meeting our prerequisites for sustainability than are others. They may be more localized, they may have greater opportunities for active management and protection, and they may have a greater history of use. Other uses will place new pressures on the forest resource that will make them either unprofitable or unsustainable.

We include personal uses of NTFPs for food, fuelwood, fodder, household goods, and medicines as **consumptive** uses. Such goods are generally produced and consumed at the home or village level, and they often have a long history of local use. They often have only a limited marketing capability, due either to the perishability of the product or the lack of an easy way to bring it to market. These limitations change, however, as new technologies for transportation and new marketing channels develop. For instance, the large, foul-smelling durian, a fruit produced in Southeast Asia, can now be purchased in specialty shops in New York City and elsewhere in the U.S. Other types of consumptive uses include NTFPs that are destined for either national or international markets—many of the items for personal use listed above, but also such goods as live animals, extracted chemicals and resins, manufactured goods, and construction materials. In many cases, such NTFPs are byproducts of another marketing venture. They are rarely the entrepreneur's primary source of income.

Nontraditional, consumptive NTFPs have recently exhibited a broad potential for market development. Marketing genetic resources for pharmaceutical and other uses is one of the more interesting of these emerging, nontraditional products. Costa Rica, for example, has a contract with a major pharmaceutical company that will give the country a royalty for any drugs that may be developed from plants found in its forests. Both the general biodiversity of a country and the potential consumptive uses of the genetic resources of the forest become marketable goods. While contracts such as Costa Rica's are laudatory, issues of common property and benefits to the local people from many of the genetic resources that are derived from the forest may make such a marketing system difficult to develop. In particular, there is likely to be a strong difference in perspective between the government, which will receive the royalties, and the people inhabiting the land, who put pressure on the values of biodiversity in the first place. Sedjo (1992) discusses a number of cases where the development of markets for the genetic characteristics of plants are associated with issues of property rights.

Nonconsumptive uses of NTFPs tend to be more diffuse and often occur in conjunction with other productive enterprises. Some of these uses—the shading used for such crops as coffee or tea or the soil improvement from alley-cropping systems—have a direct market value and are part of a landowner's production plan. Other uses, such as watershed protection, property buffers—either property demarcation or physical barriers to intrusion —carbon storage, and biodiversity are important for local and global production, but they are not valued directly, since the output is regarded as an externality. Finally, there are nonconsumptive uses, such as ecotourism, which attempt to market the uniqueness and quality of the forest resource, while leaving it intact (Lindberg 1991, Boo 1990).

Whether these nonconsumptive goods are produced in sufficient or sustainable quantities depends on the product and the particular situation. Those outputs which are produced as positive externalities are likely to be severely undervalued in the marketplace. For example, insufficient watershed protection, as evidenced by the severe levels of degradation of forested watersheds that are found in many countries, has important economic costs for economies that are dependent on them for agriculture, energy, and drinking water. The marketed uses, particularly ecotourism, may also have difficulty in maintaining their value because uniqueness and access, two key factors in determining the value of a site, may be incompatible for maintaining the long-term qualities of a particular site (Healy 1988).

ISSUES IN THE MARKET DEVELOPMENT OF NTFPs

We need to consider five broad categories when discussing the development of markets for NTFPs and their ability to provide rural development *and* forest protection:

- Sustainability and the NTFP's particular requirements for production
- Institutional structure of the market and the decision-making authority
- Market characteristics and the ability of producers to profit from their enterprises
- Externalities derived from the production and market development of the NTFP
- Government policies and their impacts, intended or unintended

Sustainability

The production, marketing, and use of NTFPs as a means to promote rural development depend greatly on the biological characteristics of the particular product. As their markets develop, the rates of extraction of these forest products will increase, and producers will be unable to manage or control increasing extraction without incurring forest degradation. Both intensive and extensive methods of forest management must be found to increase production without degradation in a sustainable manner. Sustainable forest management systems for many fruits and edible products have developed over many years, and their production has become institutionalized within the local culture. However, traditional systems may not include scale economies that would allow increasing output—and the associated decreases in production costs—without harming long-term sustainability. It is a fundamental paradox that many researchers of NTFPs encounter, namely that the creation of a market, while placing a value on the nontimber aspects of the forest, may in fact put new extractive pressures on the forest that could be equally as damaging to the forest ecosystem as harvesting it for timber (Browder 1992, Anderson 1990).

Institutional Structure of the Market and Property Rights

A substantial body of research has shown that property rights are crucial in the development of NTFP markets (Sedjo 1992, Jessup and Peluso 1986, Randall 1978). The development of long-term markets requires some form of tenure over the resource, whether it be individually or community-based. Unfortunately, many NTFPs are openly accessible, so that their production may preclude the possibility of effective management and sustainability. Investments in capital and labor to improve production in more managed systems and opportunity-cost investments to leave production units—trees or plants—in place while they mature will become less likely without clear tenure. Community-based mechanisms that deal with these problems of ownership can be effective in many situations. To date, however, successes have generally occurred in relatively stable, sheltered societies (Runge 1981).

As markets develop and human populations increase in regions where NTFPs are produced, it is unclear whether controls on common property can continue to operate as existing assurance mechanisms break down. Extractive reserves—forest reserves set aside for extraction of NFTPs—attempt to deal with this situation of increased markets and population by granting communities title to large areas of production and by employing existing organizational structures within the community to deal with issues of use (Fearnside 1989). However, there is substantial concern whether in fact these organizations can implement such assurance mechanisms and whether the reserves can be effectively separated from outside pressures for extraction, commercial exploitation, and poor migrants (Browder 1990, 1992).

Market Characteristics

Most research on market development of NTFPs has been based on case studies which focus on biological characteristics, with a minimum number of sample points in both time and space, as well as on the expansion of current market structures. Since the range of inquiry has been limited, the results of such research are highly impressionistic. In order to ascertain if an NTFP market will develop, even for existing products, it is necessary to assemble and evaluate specific information on the characteristics of the NTFP and on the markets in which it will be traded.

First, we must know elasticities of demand as well as supply for each product. There may be sufficient production capability only for existing markets, but none for expanded markets. In addition, demand may already be saturated, leaving little opportunity for expansion, or it may be extremely dependent on income. Second, we must know the availability of such production inputs as labor and capital, since their availability at specific harvesting or processing times may be critical in order for any attempt at expansion of output to be successful. Third, we need to know the availability of existing and potential substitutes for an NTFP. Substitutes for such products as ecotourism may exist not only within a specific country but also internationally. The market may already be saturated so that expanded tourism in one region may only divert visitors away from another area. Finally, we must know how to employ marketing networks to increase the amount of marketing information available to producers so that they may capture more profits. This information can be extremely difficult to gather, but it is essential if an expanded market is to be viable and sustainable. As the number and extent of NTFP markets increase and develop, and as we gain more experience, networks of NGOs and other organizations that process this information should help to analyze and market new, emerging products.

Production Externalities

Much of our discussion of NTFP market development has implied that inhabitants in the forest could live their lives unchanged as they make extra profits from selling their wares in national and international markets. This will most likely not be the case. The development of markets implies a series of changes to peoples' lives and to the environment around them. In order to assure supplies or to gain greater access, developers may have to build roads and other forms of transport. New transportation networks will allow increased use of the forest in an area and may also lead to an increase in extractive or farming activities beyond those intended by the original producers. If profits increase from an NTFP market, the local people will demand better social services, such as medical aid and schools. Such services and other improvements in the quality of life are the desirable outcome of regional development, but they definitely imply a change in the fundamental character of a region.

Impacts of Government Policies

A substantial body of research has examined the impact that government policies may have on the forest sector, either directly or indirectly (e.g. Hyde and Newman 1991, Mahar 1989, Repetto and Gillis 1988). One of the great appeals of NTFPs is that they can meet some objectives of rural development and forest protection without the need for government intervention. However, some of the more promising NTFPs are already influenced by government activities in both positive and negative ways. For instance, the Brazilian government heavily subsidized the production of natural rubber, and this policy helped to fuel the rubber tappers' desire to create extractive reserves. At the same time, government policies on land disposal and tax incentives caused increasing pressure on the land that was used by the tappers. These conflicting policies increased the tension between competing users and led to highly antagonistic responses from both sides (i.e. the murder of Chico Mendes). That the government has abolished many of its policies of land disposal does not obviate the impact that other policies may have on other products.

CONCLUSION

Let us remember two of the fundamental tenets of welfare economics. First, people are rational. Second, if an enterprise yields rents or profits then rational people will search them out. The upshot from these two principles is that one should carefully examine the arguments of those who say that people in tropical forests are acting irrationally by not engaging in the profitable enterprises that can be developed from NTFPs. Either the activities that these people are already doing in the forest are more immediately profitable for them or there is some factor that prevents the occurrence of profits that could be derived from the production of an NTFP. Thus, while we have discussed a few of the issues which we believe may limit the broad application of NTFPs as tools for regional development and environmental protection, we must turn our attention to the potential of profitability of NTFP activities relative to current extractive activities. If "sustainable" NTFP activities do not pay, then private individuals cannot be expected to engage in their production.

Most importantly, NTFPs are not a panacea for the problems of tropical deforestation and rural poverty. They are an option that may be appropriate under certain conditions and for certain products. In some situations, selling NTFPs may meet some objectives for forest protection, but conflicts between uses will most likely develop. In other cases, objectives of rural development may be met, but it is likely that outside help, inputs of capital and management, will be needed and that conflicts over forest protection may occur. Attempts to force the adoption of NTFP markets in inappropriate areas are likely to be self-defeating and ultimately harmful to the causes that NTFP proponents are trying to pursue.

Literature Cited

Allegretti, M.H. 1990. Extractive reserves: an alternative for reconciling development and environmental conservation in Amazonia. In *Alternatives to deforestation: steps toward sustainable use of the Amazon rain forest.* (A.B. Anderson, ed.). New York: Columbia University Press.

Allegretti, M.H. and S. Schwartzman. 1989. Extractive reserves: a sustainable development alternative for Amazonia. Unpublished report (US-478) to the World Wildlife Fund-U.S., Washington, D.C.

Anderson, A.B. 1990. Extraction or agroforestry? *International Agricultural Development* 10(5):13–16.

Boo, E. 1990. *Ecotourism: the potentials and pitfalls.* Volume 1. Washington, DC: World Wildlife Fund.

Browder, J.O. 1992. The limits of extractivism: tropical forest strategies beyond extractive reserves. *Bioscience* 42:174–182.

Browder, J.O. 1990. Viewpoint: extractive reserves will not save tropics. *Bioscience* 40:626.

Caldecott, J. 1988. Climbing towards extinction. *New Scientist* (June 9):62–66.

de Beer, J.H. and M.J. McDermott. 1989. *The economic value of nontimber forest products in Southeast Asia.* Amsterdam: Netherlands Committee for IUCN.

Falconer, J. 1990. *The major significance of 'minor' forest products.* Rome: Food and Agriculture Organization of the United Nations, Community Forestry Note #6.

Fearnside, P.M. 1989. Extractive reserves in Brazilian Amazonia. *Bioscience* 39:387-393.

Healy, R.G. 1988. Economic considerations in tropical forest tourism. Raleigh: USDA Forest Service, Forest Private Enterprise Initiative Working Paper #34.

Hyde, W.F. and D.H. Newman. 1991. *Forest economics and policy analysis: an overview.* Washington, D.C.: World Bank Discussion Paper #134.

Jessup, T.C. and N.L. Peluso. 1986. Minor forest products as common property resources in East Kalimantan, Indonesia. In *Proceedings of the conference on common property resource management.* (F.R. Ruslin, ed.). Washington, D.C.: National Academy Press.

Laarman, J.G. 1988. Export of tropical hardwoods in the twentieth century. In *World deforestation in the twentieth century.* (J.F. Richards and R.P. Tucker, eds.). Durham: Duke University Press.

Lindberg, K. 1991. *Policies for maximizing nature tourism's ecological and economic benefits.* Washington, D.C.: World Resources Institute.

Mahar, D.J. 1989. *Government policies and deforestation in Brazil's Amazon region.* Washington, D.C.: World Bank in cooperation with World Wildlife Fund and the Conservation Foundation.

Myers, N. 1984. *The primary source: tropical forests & our future.* New York: W.W. Norton & Co.

Myers, N. 1988. Tropical forests: much more than stocks of wood. *Journal of Tropical Ecology* 4:209-221.

Pendelton, L.H. 1992. Trouble in paradise: practical obstacles to nontimber forestry in Latin America. In *Sustainable harvest and marketing of rain forest products.* (M. Plotkin and L. Famolare, eds.). Washington, D.C.: Island Press.

Peters, C.M., A.H. Gentry, R.O. Mendelsohn. 1989. Valuation of an Amazon rainforest. *Nature* 339:656-657.

Plotkin, M. and L. Famolare, eds. 1992. *Sustainable harvest and marketing of rain forest products.* Washington, D.C.: Island Press.

Randall, A. 1978. Property institutions and economic behavior. *Journal of Economic Issues* 12(1):1-21.

Repetto, R. and M. Gillis. 1988. *Public policies and the misuse of forest resources.* New York: Cambridge University Press.

Runge, C.F. 1981. Common property externalities: isolation, assurance, and resource depletion in a traditional grazing context. *American Journal of Agricultural Economics* 63:595-605.

Ryan, J.C. 1991. Goods from the woods: managing tropical forests for preservation and profit. *Journal of Forestry* 90(4):25-28 (Excerpted from 1991 *World Watch* 4(4):19-26).

Sedjo, R.A. 1992. Property rights, genetic resources, and biotechnological change. *Journal of Law and Economics* 35(1):199-213.

Wickens, G.E. 1991. *Nonwood forest products: the way ahead.* Rome: Food and Agriculture Organization Forestry of the United Nations Paper No. 97.

CHAPTER 18

THE WOODLANDS OF TAE: TRADITIONAL FOREST MANAGEMENT IN KALIMANTAN

Christine Padoch

In September 1991, Indonesia's then Minister of Forestry, Dr. Hasjrul Harahap, and several members of his staff reviewed the situation of Indonesia's forests and announced their new forestry policy at a seminar in Washington (Harahap 1991). In the talks, the minister and his colleagues attributed the rapid rate at which much of Indonesia's old forests have been cut to several causes. However, they placed much of the blame for the rampant destruction of Indonesia's forests on the persistent and pervasive practice of shifting cultivation. A spokesman for the ministry expressed the hope that timber concessionaires—usually Indonesians who are not from Kalimantan—would eventually teach improved agricultural practices to the slash-and-burners (see also Ministry of Forestry 1991a, 1991b).

Forest management was the focus of the second part of the seminar, during which the minister announced a new program for Indonesian forestry. Management of forests would assume a more important role in forestry in the outer islands of Indonesia, that is, in forests outside Java, even though, he said Indonesians have little or no experience in managing the forests of Kalimantan—Borneo—or the other outer islands.

For the last several years, I have studied the resource management practices of a Dayak community in West Kalimantan. These people are shifting cultivators, the villains of much of Dr. Harahap's analysis. However, I have observed that indigenous farmers in this community manage their resources in a complex manner. Far from being agents of rampant deforestation, the Dayaks of Tae in Sanggau District have preserved large stands of trees. Indeed, they do precisely what the ministry claims that no Indonesians know how to do—they employ a multiplicity of management techniques and regimes to skillfully manage the forests of Kalimantan.

In this chapter, I will briefly describe the managed forests of Tae and the Dayaks who use and manipulate them. This description will concentrate on the variety of managed plots that I observed. Next, I will discuss why the management of these forests has been virtually invisible, not only to the Ministry of Forestry, but to most government agents and researchers. Finally,

I will comment on the importance of managed vegetation in West Kalimantan and in many other tropical forest areas.

A MULTIPLICITY OF MANAGED WOODLANDS

A few ethnographic reports written during the past several decades describe management of tropical woodlands by indigenous peoples (e.g. Conklin 1957, Gordon 1969). However, until about the last ten years, researchers rarely focused on these significant and widespread resource management practices. Recently, traditional forestry and agroforestry have attracted much attention. Many published reports confirm the diversity and wide prevalence of such technologies in the Neotropics (Alcorn 1984, Gomez-Pompa et al. 1987, Denevan and Padoch 1988, Hecht 1982, Posey 1984, 1985) and in Southeast Asia (Kunstadter et al. 1978, Michon 1985, Fox 1977, Torquebiau 1984, Wiersum 1982).

Many descriptions of the way indigenous agriculturists of Borneo use their resources mention the cultivation of trees, principally rubber, as an important economic activity (e.g. Freeman 1970, Dove 1988, Padoch 1983). A few notable studies have also described complex agroforestry and forest management systems (Chin 1985, Morris 1953, Weinstock, 1983).

Some people still imagine that Borneo is a heavily forested island that is sparsely inhabited by shifting cultivators. To a limited degree, this image describes a few areas in the far interior of the island, but it does not even approximate an accurate portrayal of the landscape and people of the Balai subdistrict of West Kalimantan (figure 18.1).

Balai subdistrict is located just slightly to the north of the Kapuas River and comprises an area of 396 km^2. As of 1989, Balai had a population of 21,206 people, most belonging to a tribal group known as the Tara'n Dayaks. As a whole, the subdistrict has a surprisingly high population density—54 people/km^2—for an interior, rural region in Kalimantan. My principal study site is the territory of Tae—a village composed of five hamlets, including fields and forests, which covers a total area of nearly 16 km^2 and supports about 88 people/km^2.

Balai subdistrict appears to be mostly wooded, although mature forests are found only on hilltops and along boundaries with neighboring regions. Patches of both older and younger vegetation on hillsides testify to the continuing practice of swidden farming, with wet-rice farming gaining ever more importance in most parts of Balai. As Tara'n farmers rely less on swiddens and extend and intensify their use of local varieties of irrigated and rain-fed permanent fields, they increasingly convert the hillsides to managed forests, forest-gardens, or agroforests. Although the Tara'n of Balai differentiate many kinds of managed forests, they extensively use three types in the area, which differ in their origins, priorities of management, and rights of access and inheritance. In the Indonesian language, these three types of managed forests are called *tembawang*, *tanah adat*, and *tanah usaha*.

Figure 18.1. Map of Study Site, inset is map of Borneo.

Tembawang originate as the fruit gardens that commonly surround villages, traditional longhouses, or farm or work huts. Herds of pigs that roam freely in villages make it inadvisable to plant easily destructible annual crops nearby. Therefore, Tara'n farmers often plant tree crops in broad swathes around both principal and subsidiary dwellings. Tara'n Dayaks report that they change house sites about every 20 years, often moving only a few hundred meters away. Productive fruit trees, however, are rarely cut. A long history of moves, with subsequent plantings of gardens, have formed a landscape dotted with *tembawang* that mark where longhouses, farm huts, or palm sugar or rubber-processing huts were located many decades, even a century, ago. These forest-gardens, up to 10 hectares or more in size, extend today over hundreds of hectares. They contain not only planted species, but plants that grew on their own and were spared when the plots were slash-weeded during fruit harvests.

The second type of managed forest, *tanah adat*, often appears to be similar to *tembawang*, but these forest reserves, which belong to the community, were created in quite different ways. *Tanah adat* are usually areas that have never been cleared, at least not in the memory of Tae residents. However, a *tanah adat* hardly resembles an untouched forest because its vegetation is heavily managed. Some trees have been planted among those that grew on their own, and other trees have been removed. The farmers selectively clear the underbrush, especially during fruit harvests. *Tanah adat* plots in Tae include two village cemeteries as well as sites that have been specifically reserved for the entire community to harvest. The village of Tae owns approximately 14 hectares of *tanah adat* divided among three principal plots.

The third category of managed forests in Tae is *tanah usaha*, or enterprise plots. Most of these areas are planted to marketable species. Rubber is by far the most important such species in Tae, as it is in other areas of western Borneo (Dove 1988, Freeman 1970, Padoch 1983). Some villagers have also planted limited areas of cacao and other fruits. These plots, or agroforests, usually form parts of cyclic swidden-fallow agroforestry systems, where swidden plots, which are dominated by annual crops, are gradually transformed into stands of trees. Perennial crops—usually rubber, sometimes mixed with fruit trees—are planted in the fields as the hill rice is ripening. Often after about 30 years, when rubber trees pass their age of prime production and pressure on the land intensifies, the villagers begin a new cycle by felling the plots for swiddens.

The managed forests that I have described are only three types among many that can be found in Tae. The total number in this one small village of Kalimantan would be much higher if small specialized plots, including managed clumps of useful species in poorly drained areas or along streams and paths, were counted.

DIFFERENCES IN REGIMES AND PRIORITIES OF MANAGEMENT

Tembawang, *tanah adat*, and *tanah usaha* differ both in the way they were created and the way they are managed. All are managed quite casually, with occasional slash-weeding the most important part of management. The timing and intensity of clearing differ somewhat with the species in each type of plot. *Tembawang* that are rich in fruits, such as durian and illipe nut, are usually slash-weeded once a year to clear the areas where ripe fruits are likely to fall. Clearing around illipe nut trees may be especially intensive, and controlled burns are sometimes employed. *Usaha* plots that include rubber trees and sugar palms are partially cleared whenever vegetation limits access to these species that must be tapped frequently. These weedings have the same effect, whether they are done to ease exploitation or specifically remove undesirable species to promote the growth of useful ones. Very casual management, such as the removal of a vine or other competing plant from the vicinity of a plant that produces a marketable product, may also play an important role in traditional management systems.

Besides their variable management schedules, different priorities of management also distinguish the three types of managed forests. *Tanah usaha* —enterprise plots, as their name suggests—produce cash crops. For example, many *tanah usaha* have been planted with rubber since the late 1930s and 1940s. The composition and even the existence of *tanah usaha* change in response to opportunities in the marketplace, and the diversity of their structure and species composition varies greatly from plot to plot. A *tanah usaha* that is closer to the village may often have greater diversity of species, resembling *tembawang* in the number and kinds of fruit trees that are present. Those *tanah usaha* that are too far away to be conveniently harvested contain fewer fruits.

On the other hand, *tembawang*, are, by definition, rich in fruits. Most of these forest-gardens include a variety of the familiar domesticated fruits of Southeast Asia—rambutan, langsat, mangosteen—and they always include durian. However, the villagers of Tae obviously also manage *tembawang* for species diversity. One 10 m by 200 m transect through a *tembawang* revealed that 44 species were represented among 224 individual trees or stands of bamboo with a minimum DBH of five centimeters. Thirty of these species had edible fruits or shoots. Many of these fruits were semidomesticated species or wild relatives of well-known market fruits, especially mango and rambutan.

It is often impossible to ascertain whether the Tara'n Dayaks have actually planted any particular, individual tree. They allow many fruits that grow on their own to remain, and they even encourage their development. Tara'n forest managers claim to have planted 74 different fruits at one time or another, and they have identified more than 100 fruit species in the managed forests of Tae. Many fruit trees have multiple values, especially for construction and medicine.

The emphasis on edibles in *tembawang* should not suggest that they lack species that produce cash. Most *tengkawang*, or illipe nuts (*Shorea* spp.), which are an important "forest product" export for West Kalimantan, are harvested from managed *tembawang*. In addition to *tengkawang*, Tae villagers have been marketing durian fruits in quantity for the last 10 years. On just one day at the height of the durian season in 1991, I counted more than 10,000 durian fruits that were brought to market from four of Tae's five hamlets. Even at the minimal price of about 15 cents apiece, the income that many families earned throughout the month-long season was substantial.

Many *tembawang* also contain species that are specifically managed for their quality as construction materials. However, the *tanah adat* reserves are usually cited as the areas from which community members get most of the timber needed to build their houses. Since many households in Tae have constructed new houses within the last five years, timber reserves are now quite depleted. Nevertheless, in a 10 m by 200 m transect through a *tanah adat* in Tae we found 55 species among 229 individual trees. Thirty of these species were regarded as particularly good for construction.

DIVERSITY IN PROPERTY RIGHTS

Patterns of ownership, access, and inheritance reflect the origin, management priorities, and use of each forest management type. Each type is inherited in a different way, and each allows very different patterns of access. *Tanah adat* belong to the entire community. All villagers have equal rights to use these resources, and the elected head of the village enforces the specific rules of access. *Tembawang* are owned by the households whose members first planted them, and the rights to use them pass on to all their descendants through both male and female lines. However, some descendants, specifically those who have been more active in managing the *tembawang*, exercise stronger rights than others. *Tanah usaha*, like agricultural fields, are distributed to specific descendants by those who owned the land and planted the trees. *Tanah usaha*, unlike the other types of managed forests in Tae, are bought and sold.

While this very brief summary indicates that rules of access differ greatly among the various types of managed woodlands, it does not convey how flexible these rules are. Nor have I discussed the complex way in which the managed forests are cared for and used, with specific rules that apply to particular species. For instance, the cutting of any durian tree anywhere is an offense, and fines must be paid to the community even if cut by the tree's owner. Felling a rambutan, langsat, or mangosteen involves no such fines. The cutting of any tree in *tanah adat* without the permission of the village head is an offense that carries an extremely heavy fine.

"INVISIBLE" FOREST MANAGEMENT

There are indeed Indonesians who know how to manage the forests of Kalimantan in various ways and for various purposes. They have established complex rules to govern the management and use of their forest plots. The type of management that I have described is obviously not the one that Dr. Harahap had in mind when he declared that Indonesians must learn to manage the forests of Kalimantan. There are many reasons why he, and most of us, have never noticed these patterns of forest management. The apparent casualness of the management regime is doubtless a factor that explains the "invisibility" of this traditional forest management. None of the professional foresters or agronomists that staff provincial or regional government offices acknowledge that residents of Tae practice forest management. Not even inhabitants of Tae would say that they manage their forests in any technical sense.

The appearance of these managed forests is perhaps the most important factor that makes traditional forest management in Tae invisible. Tae's managed woodlands, especially the *tembawang*, do not **look** managed. The plantings are neither even-aged nor evenly spaced. They appear haphazard. Doubtlessly, the most important cause of this invisibility is the extreme diversity of the traditional stands. The villagers of Tae not only tolerate but encourage spontaneous vegetation in their plots. Most plots are managed for a variety of products, for subsistence and the market. The woodlands of Tae will be appreciated only when *species diversity* and *multiple uses* become desirable characteristics in agroforestry and forestry. Now, these forests are far too managed for those interested in conservation, and they are not managed enough for those interested in production. To most outsiders, they are simply invisible.

Literature Cited

Alcorn, J.B. 1984. *Huastec Maya ethnobotany*. Austin: University ot Texas Press.

Chin, S.C. 1985. *Agriculture and resource utilization in a lowland rainforest Kenyah community*. Kuching: Sarawak Museum Journal Special Monograph No. 4.

Conklin, H.C. 1957. *Hanunoo agriculture*. Rome: FAO.

Denevan, W.M. and C. Padoch, eds. 1988. *Swidden-fallow agroforestry in the Peruvian Amazon*. New York: Advances in Economic Botany 5.

Dove, Michael. 1988. *Swidden agriculture in Indonesia*. Mouton.

Fox, J.J. 1977. *Harvest of the palm: ecological change in eastern Indonesia*. Cambridge: Harvard University Press.

Freeman, D. 1970. *Report on the Iban*. London: Athlone Press.

Gordon, B.L. 1969. *Anthropogeography and rainforest ecology in Bocas de Toro Province, Panama*. Office of Naval Research Report. Department of Geography, University of California, Berkeley.

Gomez-Pompa, A., J.S. Flores and V. Sosa. 1987. The "Pet-Kot": a man-made tropical forest of the Maya. *Interciencia* 12:10–15.

Harahap, H. 1991. Keynote Address by the Minister of Forestry of the Republic of Indonesia on the Occasion of the Symposium on Sustainable Forestry Development in Indonesia, Sept. 9, 1991, at the Smithsonian Institution. Unpub. ms.

Hecht, S.B. 1982. Agroforestry in the Amazon Basin: practice, theory and limits of a promising land use. In *Amazonia: agriculture and land use research*. (S.B. Hecht, ed.). Cali, Colombia: CIAT. pp. 331-71.

Kunstadter, P., E.C. Chapman, and S. Sabhasri, eds. 1978. *Farmers in the forest: economic development and marginal agriculture in northern Thailand*. Honolulu: University Press of Hawaii.

Michon, G. 1985. De l'homme de la foret au paysan de l'arbre: agroforesteries indonesiennes. Ph.D. thesis, University of Montpellier, France.

Ministry of Forestry. 1991a. The policy and implementation of environmentally sound sustainable development and utilization of forest resources in Indonesia. Unpub. report.

Ministry of Forestry. 1991b. Industrial forest plantation and forest rehabilitation. Unpub. report.

Morris, H.S. 1953. *Report on a melanau sago producing community in Sarawak*. London: Her Majesty's Stationery Office.

Padoch, Christine. 1983. *Migration and its alternatives among the Iban of Sarawak*. Instituut voor Taal-, Land-, en Volkenkunde, Verhandelingen 98. Leiden: Martinus Nijhoff.

Posey, D.A. 1984. A preliminary report on diversified management of tropical forest by the Kayapo Indians of the Brazilian Amazon. In *Ethnobotany in the neotropics*. (G.T. Prance and J. Kallunki, eds.). Advances in Economic Botany 1.

Posey, D.A. 1985. Indigenous management of tropical forest ecosystems: the case of the Kayapo Indians of the Brazilian Amazon. Agroforestry Systems 3:139–158.

Torquebiau, E. 1984. Man-made dipterocarp forest in Sumatra. Agroforestry Systems. 2:103–127.

Weinstock, J.A. 1983. Rattan: ecological balance in a Borneo rainforest swidden. Economic Botany 37(1):58–68.

Wiersum, K.F. 1982. Tree gardening and taungya on Java: examples of agroforestry techniques in the humid tropics. Agroforestry Systems 1:53–70.

CHAPTER 19

RETHINKING RESOURCE AND ENVIRONMENTAL POLICY FOR THE BRAZILIAN AMAZON

A. C. de Albuquerque dos Santos

FORESTS ARE MULTI-DIMENSIONAL NATURAL RESOURCES

Throughout history, man has valued forests for the multiplicity of products and services, both material and environmental, that they provide. Forests yield food, medicines, spices, resins, gums, latex, timber, and fuelwood. Coupled with management of water resources, forests sustain important biomes by preventing erosion, controlling water runoff, and regulating rivers and streams. They also provide the geophysical environment for millions of plant and animal species, many that have yet to be studied (Goulding 1980, 1990).

The positive values of forests in the Brazilian Amazon have been ignored because the Amazon's role in Brazilian thought, particularly in resource policy, has been similar to that of the forest in Western culture (see Harrison 1992, Cronon 1983, and Pinto 1992). The forest has often played negative roles throughout history and in Western imagination (Harrison 1992). It has been a place of lawlessness, profanity, darkness, and obscurity. It is no accident that similar negativity is at the core of resource and environmental policy in Brazil. The Europeans who discovered and explored the area brought their cultural preconceptions with them (Pinto 1992, Oliveira 1983).

From the time of its discovery, the region was the object of European scientific expeditions to tap and exploit its natural resources. The English, Dutch, French, and Portuguese vied for economic and political control of Brazil. Finally, the Portuguese expanded and consolidated their geoeconomic interests in the New World by using the Amazon Basin, especially the Amazon river estuary, as a strategic base. Paradoxically, the first expedition to demonstrate that scientific curiosity and knowledge could play an essential role in shaping resource policy in the Amazon was led by Charles Marie de La Condamine of France.

La Condamine had left for Quito in 1735. Once the mission in Quito had been completed in 1743, he became fascinated with the possibility of returning home via the Amazon. La Condamine was the leader of a new breed of people forever ready to observe, measure, take notes, examine, and try to understand context. As Voltaire observed, there were no limits to his *curiosité ardente*. His primary interest while in the Amazon was to seek **information** of every possible kind, not gold, spices, conquest, or honor. The information that he recorded and the **knowledge** that he acquired became crucial for the emerging universal interest in the Amazon in 18th-century Europe. The scientific history of the Amazon, and of South America, had begun (Smith 1990).

We have wasted most of our unprecedented opportunities to discover and develop beneficial products from the Brazilian Amazon because resource and environmental policies have focused on intensive, short-sighted extraction of natural resources. The prevailing strategies promote a culture of extraction and resource dependency which destroys forest ecosystems. We now recognize, however, that knowledge is vital to the sustainable development of natural resources. Acquisition of new knowledge will allow formulation of policies for the Amazon that will maintain ecosystem integrity.

I believe that the greatest potential to develop social and economic value in the ecosystems of the Amazon lies in what waits to be discovered within them. However, we can only realize this potential if we adopt resource and environmental policies that preserve the integrity of ***whole*** ecosystems and that create ways to discover what these ecosystems contain. In this discussion, I will explore some of the theoretical foundations for such policies and practical guidelines that allow policy-makers to shift their focus from one of resource dependency to a future built on the acquisition of knowledge.

REDEFINING FOREST RESOURCE PRODUCTS AND SERVICES

Today, an increasing number of researchers are attempting to redefine what forest products and services are (FAO 1991, Wickens 1991a, 1991b). Historical accounts and reports of numerous scientific expeditions to the Brazilian Amazon have already recorded most forest products and services that have been mentioned in recent literature. Contemporary planners and policy analysts for the region have not used these historical sources as tools to formulate a policy for natural resources. They have neglected to interpret the past, which could add a fundamental dimension to their current resource policy (Homma 1989, 1992). This knowledge-base could provide the medium through which policymakers might communicate confidently, for it provides a range of reasonably stable expectations. Without this baseline as a guide, planners are not able to steer the policy-making process—it becomes a rudderless ship.

The history of the Brazilian Amazon recounts the systematic extraction of its natural resources over the past four centuries, with bleak sociocultural and environmental impacts and little economic benefits. This history illustrates a classic case of "limits to method." Extraction is unsustainable and causes resource dependency. Freudenburg (1992) explains that:

> Residents of extractive regions need to be aware of potential risks of simply **assuming**—or hoping—that development implications of extractive enterprises will be good ones. Despite general readiness of rural residents to accept vague claims about the benefits of extractive industries, **there is a striking absence of empirical support for the claims**. At the same time, there is growing reason to conclude that it may not be simply a matter of linguistic accident that these communities are commonly called extraction-dependent [emphasis added].

Yet, with increasing frequency, planners and researchers propose the harvesting of natural resources as a viable alternative for the management of tropical ecosystems and attempt to demonstrate its potential economic advantages (i.e., Peters et al. 1989). These studies have ignored the long-term implications of such alternatives for on-the-ground management as described by Bates (1864), Wallace (1853), Da Cunha (1976), and Goulding (1980, 1990).

Mendelbaun (1991) argues that constructing an appropriate and valid history is an **integral** part of the policymaking process, not a preface to the formulation and choice of policy. One chooses a past while one chooses a future. Mendelbaun (1991) observes further that:

> Currently, policy-makers, when they reach out to historical scholars, start the conversation by asking about the past. 'I have a plan for the city,' they say. 'Now tell me about the city's history so that the plan is placed in perspective.' The appropriate starting point lies, however, in the opposite direction. 'Tell me about the future you intend,' the historian should begin. 'What history—and why—is relevant to that future?' From that point, past and present are truly in service to one another.

KNOWLEDGE PRESENTS AN OPPORTUNITY FOR THE FUTURE

In feudal times, forests were valued primarily for wildlife and hunting grounds for the nobility, with fuelwood, mast, fodder, and raw materials of secondary importance. More recently, with industrialization, forests were valued primarily for timber and wood products. Renewed interest in the concept of ***forest ecosystems*** stimulated us to reexamine strategies for research, resource policies, and management practices. A more holistic scenario is emerging where forest-based services, such as flood protection and stabilization of the climate, are becoming more valuable than timber. The whole—the forest ecosystem—is actually greater than the sum of its parts—trees.

Current literature suggests new ways to assess the **value** of forest ecosystems. The first step is to focus on forests' potential to provide nonwood products. Hopefully, the next step will be to focus on the socioeconomic role and services of forests, especially the phenomenal potential to gain knowledge from the Brazilian Amazon. Reorientation of resource policy is a major step toward integrating an improved understanding of forests and their functions, services, roles, and values. The impact of land-use and environmental policies and development projects on the roles and services of forest ecosystems and on their potential benefits for the wider, rural economy should be the **primary** concern of policy-makers.

Merely attaching costs to specific products and services of the forest is not enough. These are artificial values that ignore such crucial elements as the effects of society and culture on management techniques and habits. They are useless as financial yardsticks.

Alternatively, we might focus on site-specific investigations to understand particular roles and services of forests. Current interest to search for the market value of forest products could benefit from a deeper commitment to:

- The acquisition of knowledge about how forest ecosystems function
- The social and cultural dimensions of resource management
- A longer time frame to refocus on values beyond the marketplace

INFORMATION AS THE FOUNDATION FOR CHANGING AMAZON POLICY

The Brazilian Amazon has no future without science and technology. To achieve its future, however, requires organizing in a way that leads to useful information and knowledge. A brief review of cooperation and information is helpful before using the ecosystem paradigm to structure our scientific and technical understanding.

Barnard (1968) provides an understanding of organizations based on cooperation. He understood organizations as cooperative systems based on people willing to work together and communicate to achieve a common purpose. Manifesting a vision of purpose is, as a consequence, critical to harnessing cooperation among scientists and organizations concerned with the Amazon. This is the role of leadership.

In his classical reexaminations of organizations and information, Arrow (1974a, b) used information as a device for overcoming the many imperfections of both the market and the individual as decision-maker. Information, in Arrow's sense, is a set of signals that reduce our uncertainty about the true state of the world. Information is limited; it is scarce; it costs money to acquire and to use. The presence of uncertainty limits the individual's ability to be rational, and it certainly limits the rationality of organizations. Information helps both move away from irrational or illogical behavior toward more rational choices and actions. The value of these movements gives information its value.

Bentley (1984) has taken a rather simple version of Arrow's exploration in defining *data* as the raw numbers and qualities we can gather from a natural resource system. *Data* which is given structure—*given form to*—is *information*. *Knowledge*, in this same pragmatic sense, is defined as information which is understood in the context of a particular applied problem. Applied science and the resulting technologies, then, can be viewed as the process of developing knowledge to solve problems.

Two points become obvious in this set of three linked definitions. One is that the results of creating knowledge have value. The more important the problem, whether defined by scale, money, or what ever measure, the higher the value of new knowledge. The second obvious point is that the gathering and structuring of data costs money. Further, understanding problems in terms of their use requires investment in the education and training of people, which is expensive. This suggests that alternative linkages among data, information, and knowledge might be more efficient than other alternatives. Indeed, this is the power of certain paradigms: they give a structure to certain kinds of problems that makes the informational linkage both more valuable and less expensive than other alternatives. The ecosystem paradigm is such a structure.

THE CONCEPTUAL BASIS FOR A CHANGE IN RESOURCE AND ENVIRONMENTAL POLICY FOR THE BRAZILIAN AMAZON

Ecosystem Management

Resource and environmental policy in the Brazilian Amazon could greatly benefit from **ecosystem management** with a focus on knowledge of the whole ecosystem and not on its parts (Gordon 1992).

> ***Ecosystem*** derives from the physical-chemical definition of ***system***—a bounded space pervious to energy but not mass, with transit and transactions of both accounted for. Mass obviously goes in and out of ecosystems in real life, but it must be accounted for. Ecosystems are usually areas with intrinsic properties. Watersheds, or lakes, are the most frequently cited examples, but any bounded space can usefully be considered an ecosystem, including Earth. Management implies a **human purpose** and the means and the will to attempt to accomplish it. Ecosystem management is at a minimum, people trying to accomplish something in a bounded space.
>
> People usually try to manage only that which they can influence, so that **ecosystem management** takes place on or from the surface of Earth; its surface is traced on the Earth's surface; and [its boundaries]

> usually extend upward and downward an unspecified but relatively short distance. The boundary may be flexible in time and with objective; that is acceptable and useful as long as changes are traced and recorded. Because of its dependence on tracking energy flows, ecosystem management is green-plant centric. [emphasis added].

Gordon observes that ecosystem management, just like other theoretical forms of management, is now only imperfectly realized anywhere. However, its theoretical nature should be no barrier to the use of its principles and the practices they imply. The technology for ecosystem management will probably become more significant than either the communications revolution or biotechnology because its use potentially guarantees a livable environment.

Principles of Ecosystem Management

Let us examine a set of essential principles for ecosystem management that Gordon (1992) proposed.

Manage Where You Are

Since ecosystem management requires both the recognition and transcendence of real boundaries, it applies to a specific "site." Although the notion of *site specificity* is old in forestry, the realization of ecosystem management will require even greater emphasis on the specific properties of the place to be managed and the objectives for which it is managed. Many of the examples of "poor" past practice are the results of applying "best practices" for certain places and objectives out of context. The parallel of this principle with Lasswell's total commitment to **contextuality** in policy (Lasswell 1971) is evident.

Manage With People in Mind

All ecosystems on earth are subject to human influence, known and unknown, overt and accidental. Management, itself, implies strong human purpose. Yet, some of the most difficult tasks that have been proposed for ecosystem management involve the *exclusion* of human use and influence from specific systems. Thus, ecosystem management begins with a careful evaluation of human desires, influences, and responsibilities. The notion of *core values* that is essential to policy sciences becomes the core of ecosystem management. We ask, "Managing for whom?" From the broad perspective of regional conservation, this question becomes an essential component of resource policy. Thus, every system that can be defined in biological and physical terms connects to and interacts with a network of human **values**, **uses**, **institutions**, and other social structures.

Manage Across Boundaries

Ecosystem management keeps track of transactions that cross boundaries of defined systems and shifts these boundaries when necessary. At the most elementary level, such accounting requires that "neighboring" influences be recognized and managed. Often fee ownership is not extended. However, a complex process, in which joint goals and **incentives** are decided, and compromises and regulations are made, is required. Indeed, in the context of ecosystem management through the development of organizations, the issue of incentives becomes crucial for ecosystem management and sustainability.

Manage With Detailed Knowledge Instead of "Rules of Thumb"

Knowledge of specific processes and interactions that are responsible for activities and outputs of the ecosystem is crucial for ecosystem management to continually improve. Systems based on the landscape are so complex that we can rarely enumerate and track all of their components. **Detailed knowledge of why things occur as they do is the only way to predict the behavior and output of an ecosystem.** For example, such knowledge requires extensive sampling throughout the system, combined with information gained elsewhere that may be tested and adapted in the system at hand.

Manage Without Externalities

For our purposes, *externalities* means "contents of a system that are extraneous to the objectives of management." At first glance, this definition seems to contradict our assertion that a complete enumeration of a system's components is seldom, if ever, achieved. However, management must include all known contents of the system to determine when decisions and manipulations should be made.

Gordon (1992) makes a final point about the nature of ecosystem management:

> It is increasingly clear that ecosystem management presents one of the only available ways to begin to resolve the paradoxical need to manage "nature." Ecosystem management contains the minimum set of activities that must in prudence be applied by people to any "natural" system, including the global one. What this "ecosystem minimum set" instructs us to do with natural systems is again five-fold, and includes: (1) keep systems parts, (2) keep accurate records, (3) make forecasts, (4) test them, and (5) look outward, as well as specifically, at the system in question.

Knowledge Instead of Labor

Inadequate channels for acquiring and processing information is the **primary** cause of present externalities in the Brazilian Amazon. This deficiency prevents regional organizations from engaging in effective cooperation at the national and international levels. Arrow's analysis of the role of the organization as a processor of information becomes crucial in this context.

The future is uncertain. With the current rate of technological transformation throughout the world, participation of humans in the market system will probably change radically. Perhaps, even capitalism, like Marxism, will perish. A new, hybrid economic system may emerge. However, we may be certain that **knowledge** will be the driving force behind any new emerging economic and market system.

Drucker (1992) envisions that the change to a society based on knowledge will precipitate further transformations:

> In this society, knowledge is **the** primary resource for individuals and for the economy overall. Land, labor, and capital—the economist's traditional factors of production—do not disappear, but they become secondary. They can be obtained, and obtained easily, provided there is specialized knowledge. At the same time, however, specialized knowledge by itself produces nothing. It can become productive only when it is integrated into a task. And that is why the knowledge society is also a society of organizations: the purpose and function of every organization, business and nonbusiness alike, is the integration of specialized knowledge into a common task.

Cooperation and information will become increasingly significant in the future knowledge-based society and could provide a wide range of opportunities to correct some imperfections of the price system. First, the price system would become more sensitive to the social structure. It would also reflect various levels of specialization within an overall context. For example, instead of isolating a specific product—timber, land, fruits, nuts—to assign a price to, we would incorporate the product into the value of the whole forest ecosystem with its multiple roles. We would focus on **context** rather than on isolated fragments.

In addition, collecting information about natural systems must inevitably rely on local, on-the-ground investigation that is sustained and long-term. For a knowledge-based society to emerge, information will have to be processed efficiently, both within and among organizations. An effective organizational framework will be needed to turn this information into knowledge. And, the organizations must know how to cooperate with each other for the knowledge to become productive, to perform its social function. *Cooperation* will become more complex, more multidimensional. The locus of cooperation will also have to change. It must become decentralized and occur at all levels of organizations to allow decisions to be made quickly. To be effective, these decisions must be made by those who are *closely in touch with* economic performance, the market, technology, the environment, demographics, and knowledge—all the myriad changes of society. This locus—closeness—cf cooperation to decision-making will provide opportunities for innovation (Drucker 1992, Perrow 1986, 1991).

We need to understand and quantify market failures to level the "playing field" (Weimer and Vining 1991). The way in which individual economic agents receive and transmit information powerfully conditions the possibilities of allocating risk throughout the market. Indeed, the fact that some markets do not exist may be a possible explanation for environmental problems. We need to make a realistic account of the costs of natural resources to reflect their true market value when we allocate resources, especially in the case of an externality such as environmental pollution such as mining and deforestation. These costs must incorporate the full range of services such as environmental protection—flood control, protection of habitats for wildlife, pharmaceutical use—and **social function** that provides economic alternative uses of natural resources, ranging from low-intensity subsistence agriculture to tourism (Arrow 1974a).

Cooperation: From Factory to Craft

Recent policies to develop the Brazilian Amazon have adopted mechanisms that focus on quantity, not on quality. Extraction, not knowledge, has been the foundation for resource and environmental policy. For centuries, all major policies for the region have been based on the **assumption** that the Amazon could be "harvested," or cultivated to be harvested.

However, this assumption made the mechanism for the price system even less effective. It prevented any formulation of policy that would capture hidden values that could be incorporated into the final cost of the "product." Essentially, one could say that static, one-dimensional policies make the region like a factory. For example, the forest was important only for timber; the soil, only for ore or agriculture; and rivers, only for hydroelectricity.

The colossal biodiversity of the Brazilian Amazon—with its large-scale heterogenous systems, cultural diversity, and extremely high costs to gather and disseminate information over a vast area—underscores the urgent need to initiate resource and environmental policy that focuses on the acquisition of knowledge. To date, policy options have not considered quality—indeed, *craft* may be a more appropriate word—and very little effort and resources have been invested in **defining** the problems, such as capturing the multidimensional nature of the Amazon.

Decree 1164 in 1971 is an extreme example of policy that is based on extraction. It federalized a large part of land in the Amazon to implement policy that designated this region as the locus for excess populations of other regions of Brazil and that sanctioned the replacement of the Amazonian forest

Table 19.1. Policy Assessment Typology of Core Social, Cultural and Biological Impacts in the Brazilian Amazon

Sources of Potential Impact	Area of Occurrence	Typology of Core Social, Cultural and Biological Impacts
1. Gold Prospecting	Rondônia, Pará, Amapá, Amazonas, and several sub-basins of the Amazon, Tapajós and Madeiras Rivers	1. Silting and erosion of waterbodies; 2. Water pollution, turbidity and heavy metals increase; 3. Chaotic formation of population centers with complex social problems; 4. Intense landscape degradation; 5. Aquatic life degradation with potentially devastating impact on fish and human populations; 6. Uncontrolled migration streams of placer miners with potentially deadly conflicts and/ or contact with Indian groups; 7. Ever decreasing-quantity and quality of human life for affected population; 8. Explosive cocaine-gold trade with unprecedented socio-cultural impact for the Amazon.
2. Extensive Agricultural and Cattle Raising	The entire Amazon near highways and big cities	9. Intensive and extensive forest burning; 10. Contamination of waterbodies with pesticides; 11. Erosion and siltation of waterbodies; 12. Destruction of native productive areas with destruction of flora and fauna; 13. Lax regulation and enforcement set off deadly land speculation process and deforestation.
3. Construction of the Transamazon Highway and Associated Road Network	Became axis of unprecedented level of land access to most of the Amazon	14. Created conditions for induced and spontaneous migration streams into the Amazon; 15. Destruction of native cultures; 16. Large farming and cattle raising projects; 17. Land speculation fueled by lax regulation and control mechanisms for the utilization of fiscal incentives; 18. Large-scale and destructive fragmentation of legal authority of Amazonian states over their land with complex implications for biodiversity (Decree 1164/71); 19. Large burnings of forest ecosystems; 20. Expansion of mining activities; 21. Propagation of endemic diseases; 22. Explosive population growth rates on project target areas.
4. Explosive Land Tenure	The entire Amazon region, especially southern Pará	23. In Brazil in general and in the Amazon in particular, there is one of the most inequitable land distribution in the world; 24. From 1960 to 1986, there have been 258 rural leaders killed in the state of Pará. For the period of 1980-86 there were 36.9 deaths/year, an increase over the 1973-79 period which registered an average of 9.3 deaths of rural leaders/year.
5. Explosive Rates of Internal Migration	The entire Amazon region	25. Complex socio-environmental problems at large urban areas due to explosive demand for services, as a consequence of population growth rates of up to 40% between 1970-80; 26. Chaotic land occupation with severe environmental impacts.

6. Poor Choice to Locate Hydroelectric Plants	Amazonas: The Balbina Project	27. Cultural impact on native population; 28. Unnecessary flooding of large areas affecting forest ecosystems, potentially productive agricultural areas and settlements; 29. Large impacts on biodiversity due to drastic modification of landscape; 30. Large socio-economic losses due to poor performance of project in relation to its financial investment.
7. Industrial Mining (e.g. manganese, cassiterite, copper, bauxite)	Amapá, Amazonas, Pará (Carajás), Rondônia	31. Intense landscape degradation; 32. Water pollution and silting; 33. Social and economic limbo imposed on the peripheral populations within the area of influence of mining operations due to sub-employment and degrading social and health conditions.
8. Pig Iron Industries	Pará Great Carajás Project	34. Created demand for charcoal becomes additional pressure to deforest forest ecosystems; 35. Production of cheap energy at high environmental cost; 36. Water, air and soil pollution.
9. Lax Regulation of Industrial Activity	Amazonas: Free Trade Zone	37. Air, water and soil pollution; 38. Increased load and disposal of toxic residues; 39. Impact on urban center due to the increased demand for services.
10. Predatory Hunting and Fishing	The entire Amazon, especially near highways and large cities	40. Extinction of aquatic mammals and decrease of turtle and fish populations in the Amazon Basin; 41. Drastic reduction of species of economic and ecological value.
11. Aluminum Industry	Pará (Belém)	42. Air and water pollution; 43. Subsidized low energy prices to the project at the expense of state government of Pará.
12. Leadership and Organizational Vacuum	The entire Amazon Region	44. The Brazilian Amazon has suffered from a <u>chronic and perverse lack of leadership</u> to muster social and political support to promote the policy shifts of her interest; 45. The density and quality of the organizational framework in the Amazon is extremely low, providing extremely favorable conditions for capture by the actions and/or policies of the federal and international levels; 46. Organizational framework weakness becomes a <u>major source of policy failure</u> because it precludes the successful conception and implementation of <u>long-term knowledge-acquisition policies</u> for the region whose primary requirement is the establishment and development of a strong and effective network of on-the-ground organizations.

by classic agricultural systems. Again, the culture of extraction and resource dependency was at the heart of resource policy. Decree 1164 federalized about 65% of Amazonian lands. At the time, the legal authority over 86 million ha of the 122 million that form the state of Pará was transferred to the federal government (Pinto 1992).

In this example, the price system failed to capture the full market value of the resources that were targeted by the policies. It devalued supply at the expense of a wide range of factors (table 19.1). The exogenous expectations of demand prices also fell. To date, policy models, having invested very little in knowledge, choose an option to increase supply, a situation that brings even lower prices and ever higher social and environmental impacts.

Options for resource policy must concentrate on the acquisition of knowledge, which will form the basis for economic and ecological assessment. Then policies, informed by specific social and cultural knowledge, can become powerful tools that will foster knowledge, perhaps combined with a reduced focus on quantity. Such policies will limit the amount of goods, i.e. mostly natural resources in the Brazilian Amazon, and also reduce their social costs while raising their quality. Raising the quality of natural resources may be achieved with management strategies that consider a range of impacts from extraction, processing, and trade of natural resources. For example, various "equally weighted" products in the market system—e.g., lumber, fish, even hydropower—can have vastly different qualities, and the price system, reflecting responsible management practices, could reduce their final values with decreased social and environmental costs. In addition, large, cumbersome federal agencies that implement policy could be decentralized into smaller and more responsive ones in the region. This reorganization could improve opportunities for the future and increase cooperation, which would immediately reduce costs of implementation and, more significantly, the present level of externalities.

Institutional Change to Embody Local Context Into Organizational Framework

We could significantly improve current policies for the Brazilian Amazon by ensuring that local organizations could learn—i.e., "read" and interpret—their environment. Morgan (1986) envisions such a scenario. Reflecting on ways to facilitate self-organization, he suggests principles of holographic design:

- Get the whole into parts
- Create connectivity and redundancy
- Create simultaneous specialization and generalization
- Create a capacity to self-organize

Envisioning organizations as "brains," or holographic systems that process information represents a major departure from the classic models in the design of organizations. Morgan (1986) argues that the holographic character of the brain is most clearly reflected in the patterns of connections through which each neuron, or nerve cell, is related to a myriad of others, allowing a system that is both generalized and specialized to function. He explains that

> to compare the brain with a hologram may seem to be stretching reason beyond the limits. However, the way a holographic plate enfolds all the information necessary to produce a complete image in each of its parts has much in common with the functioning of the brain. And it is possible to extend this image to create a vision of organization where capacities required in the whole are enfolded in the parts, allowing the system to learn and self-organize, and to maintain a complete system of functioning even when specific parts malfunction or are removed.

Morgan (1986) summarizes the principles of holographic design as follows:

- **Redundant Functions**, which show a way of incorporating wholes into parts by creating redundancy and connectivity—a simultaneous specialization and generalization
- **Requisite Variety**, which provides practical guidelines for the design of part-whole relationships by showing exactly how much of the whole needs to be built into a given part
- **Learning to Learn**
- **Minimum Critical Specification**, both of which show how to enhance the capacity for self-organization

The holographic principles listed above underlie Drucker's (1992) vision of knowledge specialists:

> Because the modern organization consists of ***knowledge*** specialists, it has to be an organization of equals, of colleagues and associates. No knowledge ranks higher than another; each is judged by its contribution to the common task rather than by any inherent superiority or inferiority. Therefore, the modern organization cannot be an organization of boss and subordinate. It must be organized as a team.
>
> There are only three kinds of teams. One is the sort of team that plays together in tennis doubles. In that team—and it has to be small—each member adapts himself or herself to the personality, the skills, the strengths, and the weaknesses of the other member or members.

> Then there is the team that plays European football or soccer. Each player has a fixed position; but the whole team moves together (except for the goalie) while individual members retain their relative positions. Finally, there is the American baseball team—or the orchestra—in which all the members have fixed positions (emphasis added).

Nelson and Winter (1982) and Simon (1976, 1991) argue further that serious analysis of policy requires detailed understanding of the organizations and mechanisms, interests, and **values** at stake. They see analysis of policy as a continuing process. Public policies and programs, like private activities, are embedded in and carried out by organizations. In a basic sense, it is the organization that learns and adapts. To a considerable extent, the design of a good policy is the design of an organization, with a structure that makes it capable of learning and of adjusting its behavior in response to what is learned. There is no doubt about Nelson and Winter's perception of the role of organizations for policy:

> To recommend reasonable policy for a particular case, it is necessary to assess the existing organizational framework in detail, to make tentative judgments about an uncertain future, to draw on the fund of experience with related problems, and—above all—to recognize that new information will be coming in as the future unfolds.

KNOWLEDGE IS THE MOST CRUCIAL VALUE

Today, the world economy is at a turning point, which is not unlike the transformation from an agricultural economy in the 19th century to an industrial economy that has endured for most of this century. The old paradigm of success was based on access to natural resources and pools of labor. Now, the greatest change will involve the reorganization of work itself. In this new paradigm, strategic resources are no longer only those from the ground—oil, wheat, and coal—but **ideas and knowledge** from the mind.

The recent rise in interest in conservation and protection of biodiversity has generated many potential ways to redefine the role of Amazonian forests, with new issues in resource policy and management. Unfortunately, there is a growing perception that once one is able to assign **economic** value for a wide range of forest products and services, these treasures will be saved by the forces of the market. In addition, many contemporary researchers have listed and defined nontraditional products of the forest. However, their focus has been on present experience or on experience from the near-past, with little effort to incorporate the lessons from history. These assessments of potential forest products have too often ignored the teachings and findings of some 400 years of expeditions to the Brazilian Amazon. The sociocultural context and specificity of the region's wide range of scenarios has been replaced by an intensive and extensive search for the market value of forest products.

Knowledge is the key to the future. Efforts to assign merely a market value to forest products and services keeps the contemporary focus narrow and sacrifices long-range perspective. In this chapter, I have proposed an alternative to the prevailing contemporary view—to shift resource and environmental policy away from extraction and resource dependency. The organizing principles to create this new policy are:

- Ecosystem management
- Acquisition of knowledge, instead of labor
- Intensive and extensive cooperation
- Institutional change that allows local organizations to become mechanisms able to learn and self-organize

Decision-makers must understand that resource and environmental policy for the Brazilian Amazon will cross the boundaries of policy that were once thought to be differentiated. It is crucial to bridge these historical gaps. We may begin by breaking down the barriers that have long existed between the agencies that regulate and implement policies for the Amazon, especially in areas such as the environment, science and technology, natural resources, education, and, of course, economy.

A final organizing principle of resource and environmental policy should be the incorporation of **values**. Etzioni and Bell provide powerful suggestions in this context:

> For a society to achieve mastery of post-modern technology it must mobilize its members to work toward an active society, gain self-control, and realize more fully its **values** (Etzioni 1968).
>
> Acquiring accurate, relevant knowledge about the way the world works, including the social world, and making presumptively accurate predictions about the future consequences of one's actions are aspects of being morally responsible and accountable. **Although intentions matter, it is moral to do right than merely to intend to do it right** (Bell 1992). [emphasis added].

The principles that I have discussed in this chapter for a policy shift away from extraction and resource dependency presents decision-makers with a tool to create the conditions that would allow knowledge and cooperation to permeate the entire range of public policy for the Amazon, while minimizing adverse impacts on the region's unique biological, cultural, and social settings.

The wealth of the Brazilian Amazon is the **information** stored in its phenomenal biodiversity. The **knowledge** that we derive from this information must become the **core** value of resource and environmental policy for the region. Understanding the role and services of the resources from the region is crucial to protect and manage its vast natural wealth for the benefit of Amazonians, Brazilians, and mankind. Neglecting this focus on knowledge to search for product and market values of resources might jeopardize this opportunity to benefit the Amazon for the long-term future. We must turn this new resource and environmental policy into action. The alternative could be too costly, especially for Amazonians.

Literature Cited

Arrow, K.J. 1974a. *The limits of organization.* New York and London: W. W. Norton & Company.

Arrow, K.J. 1974b. Limited knowledge and economic analysis. *The American Economic Review.* Vol. 64, #1, March, 1974. pp 1-10.

Barnard, Chester. 1968. *The functions of the executive.* Cambridge: Harvard University Press. Originally published in 1938.

Bates, H.W. 1864. *The naturalist on the rivers Amazons.* London: Murray.

Bentley, W. R. 1984. Value of information in context of forest management and policy. Report to USDA Forest Service. The University of Connecticut Dept. Agri. Econ. AERS Staff Paper 84-4.

Bell, Wendell. 1993. *Bringing the good back in: values, objectivity, and the future.* Vol. 137, August. pp 333-347.

Cronon, William. 1983. *Changes in the land: Indians, colonists, and the ecology of New England.* New York: Hill and Wang.

Da Cunha, Euclides. 1976. Um paraíso perdido: Reunião dos ensaios Amazônicos. In *Coleção Dimensões do Brasil.* Rio de Janiero: MEC & Vozes Ltda.

Drucker, Peter F. 1992. The new society of organizations. *Harvard Business Review,* September-October, 1992. pp 95-104.

Etzioni, Amitai. 1968. *The active society: a theory of societal and political process.* New York: The Free Press.

FAO. 1991. Nonwood forest products: the way ahead. Rome: FAO. No. 97.

Freudenburg, William R. 1992. Addictive economies: extractive industries and vulnerable localities in a changing world economy. In *Rural Sociology* 57(3):305–332.

Gordon, John C. 1992. Ecosystem management: an idiosyncratic overview. Paper prepared for *Defining Sustainable Forestry*, Reston, Virginia, (January 13-15). Unpublished.

Goulding, M. 1980. *The fishes and the forest: explorations in Amazonian natural history.* Los Angeles: University of California Press.

Goulding, M. 1990. *Amazon: the flooded forest.* London: BBC Books.

Harisson, R.P. 1992. *Forests: the shadow of civilization.* Chicago: University of Chicago Press.

Homma, Alfredo K.O. 1989. Reservas extrativistas: uma opçao de desenvolvimento viável para a Amazônia? In *Pará Desenvolvimento*, No. 25, jan/dez. pp 38-48.

Homma, Alfredo K.O. 1992. Opportunidades, limitaçoes e estratégias para a economia extrativista vegetal na amazônia. In *Desenvolvimento sustentavel: um novo caminho*. (Juan L. Bardalex Hoyos, organizer). Federal University of Pará, Núcleo de Meio Ambiente (NUMA). pp 67-78.

Lasswell, Harold D. 1971. *A preview of policy sciences*. New York: American Elsevier Publishing Company, Inc.

Mendelbaun, Seymour. 1991. The past in service to the future. In *History and public policy*. (D.B. Mock, ed.). Malabar, Florida: Krieger Publishing Company. pp 39-53.

Morgan, Gareth. 1986. *Images of organizations*. Newbury Park, California: Sage Publications. 13th Printing, 1991.

Nelson, R. and Sidney G. Winter. 1982. *An evolutionary theory of economic change*. Cambridge, Massachusetts and London: The Belknap Press of Harvard University Press.

Oliveira, Adélia E. de. 1983. Ocupação humana. In *Amazônia: Desenvolvimento, Integração e Ecologia*. (E. Salati, O. Schubart, W. Junk, and A.E. Oliveira, ed.). Brasília: Editora Brasiliense e Conselho de Desenvolvimento Científico e Tecnológico. pp 144-327.

Perrow, Charles. 1986. *Complex organizations: a critical essay*. New York: McGraw-Hill Publishing Company. 3rd Edition.

Perrow, Charles. 1991. A society of organizations. *Theory and Society 20(6):725-762.*

Peters, Charles M., A.H. Gentry, and R.O. Mendelshon. 1989. Valuation of an Amazonian rainforest. *Nature* 339:655-56.

Pinto, Lúcio Flávio. 1992. *Amazônia: a fronteira do caos*. Belém: Gráfica Falangola.

Simon, Herbert A. 1976. *Administrative behavior: a study of decision-making processes in administrative organizations*. New York: Free Press. 3rd edition.

Simon, Herbert A. 1991. Bounded rationality and organizational learning. In *Organization Science*. Vol. 2, No. 1, February. pp 125-34.

Smith, Anthony. 1990. *Explorers of the Amazon*. New York: Viking Penguin, Inc.

Wallace. A.R. 1853. *A narrative of travels on the Amazon and Rio Negro*. London: Reeve.

Weimer, D. L. and Aidan R. Vining. 1991. Rationales for public policy: market failures. In *Policy analysis: concepts and practice*. Englewood Cliffs: Prentice Hall. pp. 30-77.

Wickens. G.E. 1991a. Management issues for development of nontimber forest products. In *Unasylva* 165, Vol. 42. pp 3-8.

Wickens, G.E. 1991b. Learning to see the forest through the trees. In *Unasylva*, 165, Vol. 42.

CHAPTER 20

COMMERCIAL FORESTRY AND RURAL DEVELOPMENT: LOOKING AT THE NEXT FIFTY YEARS

R. E. Evenson
W.F. Hyde

FACTORS OF ECONOMIC DEVELOPMENT INTERACT

The general process of economic development entails powerful interactions among factors such as population growth, land use, natural resource management, and the availability of both agricultural and industrial technology. One important subprocess of this development is the conversion of forest and grasslands to crops or managed pastures as the global population and international markets for agricultural products grow. A second vital subprocess is the harvesting of native forests and wood processing, which often accompany the expansion of croplands. A third subprocess is the development of forest plantations.

These subprocesses of economic development generate opportunities for employment and higher income. Consequently, policies that affect land use, particularly harvesting of forests, should be designed to maximize opportunities for employment and higher income in each developing country. Population growth, general economic growth, relative prices, and availability of technology critically determine the appropriateness of these policies. Technology, especially agricultural technology, is highly specific to a geographical region and cannot be easily transferred from place to place. Most research and development must be adapted to a specific location before a suitable technology for that region can be applied successfully by local farmers and foresters. Availability of technology, therefore, implies that local farmers and foresters have access to useful and applicable technology for their specific locations.

Population growth is a critical factor in economic development because it is accompanied by a growth in the demand for food. As population has grown, almost every country has historically expanded its use of cropland and managed pastures to its limits, i.e. until its "frontier" is closed. Deforestation, caused by population growth and increases in real income, occurs as all land that is relatively well suited for agriculture is cleared for cropland or pastureland.

Once the "frontier" is closed, each country must have access to water for irrigation and to agricultural technology that will increase crop yields. Expansion of agriculture on fragile or poor land ceases in countries that have invested in agricultural research, extension, and irrigation. These countries increase their food production with multiple cropping and higher yields per hectare. However, in countries where these investments have not been made, population growth accelerates agricultural expansion, with consequent deforestation, onto land that is unsuitable for crops.

COMMERCIAL FORESTRY IN ECONOMIC DEVELOPMENT

Commercial forestry plays two roles in economic development. First, it efficiently harvests the forestland that is cleared for cropland and managed pastures, while providing employment and income. It harvests the natural forests that are not cleared for agricultural purposes. Harvest policies that allow high levels of cutting from natural forests will depress world timber prices. Technology, however, can either depress or increase timber prices. Improved technology for harvesting and transportation may increase the quantity and quality of the harvest and its marketability, but these benefits to owners of forest resources could be offset by a harvesting policy that permits high cutting levels. Improved technologies for wood processing may use more harvested timber from a given forested area to yield more valuable wood products at lower processing costs. Availability of substitutes for forest products—fossil fuels, nonwooden building materials—will have a tendency to depress timber prices.

The economic viability of the third subprocess—plantation forestry—obviously depends on plantation technology and relative prices for wood products from plantations. For this purpose, we can consider agroforestry to be a special type of plantation forestry. As stumpage prices rise and plantation technology improves, more and more plantation projects will become viable.

Presently, plantation forestry is not contributing a large share of the world's forest products (see Gowen et al. and Sedjo-Lyon, this volume). Although there are some special exceptions, its technology has not advanced at a rapid rate. Stumpage prices, while rising in real terms for quality material, remain low relative to the investment costs. Many plantation projects require some form of subsidy to be economically viable. Some developing countries rely on the conversion of forestland to cropland for their forest products. However the bulk of our forest products come from harvesting natural forests.

THE OUTLOOK FOR PLANTATION FORESTRY

Asking what the next 50 years will bring to commercial forestry and rural development provokes more questions. What will happen to world stumpage prices? Are present rates of harvesting natural forests sustainable? Will the prices of substitutes for wood—oil, cement, etc.—rise? Will plantation technology improve? Will agricultural technology improve enough to halt the conversion of forested land to crop lands? How can economies of developing countries obtain maximum value from the forest resources that will be available to them over the next 50 years?

We can't pretend to have the answers to these questions. At best we can offer judgments based on recent history.

Technological Change

Let us first consider technological change that might occur in projects to develop forest plantations. Without considerable technological change, we do not expect the plantation sector to grow rapidly, unless stumpage prices—and concomitant prices for forest products—rise. W.R.J. Sutton (this volume) documents an impressive case of plantation forestry with radiata pine in New Zealand. This is a good example of the economic contribution that plantation forestry can make. The technology may at first appear to be simple and easily replicable in other parts of the world, but closer examination reveals that complex thinning and stand management are required. The New Zealand experience also required testing, or screening, many species in addition to marketing research. And New Zealand offered soil and water conditions not often found elsewhere. Thus it is not easy for other countries to replicate this experience.

Sutton (this volume) estimates that there are now 100 million ha of forest plantations worldwide, of which 14 million ha are planted with fast-growing hardwoods and softwoods. He estimates that plantations now supply about 10% of the world's industrial wood and less than 5% of sawlogs and peeler logs.

Is new technology developing fast enough to stimulate rapid growth in the forest plantation sector? Mergen et al. (1988) reports that investments in research are low relative to comparable investments for agricultural production. Investments of a substantial nature, especially time, are required to produce better technology. The development of improved genetic stocks is dependent on the long growth cycle of trees. Even though there may be considerable scope to make progress in biotechnology and management in plantation forestry, these gains are many years away. Trees, like agricultural crops, are sensitive to soil and climatic conditions. Technology must be adapted to regional environments. As Mergen et al. note, little capacity for research and development exists at the present time in most of the tropics.

It does not appear, then, that plantation forestry will expand through a significant contribution from technology in the near-term, although we may expect some growth over the next 20-25 years. Where investments in research are being made, especially in temperate and subtropical regions, we may expect this sector to advance significantly during the next quarter of a century. On the whole, however, it seems that plantation forestry is not on the verge of a green revolution.

Economics

We are less convinced than Sutton (this volume) that a significant rise in stumpage prices will stimulate an expansion in the forest plantation sector in the near future. Sutton argues that demand for wood products will grow. Consequently, the stumpage prices and the relative prices of forest products will rise to induce sufficient harvesting and production to meet this demand. He argues that present rates of harvesting from natural forests cannot be sustained. In addition, he argues that we are not producing substitutes for forest products rapidly enough to meet the growing demand for wood products. We do not share this judgment. We base our conclusions on the following observations:

- Many poor countries, with rapid rates of population growth, will continue to obtain timber as a byproduct of clearing forestland to grow crops because they have not applied modern technology to the production of food. This practice, however, will not be a large source of timber supplies, except in parts of Africa.
- Technology for harvesting and processing trees is progressing at a reasonably rapid rate so that more forest products per hectare of natural forest will be produced.
- Improved forest management will accelerate regeneration in national forests.
- Good substitutes for most wood products exist, and their numbers are growing.
- Few indications suggest that oil prices will rise sufficiently to increase demand for fuelwood.

Taken together, these factors suggest that forest plantations will expand modestly over the next few decades and that there will be more harvesting of natural forests than many policymakers and environmentalists, particularly in developed countries, see as desirable. Sedjo and Lyon (1990) argue that natural forests will provide at least 50% of the global timber harvest until about 2050 without any significant rise in prices. However, we believe the conflict of interest between what rich countries want poor countries to do and what poor countries perceive as being in their own self-interest on the issue of harvesting their forests will continue. We cannot provide a resolution to that basic conflict, but we do think that the chapters by Sedjo and Sutton in this volume suggest several noncontroversial directions for policy-making.

DIRECTIONS FOR FUTURE POLICIES

The first direction is the issue of property rights. Sedjo documents cases where forest resources are used inefficiently when property rights, either public or private, are not well developed. In both agriculture and forestry, private owners of land, particularly those who own small plots, generally manage their land well and use their resource to best meet their families' needs. When environmental problems arise, they are caused by factors that are beyond the control of these small landowners, and they require action by the government. When farmers and foresters do not have title to their land, they often do not manage resources well.

The second direction is the need for governments to recognize the consequences of mismanagement more fully. Many researchers, including Sedjo, document cases in which it seemed that government agencies had failed to collect adequate incomes from public revenues. In some cases, it may be difficult to determine the fair market price for a forest concession, but where there is evidence of outright corruption in public management, the political damage is considerable.

We need to note that governments of developing countries will generally determine what is, for them, an appropriate harvest policy for their natural forests, while interest groups in wealthy countries, often poorly informed, may think these policies inappropriate. The poor country will wish to serve its own interests by implementing a harvesting schedule, with associated rents for concessions, that maximizes the present value of the forest resource. It will assess environmental values, such as diversity and greenhouse warming effects, from the perspective of its own citizens and build these into its decisions regarding the amount of harvest, whether to export logs or lumber, and other policy decisions. It will not consider the interests of the international community. Indeed, in many cases, it cannot afford to do so. Unless rich countries are prepared to bear the costs resulting from decisions which depart from national sovereignty, we believe that such decisions of developing countries should be respected.

A further direction for policy is investment in the development of appropriate technology, and we believe that a number of developing countries could expand their forest products sectors in this way. Mergen et al. (1988) showed that most developing countries invest little in forestry-related research and development. Indeed, the forestry sector in these countries has underinvested in technology that is suited to a specific country to better meet its national goals. It is quite likely that more investment in research and extension programs for the public sector will lead to better use of forest resources. There is opportunity for research in production and management. Hyde et al. (1992) shows that the opportunity for research in forest products for better use of tropical species is most promising.

Efficient plantation forestry will require agronomic and genetic research. Efficient harvesting of natural forests and wood processing will require adaptive invention for improved use of local products. Modern technology in transportation and other harvesting and wood processing activities does not need to be reinvented in developing countries since much of it can be transferred from developed countries. However, it is not always appropriate to transfer the technology without adapting it to local materials and developing wood products.

BEYOND ECONOMICS

The experience with well-managed harvesting of natural forests and plantations —southern pine in the U.S. South, radiata pine in New Zealand, and some fir mixes in Scandinavia—is instructive. It demonstrates the scope for substantial improvements in management. But it does show that prices matter. When prices are high enough to justify harvesting, but not high enough to justify extensive forest plantations, which has been the case for most forested areas of the world over the past century, it is inevitable that the forestry sector will fail to meet goals that have been set for it by many of today's environmental groups. If costly schemes for reforestation are not justified by timber prices, we need a careful evaluation of their other benefits and costs. This type of evaluation is increasingly important, but it demands economic actions that go beyond commercial forestry.

Literature Cited

Hyde, W.F., D.H. Newman, and B.J. Seldon. 1992. *The economic benefits of forestry research.* Ames: Iowa State University Press.

Mergen, Francois. 1901. *Tropical forests: utilization and consumption.* New Haven: School of Forestry and Environmental Studies.

Mergen, F., R.E. Evenson, M.A. Judd, and J. Putnam. 1988. Forestry research: a provisional global inventory. *Economic Development and Cultural Change.* pp 149-171.

Sedjo, R.A. and K.S. Lyon. 1990. *The long-term adequacy of world timber supply.* Washington, D.C.: Resources for the Future.

Part Five

Overview and Summary

CHAPTER 21

SUMMARY PERSPECTIVES: RECOMMENDATIONS FOR RESEARCH AND ACTION

William R. Bentley
Marcia M. Gowen

The February 1992 workshop at Old Saybrook, Connecticut, on *Forestry and Wood-Based Biomass Energy as Rural Development Assets* was a fascinating session. It brought together engineers, foresters, ecologists; professionals working for private industry, public agencies, non-profits and universities talked with one another for almost a week about several complicated relationships among biophysical and social-economic factors. The presentations generally were integrative, usually around the themes of rural development and environmental protection during the development process. The discussions were yet more integrative, often synthesizing new ideas by bringing together relationships from widely different disciplines and professions. The results, in turn, led us to request some additional papers for this volume by authors who were not present but have much to contribute to our dialogue.

Two results stand out from the perspective of time. First, The expected consequences of using forests and wood-based biomass energy in rural development strategies have several possible twists which are difficult to understand and predict. Everyone agreed both forest resources and wood energy are development assets. Almost everyone also agreed that adding wood-based biomass energy to the mix, even if limited to use of residues from processing, could have both positive and negative consequences. The interesting point of agreement in this ambiguous conclusion is that social causal chains are the key relationships, not technology as such.

Second, considerable applied research is needed to refine our understanding of the likely consequences in specific circumstances. "Do more research" is often the recommendation of such meetings, especially if hosted by a university and a non-profit which have applied research as center pieces of their strategies, but these recommendations focus on integrative thinking and synthesis. No one argues against disciplinary research, but this group of specialists came down strongly for work that is problem oriented and client directed.

GENERAL ISSUES AND RESEARCH DIRECTIONS

Following the workshop presentations, participant discussion focused on the linkages among several issues—resource supply, harvesting, and utilization; energy production; influence of cultural diversity on resource management; and responses to harvesting and other human-caused perturbations in forest ecosystems. Wood-based energy was the integrative casual factor that drove most of the discussion. One primary conclusion was the clear need to consider all the "upstream" and "downstream" issues related to wood waste utilization at a wood processing plant. These include the types and impacts of resource production activities and the end-use impacts of rural energy production on human activities and ecosystem function.

Rural Development Issues and Priorities

Discussion of rural development led to the following definition and understanding. Rural development is a dynamic concept, and the analyst and social scientist at best can hope to know the consequences and impacts on different sectors of the population and ecology from introducing changes into rural areas. In general, the three key goals of rural development are to:

- improve the distribution of benefits from development toward the poor
- increase the productivity of rural people's assets—such as human and natural resources
- move resource management schemes toward sustainability.

Several research questions concern the local benefits and costs versus external net benefits and costs created by wood waste utilization. The major concern is, "how effective is wood-based energy in promoting rural development?" It is critical to know who benefits from wood energy? The distinction is between the local and non-local (regional and urban) beneficiaries from the wood energy. Each potential beneficiary group sees a variety of possible benefits and costs associated with the energy production and consumption stages (see figure 21.1). Research is needed that measures the actual realized benefits for each group. Research on this issue should include the importance of local empowerment, distribution, productivity, and sustainability.

At the resource production level, supply sources for feedstock timber producers can be categorized as natural forests, degraded lands (potential plantation sites), and marginal cultivated lands suitable for farm forestry. These distinctions reflect patterns of forest land uses by local people, but also incorporate relative price movements and political changes that may override local practices.

Beneficiary	*Processing State*	*Benefits*	*Costs*
Local	Resource Supply	Income, jobs	Lost alternative uses Lost tenure
Local	Production	Income, jobs, Environmental	Alternative uses
Local	Consumption	Energy products Small industries Environmental	Lost subsidies
Non-local	Consumption Taxes/revenues	Energy products Environmental	

Figure 21.1. Beneficiaries of Wood-Based Energy Production at Forest Processing Industries.

Key Research Needs and Priorities

The upstream issues related to resource supply, harvest and utilization are important. They are most important where the impacts and benefits to rural people from wood energy production for the large part are uncertain and undocumented. The technological issues associated with wood waste utilization at forest products industries are reasonably well understood. However, resource supply concerns are both more important and less well understood.

A set of nine cross-cutting research questions were raised by the group, and developed into the following general research agenda:

1. ***Policy in terms of political and institutional relationships.*** The participants, while a diverse mix of professional interests, was not a typical policy or policy analyst meeting. It is, as a consequence, even more significant that they highlighted the many market and policy failures that adversely affect the management of forest and energy resources. Bad pricing policies for publicly-owned resources, subsidies that distort markets and technological development, and political economies that are biased against both the rural and poor in their decision making are but a few of the examples noted.
2. ***Ownership or property and tenure rights as they affect both function and distribution of benefits.*** The most important policy failure noted was the lack of firm, enforceable property rights or tenure rights in many forest, range, and marginal farming contexts. In many cases, this failure is simply a slow response to rapid social change over the past five or more decades, especially in situations where "fee simple" ownership rights are not

appropriate. In other cases, however, it is a symptom of the break down of "law and order," often with leaving a vacuum filled by a mix of warlords, mafia, or anarchy. The uncertainty and ambiguity created by this vacuum leads to over-harvesting forest resources and underinvesting in either natural or plantation systems for producing needed goods and services.

3. ***Inventories: stocks and flows of all resources in physical account terms.*** The lack of definition or good measures of tropical deforestation volumes, areas and rates points to the broader needs for better physical inventories. Long taken for granted in the United States and most of Europe, unbiased, low-variance estimators of land area, timber resources, and other resource inventories are difficult to obtain and often simply not available in the tropical world. Use of satellite imagery with ground controls will help, but high-quality information requires a commitment to unbiased estimators and the continuous measurement of resources to establish both baselines and trends.
4. ***Resource management systems.*** The importance of cultural perspective was an unexpected point of agreement among the participants. Given a large number of engineers and economists, it is not surprising that considerable attention was paid to efficiency as a criterion for judging the value of development assets. However, it is surprising that the groups so readily understood that the basic values which drive how we measure efficiency can vary so strongly among cultural groups. This includes different professions as well as different indigenous and exogenous peoples. Well developed resource management plans that consider multiple-use goals and resource bases are necessary to ascertain how the introduction of new resource uses, such as wood energy production, will strain or improve local resource use patterns.
5. ***Preference ordering.*** Knowledge of local people's historic and future preferences, with some sense of priority or order, provides information useful to project planners about the potential consequences of adoption. If higher value alternative uses for timber and non-timber forest products exist, information about these values is important to firms and development planners in decisions about new plant locations or adding biomass energy utilization to existing plants.
6. ***Time and space.*** The basic development framework presented by the BEST team suggests a rational over space that uses rural biomass energy systems to both extend the grid and reach beyond the grid in bring electrical energy to rural areas. Considerably more attention needs to be given to the role small-scale biomass energy systems can ply in filling in the grid, making it more

stable, and reducing the costs of providing power to dispersed rural people and their industries. These spatial dimensions then need to be integrated with the more temporal concerns of foresters, resource economists, and investment specialists.

7. ***Markets in both the formal and informal senses.*** Fuelwood markets are largely informal, and as a consequence our knowledge is meager. A historic perspective of both America and England's energy development, however, suggests that understanding both formal and informal markets is necessary to develop future energy strategies and resource policies.
8. ***Existing and potential conflicts from intervention.*** Not everyone will benefit from improved tropical forest management or more encouragement of small-scale biomass energy systems. Understanding the rational of groups in conflict is important because it helps design policy, management and technical solutions that can minimize opposition.
9. ***Ethnic and gender relations.*** The attendees at this workshop had little expertise in these issues, but the aggregate development experience of this group led most to favor consideration of both ethnic and gender issues in context of forestry and wood-based energy.

POLICY FAILURES AND RURAL RESOURCE DEVELOPMENT

The participants were most concerned with the first four research issues. Their comments are distilled here to help guide researchers in future attempts to improve our knowledge base for rural development.

The use of wood from different sources—natural forests, plantations, agroforestry or farm forestry, or processing residues—causes different biophysical impacts. Different tenurial issues also arise with each source. Natural forests often have local usufruct rights. Removal of material primarily for wood energy may reduce the local fuelwood resources or impact on non-timber product and service values from the forest. Agroforestry on common lands produces a quite different product and service mix than the mix observed on adjacent and otherwise similar private farm land. Residues in wood scarce regions, such as South Asia, traditionally have been allocated as part of the pay package or given to poorer members of the local community. Using these resources to produce electrical power, while often representing a social efficiency gain, impacts unfairly on the poor. In many countries, exploration of forest resource supply issues and constraints is required before estimating the long-term supply of biomass wastes from timber and wood utilization sources. The many local "ownership" issues, such as those described, often mean that an apparent abundant local supply may not be sufficient to fuel a new rural energy industry.

Workshop participants were concerned with the future supply of timber from natural forests. They assumed that harvesting should be constrained to sustainable levels. Harvesting of the natural forests, whether through clearcutting or selective cutting, is not a proven sustainable practice in many tropical ecosystems. Even where information on ecological relationships exists, social factors may preclude sustainable management of the forest as a commercial timber resource. Roads, for example, when coupled with human population pressures, may lead to fuelwood and non-timber product harvesting that is, in aggregate, not sustainable even though the initial timber harvest levels are silviculturally sound.

Many countries and local regions have substantial areas of degraded lands. In India, for example, nearly half the rural landscape is seriously degraded from overgrazing and other causes. These lands offer a marvelous opportunity to increase resource productivity with rehabilitation based on planting tree and perennial grasses. Because these lands currently produce little, it is easier to skew the benefits toward the rural poor, and any successful rehabilitation is a movement toward ecological sustainability.

Using degraded lands for industrial wood or biomass energy plantations is a sensible long-term solution use of intensive timber management to promote rural development. Plantation costs, however, may not be fully recovered by intermediate and final harvest returns, especially given the higher risk premiums and consequent interest rates generally observed. Agroforestry and farm forestry offer potential wood processors another source of raw materials, but the volumes obtainable and opportunity costs of wood for various industries must be considered.

Systems of wood energy use should benefit local people, making them stakeholders in the systems. Granting this is desirable, several participants expressed skepticism about successfully achieving this result from wood energy use. Evidence from the America's Southeast, for example, suggests that higher net returns from adding higher value products in the wood processing industry tends to stay with the firms rather than increase the prices paid to the producers. Consequently, a useful research topic will be estimating the impact of wood energy production on the value added of the industry and on employment and income multipliers. Many participants hypothesize that the rural benefits from wood energy use are greater on the downstream end rather than upstream. Producers may not reap many direct benefits, but rural families may receive indirect benefits from the stability and lower cost of rural energy production from indigenous sources.

A general consensus developed that the vast diversity of institutional arrangements and ecological paradigms governing resource use throughout the developing world makes regional and country specific research over the short and long terms necessary. Greater stress on the integration and recording of the ecological and social consequences of wood-based energy use by the forest products industry in all areas of the world are needed. The results of the working groups on three areas—rural development, ecology, and energy